Table-entered Variables

The size of a table is halved when an input variable is treated as a **table entered variable** instead of as an input. Furthermore, the table display is significantly improved in most, if not all, cases.

Normally the coefficient of a product term is the implied "1". Table-entered variables are based on the idea that one or more of the literals in a term may be defined as the coefficient of the term instead of the implied "1". This means the term is reduced to the literals not included in the coefficient. For example $1(xyz)$ becomes $x(yz)$. The new coefficient is x and the term reduces to yz. Consider the equation $q^+ = jq' + k'q$. Let the input variables be j and k, while treating q as a table entered variable. Construct the table by evaluating q^+ for all j, k input variable combinations. Each evaluation result is q, 0, 1, or q'.

row	j	k	q^+	Function
0	0	0	q	hold
1	0	1	0	reset
2	1	0	1	set
3	1	1	q'	toggle

Karnaugh Map Concept

The Karnaugh map (K map) is based on the concept of adjacent states. Two states are adjacent if they differ in the value of only one variable. This concept is important because the algebraic combination of two adjacent states results in the elimination of one variable. A K map is a graphical aid for function simplification. One way to use the map is to map or write a function g onto the map so that you can read a simplified equation for the function g from the map.

Map-Entered Variables

The K map becomes unwieldy even with five variables. With more than five variables the K map rapidly becomes unmanageable. The map-entered variable reduces the required map size, thereby extending the K map's practical usefulness. The number of function variables equals the number of K map dimensions plus the number of map-entered variables.

Example Illustrating the Power of Map-Entered Variables

Write to a K map the function $g = wx'y'z' + x'yz + yz' + xz' + w'xz$ (see Example 3-33).

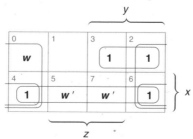

Read from the K map the simplified function $g = wz' + w'x + x'y$.

DIGITAL DESIGN

Nicholas L. Pappas, Ph.D.
San Jose State University

West Publishing Company
Minneapolis/St. Paul   New York   Los Angeles   San Francisco

Copyediting: Elliot Simon
Art: Christine C. Bentley and Edward M. Rose, Visual Graphic Systems
Composition: Bi-Comp
Cover image: Kristopher Hill, Final Copy
Production, prepress, printing, and binding: West Publishing Company

COPYRIGHT ©1994   By WEST PUBLISHING COMPANY
610 Opperman Drive
P.O. Box 64526
St. Paul, MN 55164-0526

Library of Congress Cataloging-in-Publication Data

 PRINTED ON 10% POST CONSUMER RECYCLED PAPER

Pappas, Nicholas L.
      Digital design / Nicholas L. Pappas.
          p. cm.
      Includes bibliographical references and index.
      ISBN 0-314-01230-3 (alk. paper)
      1. Digital electronics. 2. Logic design. I. Title.
TK7868.D5P36 1994
621.39′5—dc20   93-9764
                                                 ∞ CIP

CONTENTS

4

COMBINATIONAL MIXED-LOGIC CIRCUITS **170**

5

COMBINATIONAL BUILDING BLOCKS **222**

11 ASYNCHRONOUS SEQUENTIAL CIRCUITS   596

12 PROJECTS   646

PREFACE

This is a textbook on the art of digital design. This is about logic design fundamentals, the theoretical basis for modern design techniques, the art of creating designs that work reliably, and thought provoking projects providing practice.

We start by not taking numbers for granted; they are evolved and discovered in an attempt to provide a basis for creative thinking and application. This is followed by a brief overview of relevant topics in electronics.

The theory of combinational logic is fully developed from Huntington's Axioms. The truth table and the Karnaugh map are presented for what they are: very useful design aids. The connection between physical gates and logic with mixed assignments is fully developed. Then, mixed logic symbology used throughout the text is shown to facilitate the design process. Applications of mixed logic are used to naturally evolve combinational circuits and the combinational building blocks available to practitioners of the art of digital design.

Next, we show how feedback-is-remembering, which provides a basis for an account of the theory and practice of clocked sequential circuit elements also known as flip-flops. After a brief look at the traditional state diagram method, we follow up with an explanation of the theory of a modern design technique known as the Algorithmic State Machine. This, together with the mixed logic technique, simplifies the state machine design process in an amazing way. With these techniques and flip-flops in hand, the standard logic family sequential building blocks are evolved for the reader. The presentation includes all of the necessary equations for registers that store, count, and shift data.

After explaining the basic attributes of memory devices a method of address decoding and elementary memory system design is presented.

Design practice using programmable logic devices is illustrated by working significant examples. The theory underlying these devices is shown to be that which we have already learned. This followed by a presentation of methods for asynchronous circuit analysis and synthesis.

We include illustrative specific designs, and projects throughout the text for the reader's initiation into the art of digital design. The aim throughout is to make the argument simple and straightforward, emphasizing understanding, and avoiding the rote.

Some traditional methods are mentioned briefly to let the reader know that they exist; they do not need a more expansive coverage because either their importance has diminished or because they can be replaced by more effective methods. For example, traditional state machine design methods are replaced by one modern method. This is the algorithmic state machine method, which can be used to design a 2 state or a 200 state machine. In each topic I have chosen to select one method rather than a multitude, one that suffices until its replacement appears on the scene.

The book as a whole is written primarily for undergraduates. A university course in digital design should, in my view, provide two things:

1. a broad base of knowledge comprising theories and theorems of continual application to digital design; and
2. the motivation and opportunity to acquire a detailed knowledge of digital design as practiced; this detailed knowledge will allow the reader to design machines that work.

My omissions to this book are many—all, I hope, deliberate. Yet, some of these omissions are covered in the problems provided—problems that provide new experiences in lieu of drill.

My indebtedness to other books and to research papers is very great. My indebtedness to my students is greater.

INTRODUCTION

Digital design is the art of converting words into digital circuits. We convert words such as, "please design an electronic combination lock," or "please design a stack computer," or "please design a flash multiplier of two 8 digit bcd numbers."

Clearly the first problem we face is a problem of vocabulary. What is this word "digital" we use at the start of the text? We deal with this problem by using new words in self-explanatory contexts that appeal to our intuition. We also note there is no substitute for a first class dictionary. Words are worth the time it takes to look them up. The process of looking up words is a conscious act, an experience. Experience is hard to forget.

First Principles

The first principles of digital design are a set of axioms from which we derive the structure of logic gates. We then show how simple circuits are designed with these gates, building up to systems of significantly increased complexity.

The key to this process is *understanding*. Understanding is much more important than memorization. Indeed, the large number of choices in modern digital design make memorization essentially useless. If you understand the basic principles, you will not need memorization.

Understanding lets you make connections between ideas and facts, between facts and concepts. These connections are to design what girders are to structure.

Learning How to Design

First, desire to learn. No words we write can overcome the lack of desire.

Second, acquire the necessary body of knowledge. What is necessary is found in this book.

Third, practice design. If you fail to practice, then the knowledge you've gained rapidly fades away. This book, through its projects, offers you significant opportunity to practice design.

In conclusion: if you desire to learn, if you acquire the body of knowledge, and if you practice, you will learn how to design.

Additional Materials

In this text we have not copied numerous schematics and tables from various data books. Therefore a required companion to the text is an industry standard TTL data book such as Texas Instruments' *TTL Logic* (part #SDLD001A). If one chooses to use a logic family other than the LS TTL family the appropriate data book should be substituted. The accompanying data book has another major purpose: it gives credibility to mixed logic.

Students find our emphasis on mixed logic makes sense because the data book makes extensive use of mixed logic.

Standard Chapter Format

Front-of-chapter material:
Chapter Table of Contents
Chapter Overview

Additional material in the text body:
Marginalia: margin notes that give the essence of each topic in a key sentence
Examples: illustrate processes required to implement the topic being discussed
Exercises: problem statements with answer

End-of-chapter material:
Summary: review of chapter topics
References: list of relevant references
Projects: project statements and demonstration reports
Exercises: useful problems

Suggested Course Outlines

Synchronization of lectures and projects is important in the suggested course outlines shown in Tables 1 and 2. The tables show the courses taught in our department (two lectures and one laboratory session per week). An introductory circuit design course is the only technical prerequisite. A following course would be a computer design course starting from an instruction set processor (ISP) description and ending with a detailed computer design including microcode in ROM.

The time devoted to topics you select may differ. This is a function of your students' capability. The varieties are endless. We leave specific decisions to you.

Nicholas L. Pappas
San Jose, California

TABLE 1. One Semester First Course

WEEK	LECTURE	PROJECT	REPORT
1	1.1	P1	
2	1.2, 1.3	Learn Chapter 2	
3	1.4	P2	P1
4	3.1 to 3.8		
5	3.9 to 3.12	P3	P2
6	3.13		
7	Review		P3
	Exam 1		
8	4.1 to 4.8	P4	
9	4.9 to 4.10	P5	P4
10	5.1 to 5.6	P6	P5
11	6.1 to 6.10	P7	P6
12	Review	P8	P7
	Exam 2		
13	7.1, 7.2		
14	7.3, 7.4	P9	P8
15	7.5		
16	7.6 to 7.9		P9
17	Final		

TABLE 2. One Semester Second Course

WEEK	LECTURE	PROJECT	REPORT
1	8.1, 8.2	P10	
2	8.2		
3	8.3	P14	P10
4	9.1, 9.2		
5	9.3	P16	P14
6	9.4		
7	Review	P17	P16
	Exam 1		
8	10.1, 10.2		
9	10.3, 10.4	P14_pld	P17
10	10.5		
11	10.6	P12_pld	P14_pld
12	Review		
	Exam 2		
13	11.1 to 11.3	P19	P12_pld
14	11.5 to 11.7		
15	11.8 to 11.10		P19
16	11.11 to 11.13		
17	Final		

Acknowledgments

I would like to express my gratitude to my reviewers. I want them to know I seriously considered every comment, acted upon almost all of them, and discarded the few for relevant reasons. In effect, I was able to integrate their years of teaching experience. Thank you very much.

I especially want to thank Joseph B. Evans, Willard Korfhage, John R. Pavlat, Christos A. Papachristou, Sherman Reed, and Peter G. von Glahn.

Aaron S. Collins	Clemson University
Paul Darlington	University of Wyoming
Darrow F. Dawson	University of Missouri at Rolla
S. P. Desai	University of Southern Indiana
Edward Doskocz	Air Force Academy
Milos B. Ercegovac	University of California at Los Angeles
Joseph B. Evans	University of Kansas
Barry S. Fagin	Dartmouth College
Robert Fujii	Purdue University
M. L. Hambaba	Stevens Institute of Technology
James L. Kirtley, Jr.	Massachusetts Institute of Technology
Harold F. Klock	Ohio University at Athens
Willard Korfhage	Polytechnic University, Brooklyn
Earl R. Laste, Jr.	University of Lowell
S. Y. Lee	Cornell University
William E. Moritz	University of Washington
C. E. Nunnally	Virginia Polytechnical
W. Ohley	University of Rhode Island
Christos A. Papachristou	Case Western Reserve University
John R. Pavlat	Iowa State University
Ghahdi Puvvada	University of Southern California
Sherman C. Reed	University of Texas at Arlington
Irwin A. Reinhard	University of Alabama, Retired
D. G. Saab	University of Illinois, Urbana
Joy Shetler	Texas A and M University
Charles Swain	Rochester Institute of Technology
C. James Tricka	Iowa State University
Peter G. von Glahn	Villanova
Edward K. Wong	Polytechnic University, Brooklyn
Charles T. Wright, Jr.	Iowa State University

1 NUMBERS AND CODES IN DIGITAL SYSTEMS

OVERVIEW

We are schooled in decimal numbers, whose radix r is ten. (Radix ten numbers use the digits 0 to 9.) However, computers are based on binary numbers, using only the digits 0 and 1, whose radix is two. How do we move to and from radix ten and radix two numbers? We need a mathematical mechanism.

A number such as 7405.321 is merely shorthand notation for the algebraic polynomial defining the number. For each digit in the number, such as 4, there is a term in the polynomial. With the polynomial in hand, we can use elementary algebra to convert any number of any radix to one of any other radix. Then we learn to count in any radix.

In this chapter we do traditional arithmetic in radix 2 as a basis for doing arithmetic using *complements*. Our paper-and-pencil "digital hardware" has no limitations; however, actual digital hardware is limited in its ability to represent numbers. Those limitations motivate us to discover negative-number representations that digital systems can use.

The chapter closes with a brief introduction to coding and some of the more widely used codes.

INTRODUCTION

Discrete has incremental properties.

Here is an analogy to help understand the distinction between continuous and discrete mathematics. When you walk up a hill you ascend *continuously* from the bottom of the hill to the top. As you ascend you can stop anywhere, and the elevation could be any value from ground level to the height of the hilltop. In contrast, if you could use a flight of stairs to reach the top of the hill, you would ascend in *discrete* increments. This time when you stop, you would not stop anywhere; rather, you would stop on a step—not on a half step, but on a step. You can only stop at the stairs' discrete intervals of elevation.

In our everyday activities we process primarily continuous information using the *decimal* number system (which can also be used in discrete systems). On the other hand, practical digital hardware systems process discrete information using the *binary* number system, which is why knowing how to use the binary number system is important. (Binary numbers are used in hardware systems, even though decimal numbers are more natural to us, because a two-state—on–off—system can be produced more readily and reliably than a ten-state system.)

Given these two number systems, we need a bridge between them that allows us to go back and forth. As we build that bridge, we will discuss other useful number systems, as well. Let us first examine why digital hardware deals with discrete information.

The number 11,115 is exact. In this five-digit system the next number is 11,116. Clearly there is a gap between the first and the second numbers. Numbers such as 11115.2, 11115.333330047, and 11115.99 do not exist in a system limited to five-digit integers. They are merely approximated by the numbers 11115 and 11116. Numbers such as these have *finite precision*.

In practical hardware realizations, numbers are represented by a finite number of digits n, such as $n = 8$. One way of thinking about this is to say a digital hardware system processes n digits in parallel, where n is the value specified for the system. This means we cannot represent all real numbers: gaps such as the one between 00011115 and 00011116 convert the continuum of numbers into a discrete number-representation system. Now let us move on to discuss numbers of finite precision.

Digital hardware supports n digits.

1.1 Position and Radix

A number may be considered to be a shorthand notation for the polynomial it represents. That's because decimal numbers such as 3.14159 are written in *positional notation*. They are built up from the ten digits 0, 1, 2, 3, 4, 5, 6, 7, 8, and 9 and the decimal point. These digits form an ordered set of symbols. The *radix* (plural *radices*) of a number system specifies which digits are used in that number system. The decimal number system, which has ten symbols, is said to have a radix equal to ten. [Note: *base* is a synonym for *radix*.]

Numbers are written in positional notation.

Given the symbols allowed in our decimal system, how do we distinguish between three units, thirty units, and three hundred units? The simple answer, which took hundreds of years to discover, lies in one word: position. As we learned in school, we represent our decimal-system numbers as follows:

NUMBER OF UNITS	POSITION WEIGHT			NUMBER
	Hundred	Ten	One	
three	0	0	3	3
thirty	0	3	0	30
three hundred	3	0	0	300

Any number written in positional notation represents a polynomial in r, where r is the radix of the number system. For instance, the number 333 is represented by the specific polynomial $(3 \times 10^2) + (3 \times 10^1) + (3 \times 10^0)$. The general form is

$$n = (a_2 \times r^2) + (a_1 \times r^1) + (a_0 \times r^0)$$

The coefficients a_j of the polynomial in r are the digits of a number, and r is called the radix of the number. Any coefficient's subscript j also represents the power of the radix in the jth term.

In everyday activities, any number, such as 835.46, is written in positional notation. By this we mean that the digits are written side by side, with or without a radix point (which is a decimal point in this case). Observe that each position has a value equal to the digit in that position times a weight dependent on that position. The weight varies as a power of the radix, which is ten in the decimal system. Ordinarily we use decimal numbers with no explicit concern for position and radix. We are discussing position and radix here because digital logic uses numbers with radix 2.

Radix   A number made the base of a system of numbering.

Positional Principle   The positional principle operates as follows. The sequence of digits

$$a_n a_{n-1} \cdots a_1 a_0 . a_{-1} a_{-2} \cdots$$

is defined to signify a number, the magnitude of which is a sum of products involving powers of a number r, where r is called the radix or the base. That is, the position of each of the digits a_j is associated with the power j of the radix r. The number of distinct numerals in this notation is seen to be r.

Radix Point   The radix point is the index that separates the digits associated with negative powers of the radix r from those associated with zero and the positive powers of r of a number system in which a quantity is represented. Examples include the decimal point and the binary point.

EXAMPLE 1.1

The number 835.46 has five positions, one for each digit, and the weight or value of a digit depends on its position. The decimal number 835.46 represents a weighted sum, which is:

$$835.46 = 800 + 30 + 5 + 0.4 + 0.06$$
$$= (8 \times 100) + (3 \times 10) + (5 \times 1) + 4/10 + 6/100$$
$$= (8 \times 10^2) + (3 \times 10^1) + (5 \times 10^0) + (4 \times 10^{-1})$$
$$+ (6 \times 10^{-2})$$

In abstract terms,

$$n = (a_2 \times r^2) + (a_1 \times r^1) + (a_0 \times r^0) + (a_{-1} \times r^{-1})$$
$$+ (a_{-2} \times r^{-2})$$

$$= a_2 a_1 a_0 . a_{-1} a_{-2}$$

1.2

Radix Conversion

We know the decimal number system and decimal arithmetic because we have been schooled in them. The only reason we calculate in decimal is that it is convenient to do so. Had we been schooled in the radix-7 system, we would be using that as the intermediate number system instead of decimal. We must be able to convert from our familiar base ten to the bases that digital systems use. An overview of the radix conversion process follows.

Convert radix *n* to radix ten to radix *m*.

We will illustrate the need to use decimal as an interim number system by finding out what we need to know to convert a number directly from radix 8 to radix 5. We will learn that to do this conversion, we have to know how to do radix-8 division by 5 (Example 1.2a). Since we do not yet know how to do radix-8 division by any number, we avoid it by doing the following: First we convert the number from radix 8 to radix ten by writing out the polynomial in decimal and then calculating the decimal sum of the terms. (We know how to do this.) Second, the radix ten number is converted to a radix-5 number by repeatedly using decimal division (which we also know how to do). Example 1.2b provides an overview. It does not show why the methods for the two steps always work. That explanation is found in the Sections 1.2.1 and 1.2.2.

EXAMPLE 1.2 **Converting radix 8 to radix 5**

a) The algorithm for direct conversion from one radix to another radix is to divide by the second radix using the arithmetic of the first radix. Thus, to convert 624 radix 8 to *n* radix 5:

$$624_8/5 = 120 + 4/5 \quad \text{(Unfamiliar radix-8 division by 5)}$$

$$120_8/5 = 20 + 0/5$$

$$20_8/5 = 3 + 1/5$$

$$3_8/5 = 0 + 3/5$$

Therefore,

$$n = 3104 \text{ radix } 5$$

b) The algorithm we are schooled in uses decimal as an interim radix. There are two steps: we convert radix 8 to radix ten, then we convert radix ten to radix 5. Thus, to convert 624 radix 8 to m radix ten (see Section 1.2.1):

$$624_8 = (6 \times 8^2) + (2 \times 8^1) + (4 \times 8^0) = 404_{10} = m$$

To convert 404 radix ten to n radix 5 (see Section 1.2.2):

$$404_{10}/5 = 80 + 4/5$$

$$80_{10}/5 = 16 + 0/5$$

$$16_{10}/5 = 3 + 1/5$$

$$3_{10}/5 = 0 + 3/5$$

Therefore,

$$n = 3104 \text{ radix } 5 \quad \text{Q.E.D.}$$

1.2.1   Converting Radix r to Decimal

A radix-r number system requires r different symbols to represent the digits 0 to $r - 1$ (Table 1.1). For example, the radix-7 system requires only the seven symbols 0, 1, 2, 3, 4, 5, and 6 and does not use the symbols 7, 8, and 9. Radices 11 through 16 require six new symbols (see Section 1.2.2).

Expand a number into a polynomial to convert radix n to radix ten.

Converting a number of any radix r to decimal (radix 10) is straightforward. We convert the number into the polynomial it represents using decimal number equivalents for the digits and radix (see Example 1.4). Then we calculate the value of each term and the sum of the terms, using decimal arithmetic.

$$n = a_n a_{n-1} \cdots a_1 a_0 . a_{-1} a_{-2} \cdots a_{-m}$$

$$n = a_n r^n + a_{n-1} r^{n-1} + \cdots + a_1 r^1 + a_0 r^0$$

$$+ a_{-1} r^{-1} + a_{-2} r^{-2} + \cdots + a_{-m} r^{-m}$$

Tᴀʙʟᴇ 1.1

RADIX	SYMBOLS REQUIRED
1	0 (Cannot use)
2	0 1 (Binary)
3	0 1 2
4	0 1 2 3
5	0 1 2 3 4
6	0 1 2 3 4 5
7	0 1 2 3 4 5 6
8	0 1 2 3 4 5 6 7 (Octal)
9	0 1 2 3 4 5 6 7 8
10	0 1 2 3 4 5 6 7 8 9 (Decimal)
11	0 1 2 3 4 5 6 7 8 9 A
12	0 1 2 3 4 5 6 7 8 9 A B
13	0 1 2 3 4 5 6 7 8 9 A B C
14	0 1 2 3 4 5 6 7 8 9 A B C D
15	0 1 2 3 4 5 6 7 8 9 A B C D E
16	0 1 2 3 4 5 6 7 8 9 A B C D E F (Hexadecimal)

The number 1010221.2102 can have a radix equal to 3 or greater because 2 is the greatest digit in the number. But until we specify the radix, we do not know what radix is being used. This is an issue for us here. In text that uses only the decimal system this is not an issue; all numbers are assumed to be decimal. In this chapter we use many different radices, so we need to show the radix explicitly.

Show the radix of a number.

One way to show the radix is as a subscript at the right-hand end of the number:

$$1010221.2102_3$$

Radix as a subscript should not be confused with other types of subscripts, such as the a_j subscripts in the previous display.

Eхᴀᴍᴘʟᴇ 1.3 **Converting a radix-3 number to decimal**

$$n = a_6 \cdots a_2 a_1 a_0 . a_{-1} a_{-2} a_{-3} a_{-4}$$

$$= a_6 3^6 + \cdots + a_2 3^2 + a_1 3^1 + a_0 3^0 + a_{-1} 3^{-1} + \cdots + a_{-4} 3^{-4}$$

$$n = 1010221.2102_3$$

$$= (1 \times 3^6) + (0 \times 3^5) + (1 \times 3^4) + (0 \times 3^3) + (2 \times 3^2)$$
$$+ (2 \times 3^1) + (1 \times 3^0) + (2 \times 3^{-1}) + (1 \times 3^{-2}) + (0 \times 3^{-3})$$
$$+ (2 \times 3^{-4})$$
$$= 729 + 81 + 18 + 6 + 1 + 2/3 + 1/9 + 2/81$$
$$n = 835.802469 \cdots \quad \text{(Radix 10)}$$

Notice in Example 1.3 that the process terminates after adding the 2/81 term. However, such fractions may not always terminate when divided out. In general a finite number in radix j (such as 3) can only be *approximated* by a finite number of digits in radix k (such as 10). An exact representation in radix k requires an infinite number of digits in the general case.

EXAMPLE 1.4 Converting a radix-2 number to radix 10

$$n = a_2 2^2 + a_1 2^1 + a_0 2^0 + a_{-1} 2^{-1} + a_{-2} 2^{-2}$$
$$n = 111.11_2$$
$$= (1 \times 2^2) + (1 \times 2^1) + (1 \times 2^0) + (1 \times 2^{-1}) + (1 \times 2^{-2})$$
$$= 4 + 2 + 1 + 0.5 + 0.25$$
$$n = 7.75_{10}$$

EXAMPLE 1.5 Converting a radix-2 number to radix 10

$$n = a_9 2^9 + \cdots + a_2 2^2 + a_1 2^1 + a_0 2^0 + a_{-1} 2^{-1} + a_{-2} 2^{-2}$$
$$n = 1101000011.11_2$$
$$= (1 \times 2^9) + (1 \times 2^8) + (0 \times 2^7) + (1 \times 2^6) + (0 \times 2^5)$$
$$+ (0 \times 2^4) + (0 \times 2^3) + (0 \times 2^2) + (1 \times 2^1)$$
$$+ (1 \times 2^0) + (1 \times 2^{-1}) + (1 \times 2^{-2})$$
$$n = 512 + 256 + 64 + 2 + 1 + 0.5 + 0.25$$
$$n = 835.75_{10}$$

EXAMPLE 1.6   Converting a radix-8 number to decimal

Octal numbers have a radix of 8.

$$n = a_3 8^3 + a_2 8^2 + a_1 8^1 + a_0 8^0$$

$$n = 1503_8$$

$$= (1 \times 8^3) + (5 \times 8^2) + (0 \times 8^1) + (3 \times 8^0)$$

$$= 512 + 320 + 0 + 3$$

$$n = 835_{10}$$

EXAMPLE 1.7   Converting a radix-10 number to decimal

Decimal numbers have a radix of ten.

$$n = a_2 10^2 + a_1 10^1 + a_0 10^0$$

$$n = 835_{10}$$

$$= (8 \times 10^2) + (3 \times 10^1) + (5 \times 10^0)$$

$$n = 835_{10}$$

EXAMPLE 1.8   Converting a radix-5 number to decimal

$$n = a_2 5^2 + a_1 5^1 + a_0 5^0$$

$$n = 432_5$$

$$= (4 \times 5^2) + (3 \times 5^1) + (2 \times 5^0)$$

$$= 100 + 15 + 2$$

$$n = 117_{10}$$

EXERCISE 1.1   Convert 10011101.1101_2 to radix 10.

Answer: 157.8125_{10} ■

1.2.2   Converting Decimal to Radix *r*

The algorithms for direct conversion of a decimal integer N_i and a decimal fraction N_f to a radix-*r* integer n_i and radix-*r* fraction n_f are outlined next. Examples will illustrate the methods. An intuitive proof for the algorithms follows the examples in this section.

Divide radix ten integer by the "to" radix.

The conversion starts with ordinary division of N_i by *r*. This generates a new quotient and a new remainder, which is the least significant digit of n_i. Successive divisions of the new quotients by *r* generate new remainders, which are the digits of n_i. The process ends when the new quotient is zero. Let $n_i = a_n a_{n-1} \cdots a_1 a_0$,

$$\frac{N_i}{r} = q_0 + \frac{r_0}{r} \qquad a_0 = r_0$$

$$\frac{q_0}{r} = q_1 + \frac{r_1}{r} \qquad a_1 = r_1$$

$$\vdots \qquad\qquad \vdots$$

$$q_{n-1}/r = 0 + r_n/r \qquad a_n = r_n$$

Multiply radix ten fraction by the "to" radix.

The process for converting the fractional part of the number or fractions uses multiplication by *r*, generating an integer plus a fraction. The new fractional part is then multiplied by *r*, until you have a fractional part of 0 or you do not want to generate any more digits. The first integer is the most significant digit of the fraction. Let $n_f = 0.a_{-1} a_{-2} \cdots a_{-m+1} a_{-m}$.

$$r N_f = a_{-1} + n_{f-1}$$

$$r n_{f-1} = a_{-2} + n_{f-2}$$

$$\vdots$$

$$r n_{f-m+1} = a_{-m} + 0 \qquad \text{(0, or the process continues without end)}$$

EXAMPLE 1.9   **Converting decimal to binary**

$$N = 835.75 \qquad N_i = 835 \qquad \text{and} \qquad N_f = 0.75$$

Integer part:

$$N_i/r = 835/2 = 417 + 1/2 \rightarrow a_0 = 1$$

$$q_0/r = 417/2 = 208 + 1/2 \rightarrow a_1 = 1$$

$$q_1/r = 208/2 = 104 + 0/2 \rightarrow a_2 = 0$$
$$q_2/r = 104/2 = 52 + 0/2 \rightarrow a_3 = 0$$
$$q_3/r = 52/2 = 26 + 0/2 \rightarrow a_4 = 0$$
$$q_4/r = 26/2 = 13 + 0/2 \rightarrow a_5 = 0$$
$$q_5/r = 13/2 = 6 + 1/2 \rightarrow a_6 = 1$$
$$q_6/r = 6/2 = 3 + 0/2 \rightarrow a_7 = 0$$
$$q_7/r = 3/2 = 1 + 1/2 \rightarrow a_8 = 1$$
$$q_8/r = 1/2 = 0 + 1/2 \rightarrow a_9 = 1$$
$$q_9/r = 0/2 = \quad \text{Stop}$$

The number $n_i = 1101000011_2$.

Fractional part:

$$rN_f = 2(.75) = 1 + 0.5 \quad \rightarrow a_{-1} = 1$$
$$rn_{f-1} = 2(.5) = 1 + 0 \quad \text{Stop} \rightarrow a_{-1} = 1$$

The number $n_f = .11_2$. Thus the full number is:

$$n = n_i + n_f = 1101000011.11_2$$

EXAMPLE 1.10 Converting decimal to octal

$$N = 835.462 \qquad N_i = 835 \quad \text{and} \quad N_f = .462$$

Integer part:

$$N_i/r = 835/8 = 104 + 3/8 \quad \rightarrow a_0 = 3$$
$$q_0/r = 104/8 = 13 + 0/8 \quad \rightarrow a_1 = 0$$
$$q_1/r = 13/8 = 1 + 5/8 \quad \rightarrow a_2 = 5$$
$$q_2/r = 1/8 = 0 + 1/8 \quad \rightarrow a_3 = 1$$
$$q_3/r = 0/8 = 0 \quad \text{Stop}$$

The number $n_i = 1503_8$.

Fractional part:

$$rN_f = 8(.462) = 3 + .696 \qquad \rightarrow a_{-1} = 3$$

$$rn_{f-1} = 8(.696) = 5 + .568 \qquad \rightarrow a_{-2} = 5$$

$$rn_{f-2} = 8(.568) = 4 + .544 \qquad \rightarrow a_{-3} = 4$$

and so forth. The number $n_f = .354 \cdots _8$. So the full number is:

$$n = n_i + n_f = 1503.354 \cdots _8$$

EXAMPLE 1.11   Converting decimal to hexadecimal

$$N = 1514.462 \qquad N_i = 1514 \qquad \text{and} \qquad N_f = .462$$

Integer part:

$$N_i/r = 1514/16 = 94 + 10/16 \qquad \rightarrow a_0 = 10 \text{ (replace with A)}$$

$$q_0/r = \quad 94/16 = \ 5 + 14/16 \qquad \rightarrow a_1 = 14 \text{ (replace with E)}$$

$$q_1/r = \quad \ 5/16 = \ 0 + \ 5/16 \qquad \rightarrow a_2 = 5$$

$$q_2/r = \quad \ 0/16 = \ 0 \ \text{Stop}$$

The number $n_i = 5EA_{16}$.

Fractional part:

$$rN_f = 16(.462) = 7 + .392 \qquad \rightarrow a_{-1} = 7$$

$$rn_{f-1} = 16(.392) = 6 + .272 \qquad \rightarrow a_{-2} = 6$$

$$rn_{f-2} = 16(.272) = 4 + .352 \qquad \rightarrow a_{-3} = 4$$

and so forth. The number $n_f = .764 \cdots _{16}$. And the full number is:

$$n = n_i + n_f = 5EA.764 \cdots _{16}$$

Note: we use a larger number, 1514, instead of 835, to illustrate the need for only one digit per position.

The remainders a_0 and a_1 are the two-digit decimal numbers 10 and 14, which must be replaced by a one-digit symbol because positional notation allows for only one digit per position. The range of digits in the hexadecimal system is decimal 0 to decimal 15, and so the "digits" 10 through 15 are replaced with the symbols A, B, C, D, E, and F, respectively. (The choice of these six letters as replacements for decimal 10 through 15 is totally arbitrary). Recapitulating: hexadecimal numbers with radix $r = 16_{10}$, the digits are 0, 1, 2, 3, 4, 5, 6, 7, 8, 9, A, B, C, D, E, and F.

Intuitive proof of the conversion process: In a radix-r number system, the number

$$n = a_n a_{n-1} \cdots a_1 a_0 . a_{-1} a_{-2} \cdots a_{-m}$$

represents the polynomial

$$n = a_n r^n + a_{n-1} r^{n-1} + \cdots + a_1 r^1 + a_0 r^0 +$$
$$a_{-1} r^{-1} + a_{-2} r^{-2} + \cdots + a_{-m} r^{-m}$$

from which the value of n can be calculated. In general, n is the sum of an integer and a fraction:

$$n = n_{\text{integer}} + n_{\text{fraction}} = n_i + n_f$$

where

$$n_i = a_n r^n + a_{n-1} r^{n-1} + \cdots + a_1 r^1 + a_0 r^0$$

and

$$n_f = a_{-1} r^{-1} + a_{-2} r^{-2} + \cdots + a_{-m} r^{-m}$$

n_{fraction}   The decimal fraction N_f is converted by multiplying by r:

$$N_f = a_{-1} r^{-1} + a_{-2} r^{-2} + \cdots + a_{-m} r^{-m}$$
$$rN_f = a_{-1} + n_{f-1} = a_{-1} + (a_{-2} r^{-1} + \cdots + a_{-m} r^{-m+1})$$
$$rn_{f-1} = a_{-2} + n_{f-2} = a_{-2} + (a_{-3} r^{-1} + \cdots + a_{-m} r^{-m+2})$$
$$rn_{f-2} = a_{-3} + n_{f-3} = a_{-3} + (a_{-4} r^{-1} + \cdots + a_{-m} r^{-m+3})$$

$$\cdot$$
$$\cdot$$
$$\cdot$$

$$rn_{f-m+2} = a_{-m+1} + n_{f-m+1} = a_{-m+1} + (0 + \cdots + a_{-m} r^{-1})$$

$$\cdot$$
$$\cdot$$
$$\cdot$$

Multiplication by r results in an integer plus a fraction. The integer is a digit of the number. The first integer generated is the most significant of the fractional digits. If a zero fraction occurs, the process ends with an exact representation in radix r; if not, an infinite number of digits is required for an exact representation.

n_{integer}   Given the decimal integer N_i and the desired radix r, a conversion of the integer N_i from decimal to radix r proceeds as follows. With ordinary base-10 division of N_i by r, we generate an integer quotient and an integer remainder as part of a fraction:

$$N_i/r = q_0 + r_0/r = \text{quotient} + \text{remainder}/r$$

and then

$$N_i/r = q_0 + r_0/r = (a_n r^{n-1} + a_{n-1} r^{n-2} + \cdots + a_1 r^0) + a_0/r$$

We equate the fractions and the integers:

$$r_0/r = a_0/r$$

so that

$$r_0 = a_0$$

and

$$q_0 = (a_n r^{n-1} + a_{n-1} r^{n-2} + \cdots + a_2 r^1 + a_1 r^0)$$

Note that a_0 is the least significant digit of the given number, N_i expressed in radix r.

Division by r provides us with our initial result. Dividing the quotient q_{j-1} by r yields remainder r_j equal to the digits a_j. This is shown by the following series of equations.

$$N_i/r = q_0 + r_0/r = (a_n r^{n-1} + a_{n-1} r^{n-2} + \cdots + a_1 r^0) + a_0/r$$

$$q_0/r = q_1 + r_1/r = (a_n r^{n-2} + a_{n-1} r^{n-3} + \cdots + a_2 r^0) + a_1/r$$

$$q_1/r = q_2 + r_2/r = (a_n r^{n-3} + a_{n-1} r^{n-4} + \cdots + a_3 r^0) + a_2/r$$

$$\cdot$$
$$\cdot$$
$$\cdot$$

$$q_{n-2}/r = q_{n-1} + r_{n-1}/r = a_n r^0 + a_{n-1}/r$$

$$q_{n-1}/r = q_n \quad + r_n/r \quad = +0 \quad + a_n/r$$

Therefore we can say that the process of converting any integer number to a number of any radix r is a sequence of division operations generating remainders, which are the digits of the converted number with radix r that we seek. The first remainder generated is the least significant digit.

EXERCISE 1.2   Convert 361.3125_{10} to radix 2.

Answer: 1 0110 1001.0101$_2$ ■

1.2.3   Special Case: Converting Binary to Octal to Hex

Conversions among numbers with radices 2, 4, 8, and 16 are very easy to do, because each radix is a power of 2: $4 = 2^2$, $8 = 2^3$, and $16 = 2^4$. The following simple rules specify how to do the conversions.

Binary to octal: group
bits by threes.

Converting binary to octal/hex
Integers   Start from the radix point and move to the left, grouping
binary digits (*bits*)—by threes to convert to octal, and by fours to
convert to hexadecimal. Next, convert each group of three (four) bits
to octal (hex) by using Table 1.2.

Binary to hex: group
bits by fours.

Fractions   Start from the radix point and move to the right, grouping
binary digits—by threes to convert to octal, and by fours to convert to
hexadecimal. Next, convert each group of three (four) bits to octal
(hex) by using Table 1.2.

Consider the binary integer 1101001011_2. To convert to octal, the
rule specifies starting from the right, moving to the left, and grouping
bits by threes (because $8 = 2^3$). In upcoming Example 1.12, the three-
bit 101_2 group is six positions to the left of the radix point, meaning the
group has the weight 2^6, which equals 8^2. For this reason, group 101_2
contributes the value 5×8^2 to the number. The conversion is done in

TABLE 1.2   Counting in Number Systems of Various Radices

		RADIX		
2	3	8	10	16
0	0	0	0	0
1	1	1	1	1
10	2	2	2	2
11	10	3	3	3
100	11	4	4	4
101	12	5	5	5
110	20	6	6	6
111	21	7	7	7
1000	22	10	8	8
1001	100	11	9	9
1010	101	12	10	A
1011	102	13	11	B
1100	110	14	12	C
1101	111	15	13	D
1110	112	16	14	E
1111	120	17	15	F
10000	121	20	16	10
10001	122	21	17	11

two steps:

$$1101001011_2 = 1\ \ 101\ \ 001\ \ 011_2 = 1513_8$$

Converting octal to binary   Convert octal digits to binary by using Table 1.2. Concatenate the digits into one binary number.

Converting hexadecimal to binary   Convert hex digits to binary by using Table 1.2. Concatenate the digits into one binary number.

Converting octal to hexadecimal   Convert octal digits to binary by using Table 1.2. Regroup the digits by fours. Next, convert each group of four bits to hex by using Table 1.2.

Converting hexadecimal to octal   Convert hex digits to binary by using Table 1.2. Regroup the digits by threes. Next, convert each group of three bits to octal by using Table 1.2.

The following examples illustrate the method.

EXAMPLE 1.12   **Converting Binary to Octal**

Apply the rule for integers to convert $n = 1101001011_2$ to octal.

1. Group by threes from the right:

$$n = 1\ \ 101\ \ 001\ \ 011_2$$

2. Use Table 1.2 and positional notation:

$$n = 1513_8$$

EXAMPLE 1.13   **Converting Binary to Hexadecimal**

Convert $n = 1101001011_2$ to hexadecimal.

1. Group by fours from the right.

$$n = 11\ \ 0100\ \ 1011_2$$

2. Use Table 1.2 and positional notation.

$$n = 34B_{16}$$

EXAMPLE 1.14 **Various Conversions**

Binary to octal:

$$110100011_2 = 1 \ 101 \ 000 \ 011 = 1503_8$$

Binary to hexadecimal:

$$10111101010_2 = 101 \ 1110 \ 1010 = 5EA_{16}$$

Octal to binary:

$$1503_8 = 1 \ 5 \ 0 \ 3 = 1 \ 101 \ 000 \ 011 = 110100011_2$$

Hex to binary:

$$5EA_{16} = 5 \ E \ A = 0101 \ 1110 \ 1010 = 10111101010_2$$

Octal to hexadecimal:

$$1503_8 = 1 \ 5 \ 0 \ 3 = 001 \ 101 \ 000 \ 011$$
$$= 0011 \ 0100 \ 0011 \quad \text{(Regroup by fours)}$$
$$= 3 \ 4 \ 3 = 343_{16}$$

Hexadecimal to octal:

$$5EA_{16} = 5 \ E \ A = 0101 \ 1110 \ 1010$$
$$= 010 \ 111 \ 101 \ 010 \quad \text{(Regroup by threes)}$$
$$= 2 \ 7 \ 5 \ 2 = 2752_8$$

EXAMPLE 1.15 **Converting Fractions**

Binary to hexadecimal: group binary digits by fours from the left.

$$n = 0.0110100011_2 = 0.0110 \ 1000 \ 11 = 0.0110 \ 1000 \ 1100$$
$$= 0.6 \ 8 \ C = 0.68C_{16}$$

Binary to octal: group binary digits by threes from the left.

$$n = 0.0110100011_2 = 0.011 \ 010 \ 001 \ 1 = 0.011 \ 010 \ 001 \ 100$$
$$= 0.3214_8$$

1.2.4   Counting in Radix r

Counting rules are derived from the following observations.

1. For radix r the allowed digits range from 0 to $r - 1$.

Radix $r = 10$ for any r.

2. The radix r is represented by the two-digit number 10_r, in any number system. (Note: 10_r is read as "one-zero base r," *not* "decimal ten.") This becomes clear when you write the polynomial representation for the number 10:

$$10 = (1 \times r^1) + (0 \times r^0) = r$$

3. In radix r we count up as follows.

For $r = r$:

$0\ 1\ 2\ 3\ 4\ 5\ \cdots\ (r - 1)\ \ r\ \ (r + 1)\ (r + 2)\ \cdots\ (2r - 1)\ 2r\ \ (2r + 1)\ \cdots$

For $r = 10$:

$0\ 1\ 2\ 3\ 4\ 5\ \cdots\ \ \ \ \ 9\ 10\ 11\ \ \ \ \ 12\ \ \ \ \cdots 19\ \ \ \ \ 20\ 21\ \ \ \ \ \ \ \cdots$

For $r = 8$:

$0\ 1\ 2\ 3\ 4\ 5\ \cdots\ \ \ \ \ 7\ 10\ 11\ \ \ \ \ 12\ \ \ \ \cdots 17\ \ \ \ \ 20\ 21\ \ \ \ \ \ \ \cdots$

For $r = 2$:

$0\ 1\ \ \ \ \ \ \ \ \ \ \ \ \ \ 10\ \ \ \ \ \ \ \ \ \ \ \ \cdots 11\ \ \ \ \ \ 100\ 101\ \ \ \ \ \cdots$

This is all restated in Table 1.2.

4. Carrying is a very important concept. When any digit in a number goes from $r - 1$ to r, the digit entered is 0 rather than r and a 1 is carried over to the next digit to the left.

1.3   Traditional Arithmetic in Radix 2

In principle, arithmetic processes are independent of the radix r. The specific differences from radix to radix arise from the range of digits used by each radix.

1.3.1   Binary Addition

$1 + 1$ has sum zero and carry 1.

Binary number addition and subtraction tables are not different in principle from decimal number tables. The binary tables are easier to construct, because there are only two digits, 0 and 1. Written out, the tables are:

$$
\begin{array}{cccc@{\qquad}cccc}
0 & 1 & 0 & 1 & 0 & 1 & 0 & 1 \\
+0 & +0 & +1 & +1 & -0 & -0 & -1 & -1 \\
\hline
0 & 1 & 1 & 10 & 0 & 1 & -1 & 0
\end{array}
$$

Binary 10 is decimal 2 ($B = 1 \times 2^1 + 0 \times 2^0 = 2$). (The -1 is a problem for hardware implementation. This is discussed in Section 1.4.1.)

EXAMPLE 1.16 **Adding a Column of Binary Numbers**

Because $1 + 1 = 10$ in binary, there are many carries.

Method 1: We simply count the 1's in a column and enter that count, in binary, as the column sum. In the example, column 0 has four 1's: enter binary 100_2.

Method 2: This method simplifies the calculation of the sum of the column sums. Mark a carry-1 in the next column for every two 1's in the present column. An even number of 1's has a 0 sum and an odd number of 1's has a 1 sum. In the example, column 0 has four 1's. The sum is 0, and there are two carries. Column 1 has three 1's, with a sum of 1 and one carry, and so forth. All of the details are shown here. With experience you will do most of the steps in your head, as you now do with decimal calculations. As you'll notice, the large number of 1's to carry makes binary addition more work than decimal addition.

	Method 1			*Method 2*	
	111			111	
	101			101	
	110			110	
	001			001	
	100			100	
	+ 011			+ 011	
Sums	100	Column-0 sum		010	Column digit sums
	11	Column-1 sum		111	Carry
	100	Column-2 sum		1 1	Carry
Sum	10010			0110	Sum
Carry	1			1 1	Carry
Total	11010			10010	Sum
				1	Carry
			Total	11010	

1.3.2 Binary Subtraction

A quick refresher on decimal subtraction is in order. Subtracting 89 from 157 highlights the issue of borrowing amounts from higher-

Borrow radix r from
next higher digit.

weight digits. For example, since 9 cannot be taken away from 7, we must add something to 7. That something is the radix 10_{10}. We borrow 10_{10} because it is greater than any digit, by definition, and we borrow it from the value 50 of the next-higher digit. So 10 take away 9 is 1, to which we add 7 to obtain 8. Moving on to the next digit, we now have 4 take away 8, which needs a borrow. So we can say 10 take away 8 is 2, plus 4, equals 6.

$$
\begin{array}{ccccc}
 & & \text{Borrows} & & \\
 & 10 & & 100 & 10 \\
157 \rightarrow 140\ + & 7 \rightarrow 0\ + & 40\ + & 7 \\
-\ 89 \quad\ -80 & -9 & -80 & -9 \\
\hline
?\qquad\ ? & +\ 8 & 60\ + & 8 = 68
\end{array}
$$

We were taught to telescope this process of developing partial sums as follows.

$$
\begin{array}{rl}
10\ 10 & \text{10 is borrowed from next position} \\
-1\ -1 & \text{Because a 1 is borrowed} \\
1\ \ 5\ \ 7 & \\
-\quad 8\ \ 9 & \\
\hline
0\ \ 6\ \ 8 &
\end{array}
$$

Now we're ready to discuss binary subtraction. Returning to the binary subtraction table in Section 1.3.1, we note that the -1 result creates the need to borrow.

$$
\begin{array}{cccc}
0 & 1 & 0 & 1 \\
-0 & -0 & -1 & -1 \\
\hline
0 & 1 & -1 & 0
\end{array}
\qquad \text{and} \qquad
\begin{array}{l}
10 \quad \text{(Borrow 10 from next higher digit)} \\
-\ \ 1 \\
\hline
1 \quad (2-1=1)
\end{array}
$$

$$
\begin{array}{ccccccccccccc}
 & & & & & & & & & & & & 1 \\
 & & & & & 10 & & & & & & & 1 \\
 & & & & -1^* & & & & & & -1^* & & \\
0 & 1 & 1 & 0 \rightarrow & 0 & 1 & 1 & 0 \rightarrow & 0 & 1 & 1 & 0 \\
-0 & 1 & 0 & 1 & -0 & -1 & -0 & -1 & -0 & -1 & -0 & -1 \\
\hline
 & & & & 0 & 0 & 0 & 1 & 0 & 0 & 0 & 1
\end{array}
$$

where binary $10 = 1 + 1$. Borrowing 10 leaves a -1, which we mark with a *. Reinterpreting the borrowed 10 as $1 + 1$ makes all columns independent, thereby simplifying the subtraction process.

This simplification is more evident in a more complex problem. Starting at the right-hand *, borrows are created as needed.

1. Pair_0 borrows 10 from pair_1, creating -1^*.
2. Pair_2 needs to borrow from pair_3 but cannot. So pair_3 borrows 10 from pair_4, creating -1^*.

3. Now $pair_2$ can borrow 10 from $pair_3$, striking 10 and leaving 1, and so forth.

Notice how a borrow can itself be borrowed to become a -1^* as we move left. Then 1 can be borrowed from a 10_2 to become a 1 over the cancelled 10^* as we move left. With experience you will be able to telescope this process.

For $pair_j$, $j = 7\ 6\ 5\ 4\ 3\ 2\ 1\ 0$

$$
\begin{array}{r}
1\ \ 0\ \ 1\ \ 1\ \ 0\ \ 0\ \ 1\ \ 0 \\
-0\ \ 1\ \ 1\ \ 1\ \ 1\ \ 1\ \ 0\ \ 1 \\
\hline
\end{array}
$$

$$
\begin{array}{r}
1\ 0\ 1\ 1\ 0\ 0\ 1\ 0 \\
-0\ 1\ 1\ 1\ 1\ 1\ 0\ 1
\end{array}
\quad \text{becomes} \quad
\begin{array}{cccccccc}
 & 1 & & 10 & & 1 & & 10 \\
 & \cancel{10}^* & & & 10 & \cancel{10}^* & & & 10 \\
 & -1^* & & & -1^* & -1^* & & & -1^* \\
1 & 0 & 1 & 1 & 0 & 0 & 1 & 0 \\
\hline
-0 & -1 & -1 & -1 & -1 & -1 & -0 & -1 \\
\hline
\end{array}
$$

Converting 10_2 to $1 + 1$ yields

$$
\begin{array}{cccccccc}
 & 1 & & 1 & & 1 & & 1 & & 1 & & & 1 \\
 & & & 1 & & 1 & & & 1 & & & 1 \\
 & -1^* & & & -1^* & -1^* & & & -1^* \\
1 & 0 & 1 & 1 & 0 & 0 & 1 & 0 \\
-0 & -1 & -1 & -1 & -1 & -1 & -0 & -1 \\
\hline
0 & 0 & 1 & 1 & 0 & 1 & 0 & 1 \\
\end{array}
$$

1.3.3 Binary Multiplication

The binary multiplication table is

$$0 \times 0 = 0 \qquad 0 \times 1 = 1 \times 0 = 0 \qquad 1 \times 1 = 1$$

which in more general terms is

$$0 \times n_2 = 0 \qquad 1 \times n_2 = n_2$$

where n_2 is any binary number. The process for multiplying binary numbers is the same as we learned for multiplying with decimal numbers. In fact, it is easier, because only digits 0 and 1 are used. However, the carries will seem to go on forever as we add the partial products.

The partial products in Example 1.17 are positioned so that each has the correct weight. Each partial product's weight is increased by a factor of 2 for each one-bit-position move to the left. A hardware designer calls this a *shift* to the left or *shift-left-logical*. In hardware parlance, position moves are shifts. Left shifts are multiplication by 2, and right shifts are division by 2. (A more complete discussion is found in Section 8.3.)

EXAMPLE 1.17 Binary Multiplication

$$
\begin{array}{rl}
0110 & 6 \\
\underline{1011} & \underline{\times\ 11} \\
0110 & \\
0110 & \text{Partial product} \\
0000 & \\
\underline{0110} & \\
0111010 & \\
\text{Carries}\ \ \underline{000100} & \\
0110010 & \\
\text{Carries}\ \ \underline{001000} & \\
0100010 & \\
\text{Carries}\ \ \underline{1} & \\
1000010 & 66
\end{array}
$$

1.3.4   Binary Division

In general, the ratio $c/a = q + r/a$, where c is the numerator of the ratio, a is the denominator, q is the quotient, and r is the remainder, as in $67/11 = 6 + 1/11$. The method learned with decimal numbers is applicable here also. One difference is that we use binary subtraction in the process.

EXAMPLE 1.18 Binary Division

$$1000011/1011 = q + r/1011$$

Start from the left as if this were decimal division.

$$1011\sqrt{1000011}$$

We ask: Is 1011 less than 10? (no) Less than 100? (no) Less than 1000? (no) Less than 10000? (yes!) Enter 1 in the quotient, above the last 0 in 1000011. Then subtract 1011 from 10000 to get 0101. Bring down the next digit to get 01011, etc.

$$\begin{array}{r} 110 = q \\ 1011\overline{\smash{\big)}1000011} \\ -1011 \\ \hline 01011 \\ -1011 \\ \hline 00001 = r \end{array}$$ Because 1 is less than 1011

EXERCISE 1.3   Use elementary arithmetic to multiply 10101111_2 by 01101_2. State the answer in radix 2, radix 8, and radix 16.

Answer: $100011100011_2 = 4343_8 = 8E3_{16}$ ∎

EXERCISE 1.4   Use elementary arithmetic to divide 10101111_2 by 01101_2. Find quotient q and remainder r. State the answer in radix 2, radix 8, and radix 16.

Answer: $q = 1101_2 = 15_8 = D_{16}$ $r = 110_2 = 6_8 = 6_{16}$ ∎

1.4 Arithmetic Using Complements

We normally write positive and negative numbers in what is called *signed-magnitude format,* as, for example, in the following numbers:

+123987   +12.3987000005   −465   −5.67321000000327

$\pi = 3.141592654$ $e = 2.718281828$ $\ln 2 = 0.693147181$

Note that we can write these out with as many digits as we please.

Can we store these decimal numbers directly in a computer? Yes, if we first convert the numbers from decimal to binary format. This is necessary because modern digital computers are designed to work with binary digits, or *bits*. We have already learned that there are two binary digits, 0 and 1, whereas our experience is primarily with the ten decimal digits: 0, 1, 2, 3, 4, 5, 6, 7, 8, 9. Reminder: the digits 2 through 9 are symbols for the number of units (e.g., 3 is a symbol for three units). We have also learned how to convert positive decimal numbers to positive binary numbers. But how do we represent negative numbers?

In this section we need specific binary numbers to work with, so we will use four-bit numbers in order to avoid having large numbers of digits obscure the principles. In positional notation we have

$0000_2 = 0$ The smallest four-bit number

$1111_2 = 8 + 4 + 2 + 1 = 15_{10}$ The largest four-bit number

Four-bit digital system.

where any four-bit number B in polynomial notation is

$$B = (b_3 \times 2^3) + (b_2 \times 2^2) + (b_1 \times 2^1) + (b_0 \times 2^0)$$
$$= (b_3 \times 8) + (b_2 \times 4) + (b_1 \times 2) + (b_0 \times 1)$$

and where, in positional notation, $B = b_3 b_2 b_1 b_0$. This formula for B allows us to construct binary numbers with prescribed decimal values. Here is the list of four-digit binary numbers from 0000 to 1111.

BINARY	DECIMAL	BINARY	DECIMAL	BINARY	DECIMAL
0000	0	0110	6	1011	11
0001	1	0111	7	1100	12
0010	2	1000	8	1101	13
0011	3	1001	9	1110	14
0100	4	1010	10	1111	15
0101	5				

1.4.1 Discovering Negative-Number Representations

We start by recognizing that as a practical matter negative-number representation must be easily implemented by computer hardware. With this in mind let us proceed.

If the digits of a number represent only the absolute value, then we need an extra "digit" for the sign, be it plus or minus. In the past, hardware designers have chosen not to allocate an extra digit for sign (because the arithmetic is too complex). In order to avoid having to carry the plus or minus sign around with a number as it is processed in a digital system, we seek a number format that is *not* a signed-magnitude representation such as -43.

If $5 + a = 0$ in decimal, then how could we represent a if we do not represent it as -5? And in binary, if $0101 + b = 0$, then how could we represent b if we do not represent it as -0101? Let's look for help by borrowing some tips from decimal arithmetic. The first time you were asked to perform a subtraction such as $100 - 17$, you were probably advised by your teacher to notice that $100 = 99 + 1$. Thus,

$$99 - 17 = 82 \quad \text{and} \quad 82 + 1 = 83 \quad \text{so} \quad 100 - 17 = 83$$

Observe that subtraction from 99 does not use borrows. This is because 9 is the greatest digit. You need only subtract individual digit pairs:

$$9 - 7 = 2 \quad \text{and} \quad 9 - 1 = 8$$

If $x + y = 9$ then x is nine's complement of y.

We call 82 the 9's complement of 17 because

$$8 + 1 = 9 \qquad \text{and} \qquad 2 + 7 = 9$$

A *complement* is the number of units that must be added to a digit to make it equal to the *diminished radix* $r - 1$ (see upcoming Section 1.4.3). That is, two digits x and y are complements of each other when $x + y = r - 1$. Thus the digits 1 and 8 are complements of each other because $1 + 8 = 9 = 10 - 1$.

If we pretend that our "decimal" computer can only handle numbers of up to two decimal digits, then $17 + 83 = 00$, not 100, because hardware for the third digit does not exist. In effect, the most significant digit is automatically dropped. Immediately we can say $83 = 00 - 17 = -17$. Therefore, in this hardware system the digits 83 represent -17, not $+83$, as you might think. We have shown that one way to calculate the negative of 17 is to form the 9's complement and add 1. This process is called forming the 10's complement. It is also known as the *radix complement*.

Ten's complement is nine's complement plus 1.

The advantage in a hardware system's use of 10's complement is accompanied by a disadvantage: the range of numbers from 00 to 49 now corresponds to $+00$ to $+49$, and 50 to 99 corresponds to -50 to -1. What happened to $+83$?

A better question is: how are numbers with magnitude greater than 49 or 50 represented? The brief answer is: with more digits. For example, in a three-digit hardware system, the range 000 to 499 corresponds to 10's complement numbers $+000$ to $+499$, and the range 500 to 999 corresponds to 10's complement numbers -500 to -1.

Nine is the greatest decimal digit, but 1 is the greatest binary digit. By analogy, let us calculate the negative of binary 0101 by forming the 1's complement and adding 1. This process is called forming the 2's complement. First, it should be pointed out that in any radix r, complements are formed by subtracting digits individually from $r - 1$. Thus, the 9's complement is formed by subtracting a decimal digit from 9, which is 1 less than the radix 10. And the 1's complement is formed by subtracting a binary digit from 1, which is 1 less than the radix 2. This creates a special case: the 1's complement is formed by complementing the binary number's bits.

Two's complement is one's complement plus 1.

Now let us see if this 2's complement works.

1's comp	*Add 1*		
1111	1010	or	0101 (+5)
-0101	$+\ \ 1$		$+1011$ (−5?)
1010	1011		10000

After dropping the fifth digit (because it is not implemented in hardware), the result is as follows (note that any carry resulting from the addition is discarded).

$$0101 + 1011 = 0000$$

$$1011 = 0000 - 0101 = -0101 = -5_{10}$$

That is to say, $1011 = -0101$: 1011 represents -5_{10} in four-bit binary. Two's complements work in this case, and we claim they work in all cases. Let us calculate the 2's complement of several numbers directly.

If $b = -7_{10} = -0111$, what is b in 2's complement?

1's comp	*Add 1*
1111	1000
-0111	$+\quad 1$
1000	1001

So $b = -7_{10} = -0111 = 1001$.

Check
$$\begin{array}{ll} 0111 & (+7) \\ +1001 & (-7?) \\ \hline 10000 \end{array}$$

Therefore -7_{10} is represented by 1001.

If $b = -1_{10} = -0001$, what is it in 2's complement?

1's comp	*Add 1*
1111	1110
-0001	$+\quad 1$
1110	1111

So $b = -1_{10} = -0001 = 1111$.

Check
$$\begin{array}{ll} 0001 & (+1) \\ +1111 & (-1?) \\ \hline 10000 \end{array}$$

Therefore 1111 represents -1_{10}.

If $b = -8_{10} = -1000$, what is it in 2's complement?

1's comp	*Add 1*
1111	0111
-1000	$+\quad 1$
0111	1000

So $b = -1000 = 1000$.

Check
$$1000 \quad (-8)$$
$$\underline{+1000} \quad (+8?)$$
$$10000$$

And 1000 represents -8_{10} or $+8_{10}$. (In what follows the $+8_{10}$ representation is discarded.)

In this way we build a list of 2's complements. Observe that 0000 and 1000 are their own complements. The 2's complements of 0000 and 1000 are 0000 and 1000, respectively. This is not a problem from a hardware point of view because the positive and negative numbers separate into two distinct groups as follows:

Signed numbers: msb is sign bit.

1. All of the positive numbers have 0 as the most significant bit.
2. All negatives of these positive numbers have 1 as the most significant bit.

The decimal interpretation of 2's complement numbers is consistent with the process generating 10's complements. The (binary) range 0_{10} to 7_{10} corresponds to the decimal range $+0$ to $+7$, and 8_{10} to 15_{10} corresponds to -8 to -1.

2's Complement Numbers

BINARY NUMBER	2'S COMP	DECIMAL INTERPRETATION OF 2'S COMP
0000	0000	+0
0001	1111	−1
0010	1110	−2
0011	1101	−3
0100	1100	−4
0101	1011	−5
0110	1010	−6
0111	1001	−7
1000	1000	−8
1001	0111	+7
1010	0110	+6
1011	0101	+5
1100	0100	+4
1101	0011	+3
1110	0010	+2
1111	0001	+1

Signed and Unsigned Four-Bit Numbers

NUMBER	SIGNED	UNSIGNED	NUMBER	SIGNED	UNSIGNED
0000	0	0 hex	1000	-8	8
0001	1	1	1001	-7	9
0010	2	2	1010	-6	A
0011	3	3	1011	-5	B
0100	4	4	1100	-4	C
0101	5	5	1101	-3	D
0110	6	6	1110	-2	E
0111	7	7	1111	-1	F

Given that the most significant bit, MSB, is 0 for positive numbers and 1 for negative numbers, there is a temptation to call the MSB the sign bit (which is fine) *and* to segregate it from the other bits (which is a mistake). Observe that any binary number, such as 1010, may be interpreted as a signed number (equal to -6 in this example) or as an unsigned number [equal to A ($+10$) in this example]. The interpretation given to any number depends on the context.

Computer hardware, however, cannot make interpretations. The good news is that the hardware does not have to make the interpretations. (Upcoming sections 1.4.4 and 1.4.5 show why this is so.) The beauty of 2's complement representation is that additional hardware is not required to treat a sign bit, as is required in the signed-magnitude case.

The magnitude range of the signed numbers (-8 to $+7$) is about half the range of the unsigned numbers (0 to 15). The previously displayed table of signed numbers is the 2's complement interpretation of four-bit binary numbers. Two's complements of larger numbers are calculated in the same way. For example, with a word length of eight bits,

$$5_{10} = 00000101 \quad \text{and} \quad -5_{10} = 11111010 + 1 = 11111011$$

Sign extend by repeating msb.

As word length increases, leading 1's are added to a negative-number representation. This process is called *sign extension*. Positive numbers are sign-extended by adding leading 0's.

WORD LENGTH (BITS)	-5_{10}	$+5_{10}$
4	1011	0101
5	11011	00101
8	11111011	00000101
16	1111111111111011	0000000000000101

To check this, calculate $-(-5_{10}) = -11111111111111011_2$. The 2's complement of the 2's complement is the original positive number.

-5_{10} 111111111111011

1's comp 0000000000000100

Add 1 $\underline{\qquad\qquad +1}$

0000000000000101 = 101 = $+5_{10}$ Q.E.D.

The eight-bit table, abbreviated, is as follows.

NUMBER	SIGNED	UNSIGNED
0000 0000	0 decimal	0 decimal
0000 0001	1	1
.	.	.
.	.	.
.	.	.
0111 1111	127	127
1000 0000	-128	128
1000 0001	-127	129
.	.	.
.	.	.
.	.	.
1111 1111	-1	255

and so on for larger and larger numbers.

Rules are simple.

Rules to form complements The rules for forming complements are rather straightforward because there are only two digits in the binary number system.

To form 1's complement, change all 1's to 0's and all 0's to 1's.

To form 2's complement, form 1's complement and add 1.

1.4.2 Radix Complement

We have discussed 10's complement and 2's complement, which are examples of radix complement for which $r = 10$ and $r = 2$, respectively. Now we generalize complements to any radix r. For n-digit numbers of radix r, the radix r complement z^* can be defined as

$$z^* = -z = r^n - z \quad \text{(Radix complement of } z\text{)}$$

Key idea: use n digit
numbers, not $n + 1$.

This equation means $z^* = -z$ in signed radix format, such as 2's complement.

The hidden logic here is that the number r^n has $n + 1$ digits, whereas all z numbers have n digits. However, we recognize that the number $r^n - 1$ has n digits! [Note: $r^n - z = (r^n - 1) - z + 1$.] Now we do not need to work with $n + 1$ digit numbers. All numbers now have n digits.

The key to simplification of subtraction is working only with n-digit numbers. In order to do this we rewrite the equation for z^* as

$$z^* = r^n - z$$
$$z^* = (r^n - 1) - z + 1$$
$$z^* = [(r^n - 1) - z] + 1$$
$$z^* = z^{\sim} + 1$$

In words: The radix complement of z is z^*, and the diminished, or radix minus 1, complement is $z^{\sim}$. When $r = 2$, we find z^* is the 2's complement and $z^{\sim}$ is the 1's complement.

$z^{\sim}$ is called the *diminished radix complement of z*. The virtue of $z^{\sim}$ is its simple relationship to z: simply complement the digits of z as individuals. Subtract, digit by digit, each pair individually, without borrows between the pairs. Thus, our 2's complement rule is generalized to read as follows.

> Rule: the radix complement of a number is formed by complementing the individual digits and adding 1.

EXAMPLE 1.19 **Using the Diminished Radix Complement**

$r = 2$ and $n = 5$

$$r^n = 2^5 = 32_{10}$$
$$= (1 \times 2^5) + (0 \times 2^4) + (0 \times 2^3) + (0 \times 2^2)$$
$$+ (0 \times 2^1) + (0 \times 2^0)$$
$$= 100000_2 \quad (r^n \text{ has six digits})$$

And

$$r^n - 1 = 100000 - 1 = 11111$$

which has $n = 5$ digits.

This is why subtraction is truly assisted when, in decimal, you say, for example,

$$r^3 = 1{,}000 = 999 + 1 = (r^3 - 1) + 1$$

EXAMPLE 1.20

Given two n-digit numbers y and z, calculate $y - z$.

$$y - z = y + (-z) = y + z^* = y + z^{\sim} + 1$$

Let $r = 2$ and $n = 8$ digits. The range of numbers is binary 00000000 to 11111111. In this 2's complement system, which numbers in the range represent negative numbers?

GIVEN:	CALCULATE:
$z = 00000000$	$-z = -00000000 = z^{\sim} + 1 = 11111111 + 1 = 00000000$
$z = 00000001$	$-z = -00000001 = z^{\sim} + 1 = 11111110 + 1 = 11111111$
$z = 01111111$	$-z = -01111111 = z^{\sim} + 1 = 10000000 + 1 = 10000001$
$z = 10000000$	$-z = -10000000 = z^{\sim} + 1 = 01111111 + 1 = 10000000$
$z = 10000001$	$-z = -10000001 = z^{\sim} + 1 = 01111110 + 1 = 01111111$
$z = 11111111$	$-z = -11111111 = z^{\sim} + 1 = 00000000 + 1 = 00000001$

Reproducing the results in a table, we have

z	$-z$	z_{10}	$-z_{10}$
00000000	00000000	$+0$	$+0$
00000001	11111111	$+1$	-1
01111111	10000001	$+127$	-127
10000000	10000000	-128	-128
10000001	01111111	-127	$+127$
11111111	00000001	-1	$+1$

From the table we can say: $-(-z) = z$, except for 10000000 and 00000000 (these two are the odd men out). There are 127 numbers in the range from $+1$ to $+127$ and also in the range -1 to -127. The other two numbers are 10000000 and 00000000, which are their own complements.

All z whose leading digits are 0 are positive numbers. All the z whose leading digits are 1 may be interpreted and used as negative numbers.

EXAMPLE 1.21   Calculating the Radix Complement of Signed and Unsigned Numbers

Let $z = 10110101_2$.

$$-z = -10110101 = z^{\sim} + 1 = (11111111 - 10110101) + 1$$
$$= 01001010 + 1$$
$$= 01001011$$

which is $+75_{10}$. Therefore, $z = 10110101 = -75_{10}$.

Interpretation of the numbers z and $-z$ can be signed or unsigned. Do not let the $-z$ designator mislead you.

$$\text{Unsigned } z = \quad 10110101 = 181_{10}$$
$$-z = \quad 01001011 = 75_{10}$$
$$z + (-z) = 100000000 = 256_{10}$$
$$\text{Signed } z = \quad 10110101 = -75_{10}$$
$$-z = \quad 01001011 = +75_{10}$$
$$z + (-z) = \quad 00000000 = 0_{10} \quad \text{(9th digit, the carry, is dropped)}$$

Start from $z = +75_{10}$ this time.

$$\text{Let } z = 01001011$$
$$-z = -01001011 = 10110100 + 1 = 10110101$$
$$\text{Signed } z = 01001011 = 75_{10}$$
$$-z = 10110101 = -75_{10}$$
$$z + (-z) = 00000000 = 0_{10}$$

The process is reversible.

EXERCISE 1.5   Find the 10's complement of $123{,}456{,}789_{10}$.

Answer: $876{,}543{,}211_{10}$ ■

EXERCISE 1.6   Find the 2's complement of 1011011110100011_2. State the answer in radix 2, radix 8, and radix 16.

Answer: $0100100001011101_2 = 44135_8 = 485D_{16}$ ∎

1.4.3 Diminished Radix Complement

The diminished radix complement was defined and used in the prior section; it is the $(r-1)$ complement $z\tilde{\ }$, which is formed by complementing individual digits. In contrast, the r complement is z^*. Since $z^* = z\tilde{\ } + 1$, we add 1 to the $(r-1)$ complement to get the r complement.

EXAMPLE 1.22

The $(r-1)$ complement of $0,123,456,789_{10}$ is $9,876,543,210$. This is the 9's complement in radix 10 because $r - 1 = 9$.

The $(r-1)$ complement of 01234567_8 is 76543210. This is the 7's complement in radix 8 because $r - 1 = 7$.

1.4.4 Overflow and Underflow

Overflow is an important consequence of the limitation on the number of digits. The addition of two n-bit numbers can result in an $(n + 1)$-bit number. The situation is referred to as *overflow* when the $(n + 1)$-bit result is positive, and *underflow* when the result is negative. In practice, the over/under distinction is not important. (You will find overflow and underflow in the literature, which is why we mention both.)

The range of our four-digit binary signed number system is -8_{10} to $+7_{10}$. When you add $7_{10} + 5_{10}$, the sum, 12_{10}, is outside the range, and we say we have an overflow outside the range. Overflow cannot occur when you add numbers of different sign. Consider the following scenario.

Overflow is a $n + 1$ digit result in an n digit system.

The result of adding two numbers of the same sign can have more digits than the four implemented digits in our example. In this situation a positive result is said to overflow and a negative result underflows. For example,

Carries

	0111			1011
$+5_{10}$	0101		-5_{10}	1011
$+7_{10}$	0111		-7_{10}	1001
$+12_{10}$	01100 Overflow		-12_{10}	10100 Underflow

The sums are in error because in the computer the fifth digit is dropped, leaving the incorrect answers 1100 and 0100, which equal -4_{10} and 4_{10}, respectively.

One way to represent overflow and underflow graphically is on a number line (the line is not to scale):

-12	-8	-7	-5	0	$+5$	$+7$	$+12$

$\leftarrow$Underflow$\dashv$ $\vdash$Overflow$\rightarrow$

In what follows we mention only overflow as we focus on the most significant bits (MSBs). Concentration on the MSBs leads to a way to detect overflow. Overflow is possible when both numbers are positive (MSBs are 0) or both are negative (MSBs are 1). Overflow has occurred when the MSB of the result differs from the MSBs of the two numbers added. Let a_m, b_m, f_m be binary variables representing the three MSBs. For the sake of completeness we show the following overflow equation, where $'$ is the NOT operator. (The NOT operator and this Boolean logic equation are explained in Chapter 3.)

$$V = f'_m b_m a_m + f_m b'_m a'_m$$

1.4.5 Multiple-Precision Arithmetic

Avoiding overflow requires multiple-precision arithmetic. The trick to avoiding overflow with any *pair* of in-range numbers is to double the word length used but not to extend the range of the original number set. In our four-bit-word computer, a double-width number is stored in two words: the least significant word (LSW) and the most significant word (MSW). We let the LSWs hold the original numbers, such as the 5 and the 7 discussed earlier. When 5 or 7 is positive (negative), we start out with 0000 (1111) in the MSW. This process is called *sign extension*.

Let us add 7 and 5 using double-precision arithmetic, which is another way to express the use of double-width numbers. The key to this process is adding the carry from the LSW sum to the MSW sum.

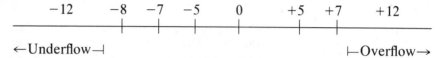

$+5_{10}$	0000 0101		-5_{10}	1111 1011
$+7_{10}$	0000 0111		-7_{10}	1111 1001
$+12_{10}$	0000 1100		-12_{10}	11111 0100

Now the MSB of the MSW, not the MSB of the LSW, determines the sign of the result. And so we have yet another interpretation of bits stored in a digital system. However, note that the digital system's adder adds according to the addition table

$$
\begin{array}{cccc}
0 & 1 & 0 & 1 \\
\underline{+0} & \underline{+0} & \underline{+1} & \underline{+1} \\
00 & 01 & 01 & 10
\end{array}
$$
(carry 0 or carry 1 and sum)

independent of the ultimate interpretation of the result. The digital system hardware does not interpret the binary numbers: we do.

Double precision is an 8-digit form of a 4-digit system.

We can extend the precision as much as needed to overcome the number-of-digits limitation imposed by using only single word lengths. Add-with-carry is used for all but LSB addition.

Perhaps further comment on not extending the range of the original number set is in order. The range of the double-precision eight-bit numbers is $+127$ to -128. If we use the full range, we can overflow for the same reasons we overflow with four-bit numbers. If we limit the range to six-bit numbers, overflow is not possible. How do we limit the range? The answer is "with difficulty," because predicting what numbers will arise in a calculation is almost impossible. By not extending the range beyond the four-bit range, our program will not overflow when double precision is used unless we are adding a long column of numbers, for example. Then the program may create numbers exceeding a double-precision range. In that case we use triple precision, etc.

1.5 Codes

Codes are used to represent events such as pressing the z key on a keyboard, encoding a shaft position, or an error in a memory-read operation. One way or another, the event information is converted, i.e., encoded, into binary for processing by a digital system. How this is done is not our concern here. Our concern is the result: the codes used to represent information.

> A code is a human interpretation of binary bit patterns. The interpretation given to any binary bit pattern depends on the context. The hardware cannot make interpretations.

1.5.1 Binary Encodes Decimal

From one point of view, we encode decimal numbers when we convert them to binary. For example, the five-digit decimal number 65,536 converts to the 17-digit binary number

$$10000000000000000$$

The practice of using commas to enhance perception of large decimal numbers is *not* applied to binary numbers. We suggest you simply insert spaces every four digits: for example, 1 0000 0000 0000 0000. Or convert to hex and move on; however, keep in mind that numbers in digital hardware are in binary.

> Reminder: given any unsigned binary number, such as 110101, we can add any number of leading zeros, such as
>
> $$0000000110101$$
>
> without changing the value.

1.5.2   BCD Numbers

BCD is decimal in binary form.

Decimal 53 encoded in binary is 110101_2. The digits 5 and 3 lose their identity when encoded: nothing in 110101_2 corresponds to the 5 or the 3. There are advantages to keeping the identity of decimal digits after encoding, such as business accounting executed in decimal with no errors and decimal numerical displays. This brings us to *binary-coded decimal* code, known as *BCD*.

BCD code is straightforward: single digits 0 through 9 are simply converted to four-bit binary. A multidigit number such as 835 is converted into a bunch of four-digit numbers.

$$835 = 1000\ 0011\ 0101 = 100000110101$$

DECIMAL	BCD
0	0000
1	0001
2	0010
3	0011
4	0100
5	0101
6	0110
7	0111
8	1000
9	1001

However, four-bit binary also represents decimal 10 to 15. This is a complication when we do arithmetic with BCD numbers, as shown shortly. (An intuitive proof for BCD addition is found at the end of this section.)

In any specific digital system, all numbers are considered to have the same number of digits. In other words, leading 0's are made explicit. The same is true of BCD numbers. In a four-digit BCD number system, 53 is represented by four separate bit groups:

$$0000 \quad 0000 \quad 0101 \quad 0011$$

This is *not* binary 0000000001010011 with spaces every four bits. Binary

$$0000 \quad 0000 \quad 0101 \quad 0011$$

equals 83_{10}, not 53. Do not allow the clarifying spaces to confuse you.

BCD addition of any two digits has two types of sums, because the sum of any two digits can range from 0 (0 + 0) to 18 (9 + 9).

1. If the sum is in the range 0 to 9, then the result is in BCD format and no action is required.

2. If the sum is in the range 10 to 18, then the sum is not in BCD format. The rule is: add 6 to restore BCD format.

For example: In BCD format, 7 plus 5 is

$$0111 + 0101 = 1100$$

This is 12_{10} encoded in binary, which is correct in binary; however, 12_{10} in BCD has the code 0001 0010. The situation is as follows: we have the binary sum $s1 = 00001100$ and we want the BCD sum $s2 = 0001\ 0010$. In this case we can convert $s1$ to $s2$ using binary arithmetic by adding 10000_2 (16_{10}), which is ten in BCD (0001 0000) and subtracting ten (1010_2). Adding 6 is equivalent to adding 16_{10} and subtracting 10_{10}:

> Add 6 to adjust sums greater than 9.

$$
\begin{array}{r}
1100 \\
\text{Add 16} \quad +10000 \\
\hline
11100 \\
\text{Subtract 10} \quad +10110 \quad \text{(five-digit 2's complement)} \\
\hline
110010
\end{array}
$$

Now we drop out the carry to get 10010, which becomes 0001 0010 in BCD as desired. We then combine the two operations by noting that $16 - 10 = 6$:

$$
\begin{array}{r}
1100 \\
\text{Add 6} \quad \underline{0110} \\
10010 \quad \text{Interpret as 0001 0010 BCD}
\end{array}
$$

Example: Let us add the two numbers 0985_{10} to $1,842_{10}$ ($= 2,827_{10}$) using BCD format. The digit sums in decimal are

$$
\begin{array}{rrrr}
0 & 9 & 8 & 5 \\
1 & 8 & 4 & 2 \\
\hline
1 & 17 & 12 & 7
\end{array}
$$

In BCD format we have

	0000	1001	1000	0101
	+0001	1000	0100	0010
	0001	10001	1100	0111
Add 6		0110	0110	
	0001	0111	0010	0111
Carry	1	1		
	0010	1000	0010	0111 $= 2827_{10}$

Conversion from BCD to binary and from binary to BCD are done with paper and pencil by using decimal as an intermediary. In hardware, direct conversions are done (these conversions are outside the scope of this discussion).

The following is an intuitive proof that adding 6 is a general solution when a BCD digit x is greater than 9 after addition. First, consider the example when the BCD sum of two BCD digits is 18. The unadjusted sum S is in binary. After we group the bits by fours,

$$S = 10010 = 0001 \ 0010$$

In binary the group 0001 has weight 16 and the group 0010 has weight 1. However, in BCD the 0001 group would have weight ten and the 0010 group would have weight 1. The binary 0010 group needs adjustment to 1000 (8) before it can represent the BCD digit 8. Observe that for sums greater than 9, the binary sum S ranges from 01010 to 10010.

$$
\begin{aligned}
S &= x \ \text{(Binary number)} \\
&= x_1 x_2 \ \text{(Converted to BCD format)} \\
&= (x_1 \times 10000) + (x_2 \times 1) \\
&= (x_1 \times 1010) + (x_1 \times 0110) + (x_2 \times 1) \\
&= (x_1 \times 1010) + [(0110 \times x_1) + x_2] \times 1 \\
&= (x_1 \times 10_{10}) + [(6_{10} \times x_1) + x_2] \times 1
\end{aligned}
$$

We interpret $x_1 x_2$ as the tens and ones digits of the sum in BCD format.

If the sum x is less than ten then $x_1 = 0$, in which case the digits of the BCD number S are 0 and x_2.

If the sum x is greater than 9, then $x_1 = 1$, and the digits of the BCD number S are 1 and $6 + x_2$.

EXERCISE 1.7   Convert $6,789_{10}$ and $3,211_{10}$ to BCD and add them. State the answer both in BCD and in radix 10.

Answer: 0001 0000 0000 0000 0000 = $10,000_{10}$ ∎

EXERCISE 1.8   Convert $6,189_{10}$ and $5,237_{10}$ to BCD and add them. State the answer both in BCD and in radix 10.

Answer: 0001 0001 0100 0010 0110 = $11,426_{10}$ ∎

1.5.3 ASCII Characters and Operators

ASCII is the acronym for American Standard Code for Information Interchange. This seven-bit code, $d_6 d_5 d_4 d_3 d_2 d_1 d_0$, is used throughout the world to encode alphanumeric characters into binary patterns. One useful point of view is the following: The three high bits, $d_6 d_5 d_4$, define eight subgroups, and the four low bits, $d_3 d_2 d_1 d_0$, define 16 elements of each subgroup.

$d_6 d_5 d_4$	$d_3 d_2 d_1 d_0$ SUBGROUP FUNCTION
000	Nonprinting control codes
001	Nonprinting control codes
010	Printing punctuation marks
011	Digits 0 to 9 and printing punctuation marks
100	Capital letters and @ mark
101	More capital letters and punctuation marks
110	Lowercase letters and punctuation marks
111	More lowercase letters and punctuation marks

The nonprinting, or noncharacter, codes—00 to 1F (hex)—are control function codes generated by typing, respectively,

$$\hat{} @, \hat{} A, \hat{} B, \ldots , \hat{} Z, \hat{} [, \hat{} \backslash, \hat{}], \hat{}\hat{}, \hat{}_-$$

They do not print symbols. They implement control actions. A discussion of these actions is beyond the scope of this discussion. [The symbol $\hat{} B$, for example, means press the control (CTRL) key and

TABLE 1.3   **ASCII Characters and Operators**

Let the two-character ASCII code number (left-hand entry in each double column) = xy, where $x = d_6 d_5 d_4$ and $y = d_3 d_2 d_1 d_0$.

00	NUL	10	DLE	20	SP	30	0	40	@	50	P	60	`	70	p
01	SOH	11	DC1	21	!	31	1	41	A	51	Q	61	a	71	q
02	STX	12	DC2	22	"	32	2	42	B	52	R	62	b	72	r
03	ETX	13	DC3	23	#	33	3	43	C	53	S	63	c	73	s
04	EOT	14	DC4	24	$	34	4	44	D	54	T	64	d	74	t
05	ENQ	15	NAK	25	%	35	5	45	E	55	U	65	e	75	u
06	ACK	16	SYN	26	&	36	6	46	F	56	V	66	f	76	v
07	BEL	17	ETB	27	'	37	7	47	G	57	W	67	g	77	w
08	BS	18	CAN	28	(	38	8	48	H	58	X	68	h	78	x
09	HT	19	EM	29	)	39	9	49	I	59	Y	69	i	79	y
0A	LF	1A	SUB	2A	*	3A	:	4A	J	5A	Z	6A	j	7A	z
0B	VT	1B	ESC	2B	+	3B	;	4B	K	5B	[	6B	k	7B	{
0C	FF	1C	FS	2C	,	3C	<	4C	L	5C	\	6C	l	7C	|
0D	CR	1D	GS	2D	−	3D	=	4D	M	5D	]	6D	m	7D	}
0E	SO	1E	RS	2E	.	3E	>	4E	N	5E	^	6E	n	7E	~
0F	SI	1F	US	2F	/	3F	?	4F	O	5F	_	6F	o	7F	DEL

Meaning of Control Codes

00	NUL	Null	10	DLE	Data Link Escape
01	SOH	Start of Heading	11	DC1	Device Control 1
02	STX	Start of Text	12	DC2	Device Control 2
03	ETX	End of Text	13	DC3	Device Control 3
04	EOT	End of Transmission	14	DC4	Device Control 4
05	ENQ	Enquiry	15	NAK	Negative Acknowledge
06	ACK	Acknowledge	16	SYN	Synchronize
07	BEL	Bell	17	ETB	End Transmitted Block
08	BS	Back Space	18	CAN	Cancel
09	HT	Horizontal Tab	19	EM	End of Medium
0A	LF	Line Feed	1A	SUB	Substitute
0B	VT	Vertical Tab	1B	ESC	Escape
0C	FF	Form Feed	1C	FS	File Separator
0D	CR	Carriage Return	1D	GS	Group Separator
0E	SO	Shift Out	1E	RS	Record Separator
0F	SI	Shift In	1F	US	Unit Separator
			20	SP	Space
			7F	DEL	Delete or rubout

hold it down, press the B key, release the B key, release the control key.]

Memory is byte oriented, so the message

Hello there!

is stored as (hex)

48 65 6C 6C 6F 20 74 68 65 72 65 21

where 20_{16} represents the space "character."

There are many ramifications to ASCII, and 20_{16} (hex) for space is one of them. Others are the nonprinting codes. Some may be considered operators controlling printers, others as operators providing means for interaction in communication protocols. These are the control codes listed in Table 1.3. These and other topics are outside the scope of this discussion.

EXERCISE 1.9

Write in ASCII "We learn by doing exercises."

Answer: 57 65 20 6C 65 61 72 6E 20 62 79 20 64 6F 69 6E 67 20 65 78 65 72 63 69 73 65 73 2E ∎

EXERCISE 1.10

Convert the number 11101_2 to an ASCII character string that prints the number in decimal.

Answer: 32 39 ∎

1.5.4 Other Codes

One useful way to characterize different codes is to specify the distance between code elements. Simply put, the distance between any two elements is the number of corresponding pairs of bits in the two elements that are different. For instance, the distance from 010 to 101 is 3; the distance from 010 to 011 is 1.

Gray code: The unit-distance Gray code is useful in encoding mechanical position into binary patterns. One-bit changes at code-change boundaries allow for more tolerance of mechanical imperfections.

DECIMAL	3-BIT GRAY CODE
0	000
1	001
2	011
3	010

DECIMAL	3-BIT GRAY CODE
4	110
5	111
6	101
7	100

One-out-of-n codes: A one-out-of-n code has 1 (0) bit, with the remaining bits equal to 0 (1). The code's virtue lies in the ease of decoding. The 10-bit code can encode the 10 decimal digits.

10-BIT 1-OUT-OF-n CODES

0000000001	1111111110
0000000010	1111111101
0000000100	1111111011
0000001000	1111110111
0000010000	1111101111
0000100000	1111011111
0001000000	1110111111
0010000000	1101111111
0100000000	1011111111
1000000000	0111111111

Excess-3 code: BCD code variations are numerous. One variation is the excess-3 code, which has the self 9's complementing property. Complement 0101 (2) to get 1010 (7). Digits 2 and 7 are 9's complements.

DECIMAL	EXCESS-3 CODE
0	0011
1	0100
2	0101
3	0110
4	0111
5	1000
6	1001
7	1010
8	1011
9	1100

Error detection and correction codes Error detection and error correction are achieved by adding extra bits to whatever code we are using. (Ascertaining what bits to add is a fascinating and very complex subject that is beyond the scope of this discussion.) One simple error-detection code is based on the idea of odd or even parity. Suppose you are using the ASCII code. The letter G in ASCII is 1000111, which has four 1 bits. An even-parity scheme adds a 0 bit so that the result has an even number of 1 bits: the code for G becomes the eight-bit word 01000111. Odd parity means the result has an odd number of 1 bits: the code for G becomes 11000111. (Further discussion is found in Section 5.6.)

SUMMARY

Position and Radix
The numbers we use every day are written in positional notation. A number represents a polynomial in the number's radix r. For instance, the binary number 1011.101_2 represents the polynomial

$$(1 \times 2^3) + (0 \times 2^2) + (1 \times 2^1) + (1 \times 2^0)$$
$$+ (1 \times 2^{-1}) + (0 \times 2^{-2}) + (1 \times 2^{-3})$$

which is composed of an integer part, $1011 = (1 \times 2^3) + (0 \times 2^2) + (1 \times 2^1) + (1 \times 2^0)$, and a fractional part, $.101 = (1 \times 2^{-1}) + (0 \times 2^{-2}) + (1 \times 2^{-3})$. The polynomials are written in radix-10 numbers.

Radix Conversion
The binary number 1101.101 is converted to a radix-10 (decimal) number by writing the polynomial in radix-10 (decimal) numbers and summing the terms. The integer

$$1101 = (1 \times 2^3) + (1 \times 2^2) + (0 \times 2^1) + (1 \times 2^0)$$
$$= 8 + 4 + 0 + 1 = 13_{10}$$

and the fraction

$$101 = (1 \times 2^{-1}) + (0 \times 2^{-2}) + (1 \times 2^{-3})$$
$$= \tfrac{1}{2} + 0 + \tfrac{1}{8} = 0.625_{10}$$

The decimal integer number 13 is converted to binary integer number 1101_2 through a sequence of divisions by 2 until the quotient is 0.

$$13 = (6 \times 2) + 1 \qquad \text{13/2 has quotient 6 and remainder 1}$$
$$6 = (3 \times 2) + 0 \qquad \text{6/2 has quotient 3 and remainder 0}$$

$$3 = (1 \times 2) + 1 \qquad \text{3/2 has quotient 1 and remainder 1}$$
$$1 = (0 \times 2) + 1 \qquad \text{1/2 has quotient 0 and remainder 1}$$

Traditional Arithmetic in Radix 2

The methods of addition, subtraction, multiplication, and division are independent of the radix. The tables used in each operation differ in the range of digits used.

Arithmetic Using Complements

The negative of a binary number is represented by a 2's complement number. The rules for forming 2's complement numbers are straightforward:

To form the 1's complement, change all 1's to 0's and all 0's to 1's.

To form the 2's complement, form the 1's complement and add 1.

Codes

BCD (binary-coded decimal) numbers maintain the identity of decimal digits. The number 985_{10} becomes the three-"BCD digit" BCD number 1001 1000 0101. BCD addition proceeds by adding BCD digit by BCD digit, as in decimal arithmetic. When the sum of any digit pair exceeds 9, add 6 to that sum to generate the sum BCD digit and carry to the next-higher digit pair.

REFERENCES

Arazi, B. 1988. *A commonsense approach to the theory of error-correcting codes.* Cambridge, Mass.: MIT Press.

Cavanagh, J. J. 1984. *Digital computer arithmetic.* New York: McGraw-Hill.

Couleur, J. F. 1958. BIDEC—A binary-to-decimal or decimal-to-binary converter, *IEEE Transactions Electronic Computers, EC-7:* 313–316.

Flores, I. 1963. *The logic of computer arithmetic.* Englewood Cliffs, N.J.: Prentice Hall.

Knuth, D. E. 1969. *The art of computer programming: seminumerical algorithms.* Reading, Mass.: Addison-Wesley.

Peterson, W. W., and E. J. Weldon. 1972. *Error-correcting codes.* 2nd ed. Cambridge, Mass.: MIT Press.

Sweeney, P. 1991. *Error control coding: An introduction.* Englewood Cliffs, N.J.: Prentice-Hall.

PROBLEMS   **1.1** What polynomials do the following numbers represent?

(a) 53.6 radix 7
(b) 1111.1 radix 3
(c) 1111.1 radix 9
(d) 101.101 radix 2

1.2 What is the radix in each of the following cases?

(a) $53 + 120 = 213$
(b) $220 + 111 = 1101$
(c) $44^{.5} = 6$
(d) $213^{.5} = 13$

1.3 Convert each of the following to decimal.

(a) 10010.111 radix 2
(b) 3321.21 radix 6
(c) AF3.BC radix 16
(d) 2222.22 radix 3

1.4 Convert each of the following decimal numbers to radix r.

(a) 313.9702 to radix 8
(b) 15,625 to radix 5
(c) 129.376 to radix 2
(d) 16,387 to radix 2

1.5 Convert each of the following to binary with a five-digit fraction.

(a) $\pi = 3.14159$
(b) $\ln 2 = 0.693147$
(c) $e = 2.718282$
(d) $\sin \pi/4 = 0.707107$

1.6 Perform each of the indicated radix conversions.

(a) radix 2 10111001 10111 to radix 16
(b) radix 16 3A7B0 to radix 2 and to radix 8
(c) radix 8 777777 to radix 16 and to radix 2
(d) radix 16 A5A5A to radix 8 and to radix 2

1.7 Count as indicated.

(a) Count from 0 to 16 decimal in radix 3.
(b) Count from 0 to 9 decimal in radix 2.

1.8 Perform the following elementary arithmetic problems in radix 2.

(a) 1000011/1011
(b) 1010 − 1111
(c) 1010 × 1010
(d) 1001101 − 1111

1.9 Calculate the 2's complements of each of the following signed numbers. Check each answer.

(a) 00000001_2
(b) 11110101_2
(c) $A736_{16}$
(d) 56021_8

1.10 Do the following problems in complement arithmetic.

(a) radix 10:   2,837 − 3,948
(b) radix 16:   1,234 − 5,678
(c) radix 2:   11000011 − 11111100
(d) radix 8:   11111 − 77777

1.11 Perform the following radix-2 operations. Does overflow or under-derflow occur?

(a) 0001 + 1111
(b) 0111 + 0111
(c) 1001 + 1011
(d) 0111 + 1001

1.12 Do the following problems, avoiding overflow/underflow. Use multiple-precision arithmetic.

(a) radix 2:   0111 + 0111
(b) radix 16:   7777 + 5555
(c) radix 2:   1001 + 1001
(d) radix 16:   A732 + CBFA

1.13 Convert each of the following to BCD.

(a) 9,705 decimal
(b) CE3 radix 16
(c) 777 radix 8
(d) 1110110001111010 radix 2

1.14 Convert each of the following decimal numbers to BCD and perform BCD addition.

(a) 3,769 + 9,512
(b) 9,999 + 9,999
(c) 9,876 + 6,789
(d) 9,876 + 1,234

1.15 Here is something new: Convert the following decimal numbers to BCD and perform BCD subtraction.

(a) 742 − 585
(b) 3,071 − 1,983

1.16 Use the ASCII code to write the following.

(a) This is an encoding process.
(b) 0 1 2 3 4 5 6 7 8 9

1.17 Answer the following questions about converting ASCII code to binary.

(a) What are the logical and arithmetic operations that convert characters 0 to 9 from ASCII code to binary numbers 0000 to 1001?
(b) What are the logical and arithmetic operations that convert characters A to F from ASCII code to binary numbers 1010 to 1111?

2 ELECTRONICS IN DIGITAL DESIGN

Digital gate circuits are assemblies of discrete analog components such as resistors, diodes, and transistors. An understanding of the characteristics of these components and the capability to analyze circuits assembled from these components are needed.

First, we apply the pn junction diode and the Shottky barrier diode equations pragmatically to build the necessary understanding of their characteristics and use. The key characteristics are turn on forward voltage, voltage drop when turned on, and essentially zero current when reversed biased.

Next, the npn junction transistor action is explained in an elementary and perhaps oversimplified, yet pragmatic way. Then we explain that the npn junction transistor is converted into a Shottky clamped transistor to avoid significant increases in device turnoff time when used in TTL (transistor transistor logic) circuits. The conversion is straightforward: a Shottky barrier diode is connected from base to collector of a npn junction transistor.

Digital gate circuits are formed by wiring together sets of discrete analog components in various ways. We show how these complex circuits can be analyzed in a straightforward way to reveal their digital characteristics.

Because digital designers deal with functions and not individual components, we show how to treat digital gate circuits as functional black boxes. An extensive discussion of TTL characteristics provides a baseline.

The baseline allows for discussion of the characteristics of TTL and CMOS gate on an incremental basis. Open collector and other types of gates are also discussed.

INTRODUCTION

This chapter is a very quick course that starts with a discussion of pn junction diode and npn junction transistor characteristics. Next, we show how resistors, diodes, and transistors are assembled to form digital gates. A discussion of gate characteristics follows. Finally, CMOS devices are discussed.

2.1 Diodes

The diode is a two-terminal device whose impedance depends upon the magnitude and polarity of the voltage across the terminals. In elementary terms, a modern forward-biased diode has a low impedance, while the impedance is very high when the diode is reversed biased. We will see how circuits are wired to take advantage of this property.

The pn junction diode is represented by a straightforward equation for the diode current as a function of the applied voltage and diode physical parameters (Figure 2.1). This equation arises from application of Schrodinger's wave equation from the quantum mechanics to the electron gas at the junction of two different blocks of silicon, p-type and n-type, in contact with each other. (We will not pursue the source of the diode equation here because diode theory is outside the scope of this text.)

$$i_d = i_s(e^x - 1) \qquad \text{where } x = eV_{pn}/KT = V_{pn}/26mv \ (300°K)$$

$$i_s = ae(n_{po} \, Ln/\gamma_n + p_{no} \, L_p/\gamma_p) = 10^{-15} \text{ amperes (estimated)}$$

$$e = \text{charge of the electron } (1.602 \ 10^{-19} \text{ coulombs})$$

$$V_{pn} = \text{voltage (v) applied to the diode junction}$$

$$K = \text{Boltzman's constant } (1.3805 \ 10^{-23} \text{ J/°k})$$

$$T = \text{temperature (°k)}$$

$$a = \text{cross-sectional area (m}^2)$$

$$L_n = \text{diffusion length of an electron (m)}$$

$$L_p = \text{diffusion length of a hole (m)}$$

$$\gamma_n = \text{average lifetime of an electron in free state (s)}$$

$$\gamma_p = \text{average lifetime of a hole in free state (s)}$$

$$n_{po} = \text{equilibrium concentration of electrons in p-type semiconductors.}$$

$$p_{no} = \text{equilibrium concentration of holes in n-type semiconductors.}$$

When the normalized voltage x is more negative than −4, the exponential term is essentially zero and the diode current is constant with value i_s (i.saturation). A typical reversed-bias diode saturation current is 10^{-15} amperes, which is a zero in TTL digital logic circuits. The normalizing voltage value is approximately 26 millivolts (mv) at

FIGURE 2.1
pn Junction diode

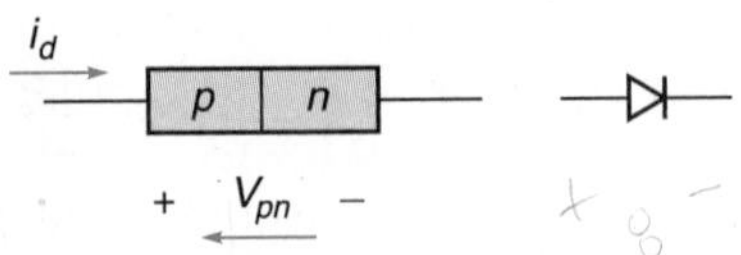

room temperature. The −4 means that a diode is off for all practical purposes when reversed bias exceeds −4 ∗26mv = −104mv. At V_{pn} = −104mv and with a current of 10^{-15} amperes, the impedance is about 10^{14} ohms!

When the normalized voltage x is positive, the diode is still off for all practical purposes in our TTL circuits until x reaches 24 as the diode begins to conduct TTL significant current. When x is 24, the forward current is still only about 26 microamperes (Figure 2.2). In elementary terms, the TTL circuits need hundreds of microamperes flowing to function. When x = 24, the forward voltage is 24 ∗ 26mv = 624mv. When x is increased from 24 to 27, the current is multiplied by e^3 = 20 increasing the current from 26 to 532 microamperes (μa). With a forward voltage of V_{pn} = 27 ∗ 26 = 702mv, the diode current is 532 microamperes. The static impedance is 702mv/532μa or 1320 ohms. The diode is turned on. As the forward voltage is increased, the diode current increases until limited by other circuit parameters. [The diode can be destroyed if the circuit does not limit the current.]

The Shottky barrier diode is formed at the junction of a metal block and an n silicon block. The Shottky diode current equation is more complex and the parametric values are different. Stated for our purposes in the most elementary and oversimplified terms, the Shottky diode behaves like a pn junction diode except that the on voltage is 0.4v instead of 0.7v.

> **PN diode turns on when forward voltage exceeds 0.7v.**

> **Shottky diode turns on when forward voltage exceeds 0.4v.**

$$i_d = i_s e^{d\sqrt{x}} (1 - e^{-x}) \qquad \text{where } x = eV_0/KT \text{ and d is a constant}$$

Analysis of TTL circuits is easier if you keep in mind that the pn junction diode is off (like an open switch) for all practical purposes when the forward voltage falls below 700 millivolts or 0.7 volts and that the diode is like a closed switch when forward voltage is above 0.7 volts. The Shottky diode is off for all practical purposes when the forward voltage falls below 0.4 volts.

FIGURE 2.2
pn Junction diode volt-ampere plot

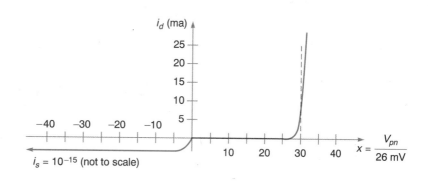

2.2 Transistors

Here is a quick study of transistor action. One way to define transistor action is the ability of the base-emitter current to modulate the collector-emitter current. Our explanation is facilitated when we think of the silicon npn block representing the transistor (Figure 2.3) as two superimposed pn junction diodes using the same p block which is very thin. In any circuit (Figure 2.3), the base-collector pn junction is reverse biased. As an isolated diode, the base-collector diode conducts i_s, which we will treat as zero microamperes.

On the other hand, the base-emitter pn junction is forward biased. As an isolated diode the current equation is

$$i_d = i_s[e^x - 1] \qquad \text{where } x = eV/kT$$

An electric field E is established in the n, p, and n blocks by the potential differences $V_{ce} - V_{be}$ and $V_{be} - V_e$ where V_e is 0v. The collector E field attracts electrons. Which electrons? The electrons from the emitter n block. Remember that relatively speaking there are no electrons in the reversed-biased base-collector (pn) junction.

FIGURE 2.3

Transistor symbol and circuit

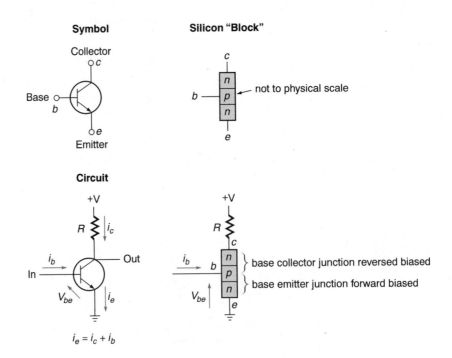

$$i_e = i_c + i_b$$

Suppose V_{be} is such that the base-emitter diode equivalent current would be $1000 \mu a$ with an open collector circuit. In that case $i_b = i_{diode} = 1000 \mu a = i_e$. Therefore, the emitter current $i_e = 1000 \mu a$.

When the collector circuit is in place, the resulting E field attracts electrons that normally would flow from the emitter to the base.

Alpha is fraction of emitter current collected.

Here is the key to transistor action. With V_{be} constant, the emitter current does not change when E collects electrons for the collector, because the emitter is unaware of the collector's action. As the collector voltage increases from zero, an ever increasing number of electrons are collected. The collector current i_c grows from $-i_s$ (zero in practice) to a fraction of the emitter current, which is known as alpha. This process reduces the base current accordingly.

$$i_c = \alpha \, i_e \qquad i_b = (1 - \alpha) \, i_e \qquad i_c + i_b = i_e$$

Because the maximum collected is αi_e, at some point a further increase in V_{ce} will not change i_c. This means i_c becomes a constant current generator. A plot of i_c versus V_{ce} is something like the plot in Figure 2.4.

Beta is base-to-collector current gain.

Let's calculate the current ratio i_c/i_b, which is known as beta.

$$\beta = i_c/i_b = \alpha i_e/(1 - \alpha)i_e = \alpha/(1 - \alpha)$$

Alpha $= 0.99$ implies beta $= 99$.

The other currents are

$$i_e = 1000 \text{ microamperes}$$

$$i_b = (1.0 - 0.99)1000 = 10 \mu a$$

$$i_c = (0.99)1000 = 990 \mu a$$

Recapitulation: V_{be} forces emitter current to be 1000 microamperes but the collector takes 990 microamperes leaving $i_b = 10 \mu a$ for the same V_{be}.

FIGURE 2.4
Transistor i_c vs V_{ce} plot for fixed i_b

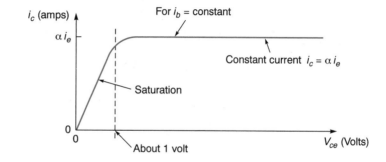

In practice we do not design to set V_{be}. For example, we design to force i_b to be 10 microamperes and let V_{be} take the value the diode equation dictates. A family of i_c versus V_{ce} with i_b as a parameter is shown in Figure 2.5.

We drive a transistor with a current square wave (Figure 2.6) and ask what happens to the collector current. When i_b is zero, i_c must be zero. With i_c zero, no voltage drop occurs across the 5K resistor and the collector voltage is at 5 volts. That sets the $i_c = 0$ (or off) operating point.

When i_b is $10\mu a$, i_c must be $990\mu a$ for sufficiently high collector voltage (Figure 2.6). If the transistor's collector is shorted to ground, the current is 1 milliampere (5v/5K). This is what a 5K load line shows. Increasing i_b increases i_c, and increasing i_c causes voltage drop ($v_r = i_c$ 5K) across the 5K load resistor. At 1ma V_{ce} is zero. The load line intersects the transistor i_b 10 microampere curve at $V_{ce} = 0.6$ volts. Consequently $i_c = 880\mu a$. Therefore, as i_b switches from 0 to 10 microamperes the collector current switches from 0 to 880 microamperes. Notice that when V_{ce} is 0.6 volts the base-collector junction is forward biased by 0.1 volts (x = 4) because V_{be} is at 0.7 volts. The base-collector junction has started to turn on. The transistor is said to have moved into its saturated region.

At this point the Shottky clamped transistor enters the picture. The Shottky clamped transistor is an npn transistor with a Shottky barrier diode connected from base to collector (Figure 2.7). The purpose is to limit x if a transistor tries to move into the saturation region. The Shottky turns on at 0.4v where x is about 15. Although the Shottky is on, the base-collector pn junction is still off. This limits the

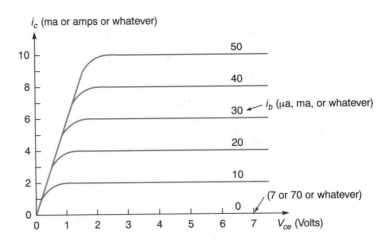

FIGURE 2.5

Transistor i_c vs V_{ce} plots for a set of i_b

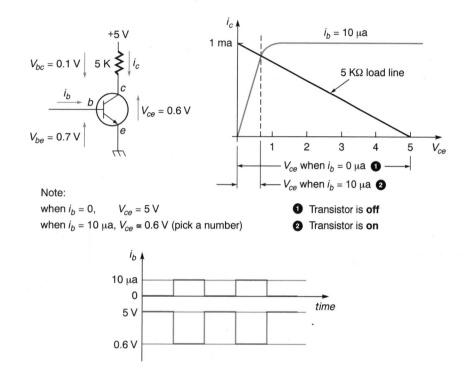

FIGURE 2.6
Switching transistor action

+5 V
V_{bc} = 0.1 V
5 K
i_c
c
i_b
b
V_{ce} = 0.6 V
e
V_{be} = 0.7 V
i_c
1 ma
i_b = 10 μa
5 KΩ load line
1 2 3 4 5 V_{ce}
V_{ce} when i_b = 0 μa ❶
V_{ce} when i_b = 10 μa ❷
Note:
when i_b = 0, V_{ce} = 5 V
when i_b = 10 μa, V_{ce} ≅ 0.6 V (pick a number)
❶ Transistor is **off**
❷ Transistor is **on**
i_b
10 μa
0
5 V
0.6 V
time

Shottky diode enhances speed.

move into the saturation region. The Shottky barrier diode clamps the transistor out of saturation by diverting current from the base to the collector (Figure 2.7). Why are we concerned about diverting current from the base to the collector? Speed of operation. A saturated transistor takes more time to turn off.

FIGURE 2.7
Shottky clamped transistor

c
i_d
i_c
b
i_b
$i_b - i_d$
i_e
e
V_{bc} = 0.1 V
c
b
V_{ce} = 0.6 V
e
V_{be} = 0.7 V

When clamped

$i_e = (i_b - i_d) + (i_c + i_d)$

$i_e = i_b + i_c$ qed

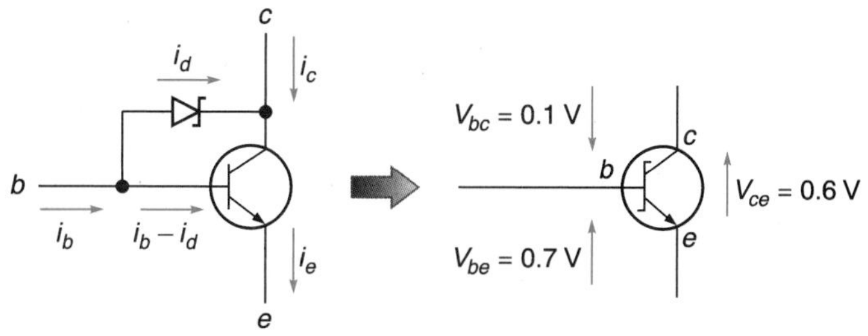

The Shottky clamped transistor symbol (Figure 2.7) is misleading. The junctions are not Shottky junctions. The Shottky diode has been added from base to collector.

2.3 TTL Circuit Analysis

The 74LS00 ''nand'' gate and the 74LS05 inverter illustrate how analog parts are used to construct digital circuits. Let us take a look at the more simple 05 circuit first. Analysis of these non-linear circuits requires us to make assumptions that will prove to be either correct or incorrect as we proceed through the analysis. This take-a-risk ap-

FIGURE 2.8
74LS05 Circuit

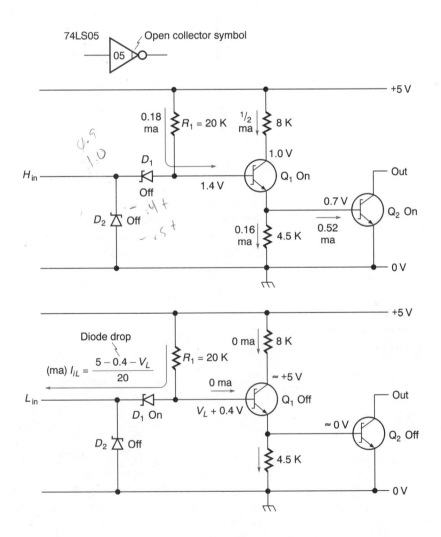

proach allows us to be guided expeditiously to the solution. Further-more, nonexact statements expedite the process without affecting the reality. A base-emitter diode, for example, needs 0.7 volts across it to support the on condition.

The Shottky diode from input to ground (Figure 2.8) has no digital function. Diode D_2 protects the input circuit from negative voltages that are connected accidentally or from voltages arising from oscilla-tions known as ringing, which is explained later in this chapter. We will ignore these input diodes in the following.

Circuit defines H and L levels.

05 input H: Diode D_1 and the 20K resistor form a standard input circuit. An input is defined as high (H) when D_1 is reversed biased and shut off. How high is high is determined as follows. Assume D_1 is off so that R_1 current flows into Q_1's base turning it full on (assumed). Consistent with Q1 on, we have Q_2 on. Working backwards from zero volts at Q_2's emitter, the voltage at the base is one diode drop up or 0.7 volts. For a similar reason the base of Q_1 is at 1.4 volts. Shottky diode D_1 is shut off when its forward voltage is less than 0.4 volts. There-fore, a high input (H) is greater than 1.0 volts.

Because Q_1 is full on, its collector is 0.4 volts lower than the base or 1.0 volts. With potentials established the currents are readily calcu-lated as shown (Figure 2.8). The open collector output is discussed in the open collector and other gates (Section 2.5, page 69).

05 Input L: When the input falls below 1.0 volts, Q_1's base voltage falls below 1.4 volts. Consequently Q_1 and Q_2 are turned off because the two transistor base emitter diodes need 0.7 volts each to support the on condition. The transistors' off currents are zero. In effect, an input is defined as low (L) when diode D_1 is turned on.

The 74LS00 circuit consists of four elements: AND gate, phase splitter, low-level driver, and high-level driver (Figure 2.9).

00 inputs H: With all inputs high, the AND condition is satisfied. This means Q_2 and Q_{3b} are on and, thus, the base of Q_2 is at 1.4v. The base of Q_{3b} is at 0.7v. Q_{3a} must be off because any current through the 1.5K resistor would make the base of Q_{3a} less than 0.7v. The role of Q_{3a} is to assist Q_{3b} in turning off. The subtle nature of this circuit is a digression that we will not pursue.

As the AND condition is satisfied, the base voltage of Q_2 rises pulling the emitter voltage up as Q_2 turns on. However, the collector current forces a voltage drop across the 8K resistor pulling the col-lector voltage down to 1.0 volts. The collector goes down as the emitter goes up. This is why Q_2 is said to perform a phase splitting function.

FIGURE 2.9
74LS00 Circuit

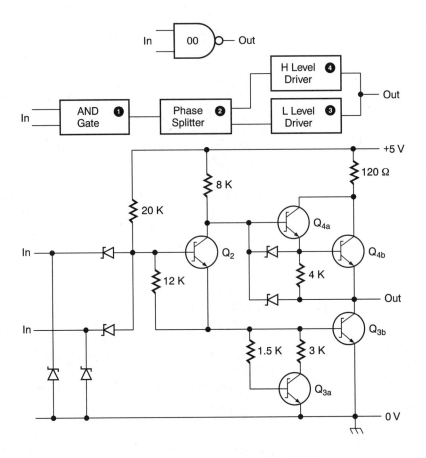

The low-level driver Q_{3b} is on and so its collector is at, say, $0.7 - 0.4 = 0.3$v. The base of Q_{4a} is at 1.0 volts because it is connected to the Q_2 collector. The 1.0 minus 0.3 volts across the Q_{4a} and Q_{4b} base-emitter diode in series does not support their being on. Thus Q_{4a} and Q_{4b} are off as they should be. Accordingly, with all inputs high, the output is low.

00 inputs L: When any input is pulled below 1.0 volts, the base of Q_2 falls below 1.4 volts shutting Q_2 and Q_{3b} off. The Q_2 collector current goes to zero, and the collector rises turning on Q_{4a} and Q_{4b}. Any low input forces the output high. The output voltage is 5v minus the 8K voltage drop minus 1.4 volts. The 8K voltage drop is proportional to the Q_{4a} base current. In turn the base current is essentially proportional to the output load current. Consequently, with any input low the output is high.

2.4 — TTL Characteristics

Any TTL gate has at least one input and one output. We need to know the rules, or criteria, for connecting outputs to inputs in order to wire together a number of gates. This implies knowing the input and output VI (voltage-current) characteristics as well as the transfer characteristic (output voltage versus input voltage). In addition to these static dc characteristics that define logical operations, a gate has transient characteristics that define computational performance.

2.4.1 DC Parameters and Fanout

There are two sets of dc parameters—voltages and currents—that define reliable operation of interconnected gates. The parameter values are stated in terms of minimum and maximum values. Stating them in terms of specific single values is precluded by process manufacturing tolerances. Thus there are high-level (H) and low-level (L) voltage ranges, not values like 3 volts and 0.6 volts.

Specific values depend upon logic family.

V_{il} Guaranteed maximum low-level (L) input voltage allowed for reliable gate operation

V_{ol} Guaranteed maximum low-level (L) output voltage allowed for reliable gate operation

V_{ih} Guaranteed minimum high-level (H) input voltage allowed for reliable gate operation

V_{oh} Guaranteed minimum high-level (H) output voltage allowed for reliable gate operation

Notice (Figure 2.10) that the dc low-level output voltage V_{ol} is guaranteed to be lower than the dc low-level input voltage V_{il}. Notice, too, that the difference $V_{il} - V_{ol}$ is called the low-level dc noise margin. Output voltage transients may venture into the noise margin region without jeopardizing reliable operation.

H and L represent voltage ranges.

Also, notice (Figure 2.10) that the dc high-level output voltage V_{oh} is guaranteed to be higher than the dc high-level input voltage V_{ih}. The difference $V_{oh} - V_{ih}$ is called the high-level dc noise margin. Again, output voltage transients may venture into the noise margin region without jeopardizing reliable operation.

Finally observe (Figure 2.10) that the H and L levels are voltage ranges defined by the output voltage specifications.

Figure 2.10
**TTL DC
Specifications**

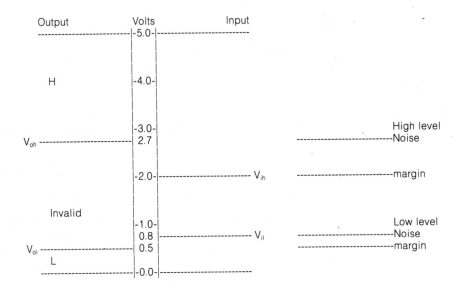

The currents define reliable system operation as well as reliable gate operation. The current specifications indicate output and input impedance levels when currents are related to the voltages. Here are the definitions and specifications.

Adequate currents allow one gate to drive many gates.

I_{il} Guaranteed maximum low-level (L) input current allowed for reliable gate operation

I_{ol} Guaranteed minimum low-level (L) output current allowed for reliable gate and system operation

I_{ih} Guaranteed maximum high-level (H) input current allowed for reliable gate operation

I_{oh} Guaranteed minimum high-level (H) output current allowed for reliable gate and system operation

These are the values for currents and voltages for 74LS TTL gates.

V_{il} is 0.8 volts maximum independent of current

V_{ol} is 0.5 volts maximum when Iol is 8 milliamperes

V_{ih} is 2.0 volts minimum independent of current

V$_{oh}$ is 2.7 volts minimum when Ioh is 400 microamperes

I$_{il}$ is 0.4 milliamperes maximum when input voltage is at low level L. Note that L may range from ground (zero volts) to 0.5 volts.

V$_{ih}$ is 20 microamperes maximum when input voltage is at high level H. Note that H may range from 5 to 2.7 volts.

Currents define fanout.

Fanout is a system parameter. Fanout is the number of input circuits one output circuit can drive. There are both a high level fanout and a low-level fanout. The smallest fanout value is the system fanout.

High level fanout: I_{oh}/I_{ih} = 400/20 = 20 for 74LS TTL
Low level fanout: I_{ol}/I_{il} = 8/0.4 = 20 for 74LS TTL

2.4.2   Input

Now we learn why input current is less than $20\mu a$ at H input voltage and less than 0.4 milliamperes at L input voltage. The input circuit is a Shottky diode in series with a normal 20K resistor connected to +5 volts. Observe that 5v/20Kohms is a 0.25 milliampere current. The specification for I_{il} is 0.4 milliamperes. In effect, this allows for a manufacturing tolerance on the 20K resistor.

H and L input currents are very different.

As the input voltage V_{in} falls from +5 volts, the current flowing into the circuit is I_{ih}, which is the sum of reversed-bias diode current plus leakage. I_{ih} decreases to zero when V_{in} reaches 1.4v (Figure 2.11).

FIGURE 2.11
**TTL Input current
vs input voltage**

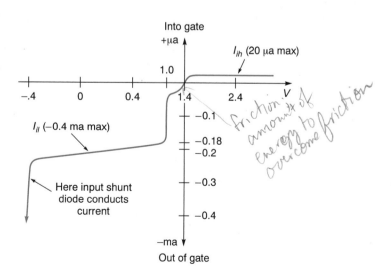

This occurs because the collector of the series input diode connected to the base of Q_1 is at 1.4 volts (Figure 2.8 on page 56). As V_{in} falls below 1.4 volts, the current reverses to become I_{il} and forward diode current begins. However, until V_{in} drops to 1.0 volts, the Shottky diode is off. At 1.0 volts the current jumps to 0.18ma as the diode takes all the base current of Q_1 as it switches off. As V_{in} falls below 1.0 volts, the current increases linearly because the diode acts like a 0.4-volt battery (See I_{il} equation in Figure 2.8.) I_{il} is at its maximum when the input voltage V_{in} is at ground (zero volts).

If V_{in} goes negative, the input shunt diode turns on when V_{in} passes −0.4 volts and only V_{in} source impedance limits the current (Figure 2.11).

2.4.3   Output

Now we learn why output current is specified to be less than 400 microamperes at H input voltage and less than 8 milliamperes at L input voltage.

Low level L: A Shottky clamped output transistor with grounded emitter is equivalent to a resistor as the collector voltage rises from zero with base current fixed (Figure 2.4 on page 53 and Figure 2.12). When the collector rises above 1 volt the turned-on transistor comes out of saturation. Then the transistor becomes equivalent to a constant current source.

L equivalent output circuit is a 0.2v battery in series with a 16.3 ohm resistor.

V_{ol} at 0.5 volts is the specified upper bound of the L region. In this region, the turned-on low-level output transistor is equivalent to a resistor connected to ground in series with a 0.2-volt battery. [The origin of the 0.2-volt battery is the output transistor's voltage drop V_{cesat}.] The value of this source resistor is taken from our simplified version of a 74LS TTL low level output characteristic (Figure 2.12). As current I_{ol} increases from 0 to 18 milliamperes the voltage V_{ol} increases from 0.2 volts to 0.5 volts. The curve is essentially a straight line in the low level L region so we reasonably can assume we are dealing with a resistor characteristic. We have dv/di equal to $0.5 - 0.2$ volts/$18 - 0$ milliamperes or 16.7 ohms.

At maximum low level fanout of 20, the voltage drop across the low level source impedance is $I_{ol}R_{sl} = 133$mv.

$$I_{ol}R_{sl} = 20I_{il}R_{sl} = 20 \cdot 0.4ma \cdot 16.7ohms = 133mv.$$

Adding 0.2 volts offset to 0.133 volts brings Vol to 0.333 volts, which is less than the 0.5 volts specification.

The different 05 and 00 output circuits (Figure 2.8 on page 56 and Figure 2.9 on page 58) result in the plots shown for higher output voltages forced upon the output transistor by the output circuit.

FIGURE 2.12
**TTL Output voltage
V_{ol} vs output
current I_{ol}**

FIGURE 2.12
**TTL Output voltage
V_{ol} vs output
current I_{ol}**

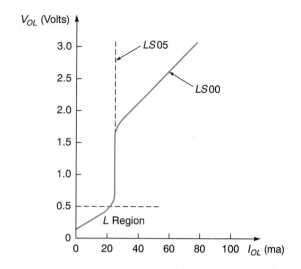

High level H: Transistors Q_{4a}, Q_{4b} (Figure 2.9) form a Darlington circuit, which has a low output impedance. The $dv/di = 3500mv/38ma = 92$ ohms (Figure 2.13). At maximum high-level fanout of 20 the voltage drop across the high-level source impedance is $I_{oh}R_{sh} = 37mv$. This is negligible compared to two diode voltage drops.

H equivalent output circuit is a 3.6v battery in series with a 92 ohm resistor.

$$V_{oh} = +5v - 2 \cdot 0.7v - I_{oh}R_{sh} = 3.6v - I_{oh}R_{sh}$$

$$I_{oh}R_{sh} = 20I_{ih}R_{sh} = 20 \cdot 0.02ma \cdot 92 \text{ ohms} = 37mv$$

$$\text{Voh} = 3.6v - .037v > 2.7 \text{ volts as specified}$$

FIGURE 2.13
**TTL Output voltage
V_{oh} vs output
current I_{oh}**

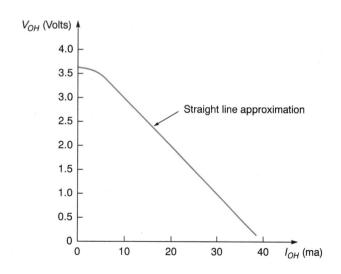

Recapitulation The high-level equivalent circuit is a 3.6-volt battery with a 92-ohm source impedance, and the low-level equivalent circuit is a 0.2-volt battery with a 16.7-ohm source impedance.

2.4.4   Transfer

A gate transfer function defines the output voltage as a function of the input voltage (Figure 2.14). Observe that the output is constant when the input is in the L or H voltage regions. The output changes state when the input traverses the invalid region from L to H or H to L. Note that the slope of the transfer function is finite as the output changes state. A useful concept is that a digital gate is a low-fidelity linear amplifier when the input is in the (digital) invalid region. (See the oscillator in Figure 11.12 on page 606.) The amplifier gain is greater than one. An estimate of this gain is $dVout/dV in = 3.3/0.2 = 16.5$ (Figure 2.14).

> Gate is a low-fidelity amplifier with a gain of, say, 16.

2.4.5   Switching Speed

Guaranteed switching speed parameters are propagation delay times from gate input transitions to gate output transitions (Figure 2.15a). The symbols are t_{PHL} for high to low output transitions and t_{PLH} for low to high output transitions. For the 74LS family, time is measured from the 1.3-volt value on the rising or falling slope. The time values are different because the circuits are not linear. The typical not guar-

> Gate output transitions are delayed gate input transitions.

FIGURE 2.14
TTL Transfer function V_{out} vs V_{in}

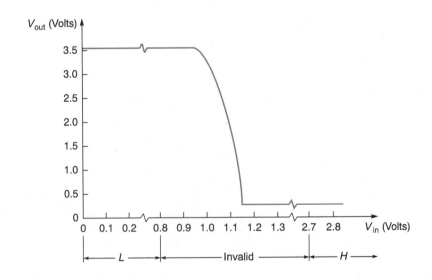

FIGURE 2.15
TTL Timing parameters

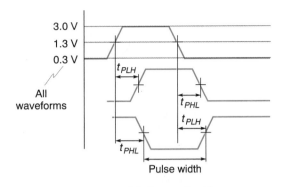

(a) Propagation Time

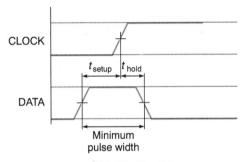

(b) Flip-Flop Trigger

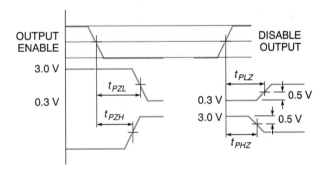

(c) Three-state

anteed value for these parameters is 9 nanoseconds (ns). The maximum guaranteed values are 15 nanoseconds for the 74LS00 gate. A designer pays attention to maximum values.

The guaranteed values are specified with respect to standard test conditions such as

$$V_{cc} = 4.5v \text{ to } 5.5v$$

$$C_{load} = 50pf$$

$$R_{load} = 500ohms$$

$$Temp = \text{in range}$$

Setup and hold time
specs must be satisfied.

A memory element such as a flip-flop (Chapter 6) has a clock input. Edge triggered flip-flops require stable data inputs for time intervals before and after the clock edge (Figure 2.15b). The setup and hold time requirements are consistent with the assumptions of fundamental mode asynchronous circuit design (Chapter 11).

A three-state gate output (Section 5.2.2) is connected to the gate load when the enable control input is asserted. Enabling a connection changes the output impedance from an open circuit (Z) to a standard gate output (Figure 2.15c). The gate is enabled after a propagation time of t_{PZL} for a low output and t_{PZH} for a high output. Conversely the gate is disabled after a propagation time of t_{PLZ} for a low output and t_{PHZ} for a high output. Disconnect is achieved by turning off the high and the low transistor drivers.

t_{PLH}　Propagation time from an input to an output switching from L to H

t_{PHL}　Propagation time from an input to an output switching from H to L

t_{SETUP}　Time that data must be stable prior to clock edge

t_{HOLD}　Time that data must be stable after clock edge

t_{PZL}　Propagation time from an input to an output switching from high impedance to L

t_{PZH}　Propagation time from an input to an output switching from high impedance to H

t_{PLZ}　Propagation time from an input to an output switching from L to high impedance

t_{PHZ}　Propagation time from an input to an output switching from H to high impedance

FIGURE 2.16
Ringing

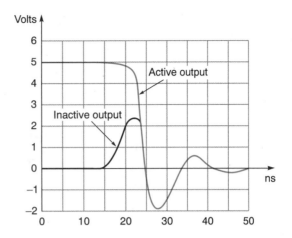

2.4.6 Ringing

Ringing is the damped sinusoidal oscillation following a high to low transition in a waveform (Figure 2.16). The ringing frequency is determined by load capacitance and the inductance of the wiring.

The damping is set by the resistive loading. Series resistors that are external to a gate output provide additional damping.

However, the waveform rebound may move the signal voltage out of the L region causing erroneous output transitions. The diodes from input to ground (Figure 2.8 on page 56 and Figure 2.9 on page 58) clip the oscillating waveform's negative excursions. Ringing can occur on the low to high waveform transition, but greater output impedance in the H state reduces the peak to peak swings. Ringing causes another

Ringing can cause false L to H transitions.

FIGURE 2.17
Package pin inductance

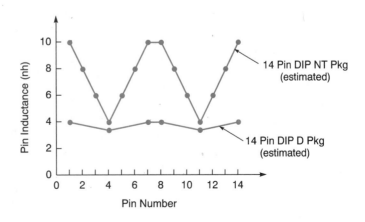

FIGURE 2.18

**Open collector
calculations**

From the data book:

$V_{ih} - \min = 2.7\,\text{V}$

$I_{ih} - \min = 20\,\mu\text{a}$

$I_{oh} - \min = 100\,\mu\text{a} \; @ \; V_{OH} = 5.5\,\text{V}$

$$\left\{ R_{max} = \frac{2.3\,\text{V}}{MI_{OH} + NI_{iH}} \right\}$$

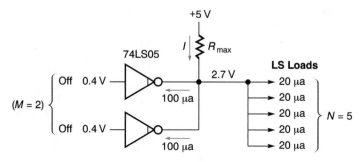

$$R_{max} = \frac{\Delta V}{I_{total}} = \frac{(5 - 2.7)\,\text{V}}{(2 \cdot 100 + 5 \cdot 20)\,\mu\text{a}} = \frac{2.3\,\text{V}}{300\,\mu\text{a}} = \frac{2300}{300}\,10^3\,\Omega$$

$$R_{max} = 7.67\,\text{K}\Omega$$

(a)

From the data book:

$V_{ol} = 0.5\,\text{V} \; @ \; I_{ol} = 8\,\text{ma}$

$I_{in} = -0.4\,\text{ma} \; @ \; 0.4\,\text{V} \;\; (< 0.4 \; @ \; 0.5\,\text{V})$

$$\left\{ R_{min} = \frac{4.5\,\text{V}}{I_{ol} - NI_{il\,\max}} \right\}$$

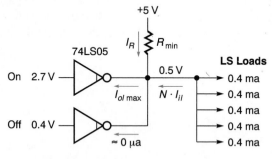

$$R_{min} = \frac{\Delta V}{I_R} = \frac{5 - 0.5}{8 - 5 \cdot (0.4)} = \frac{4.5\,\text{V}}{6\,\text{ma}}$$

$$R_{min} = 750\,\Omega$$

(b)

problem. It limits the switching rate prematurely because the ringing must die out before the next switching transition occurs.

A more important negative consequence is the simultaneous switching of inactive outputs (Figure 2.16). The voltage induced by the ground lead inductance raises the potential of inactive output pins. Packages with shorter leads externally and internally significantly reduce inductance (Figure 2.17).

2.5 Open Collector and Other Types of Gates

The 74LS05 (Figure 2.8 on page 56) is an example of a gate with an open collector output. One application of this type of gate is the wired AND connection (Figure 2.18). The AND condition is true when all 05 gates are off. In that case (Figure 2.18a) the output voltage is specified to be V_{oh} minimum or 2.7 volts (Figure 2.10). The load input currents are I_{ih} for each gate and the off 74LS05 device leakage current is 100 microamperes maximum. The equation for R_{max} is derived by applying Ohm's law to the output node. It is a maximum R because 2.7 volts is the minimum voltage.

You can make wired OR or AND circuits.

When all legs of the AND gate except one are off, the current sink capability is minimum while the AND condition is false. In this case (Figure 2.18b), the output voltage is 0.5 volts maximum. The equation for R_{min} is derived by applying Ohm's law to the output node once more. It is a minimum R because 0.5 volts is the maximum voltage.

The 74LS132 is an example of a circuit with Schmidt trigger capability. The transfer function exhibits hysteresis as input voltage rises from L to H and then falls from H to L. The gain is very high so that a small voltage delta across either threshold forces a full H-L transition of the output (Figure 2.19).

FIGURE 2.19
74LS132 Transfer function

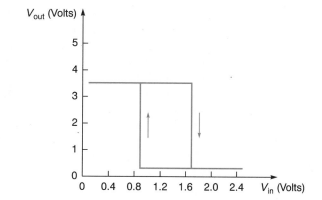

FIGURE 2.20
**Schmidt trigger
"squares up" a
waveform**

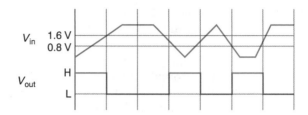

Schmidt trigger converts a waveform to a two-level H, L waveform.

Positive feedback in the circuit causes rapid H to L and L to H transitions of the output when thresholds are crossed by a slowly slewing waveform (Figure 2.20).

2.6 CMOS Characteristics

CMOS stands for "complementary metal oxide semiconductor." The performance of modern MOS transistors qualifies them for the major role in digital logic that they now have. In addition, a CMOS circuit uses essentially zero power when not switching (i.e., while standing by). The four major CMOS logic families are designated HC, HCT, AC, and ACT. HCT and ACT parts are designed with TTL compatibility. The AC/ACT families are significantly faster than the HC/HCT families that appeared first. CMOS is the dominant technology used in higher levels of integration.

MOS transistors are field effect transistors (FETs) with insulated input gates. The circuit symbols for p-channel and n-channel MOS transistors are shown in Figure 2.21. Assume that these MOS transistors are designed to be off when the gate-source voltage V_{gs} is zero. The drain-to-source resistance of an MOS transistor that is off is very high. On the other hand, the resistance is very low when V_{gs} is, say, 3 volts.

The elementary CMOS circuit is the inverter in Figure 2.22. When V_{in} is at low voltage L, the p-channel transistor V_{gs} is several volts, which turns it on. At the same time the n-channel transistor V_{gs} is "zero," which turns it off. Conversely, when V_{in} is at high voltage H, the n-channel device is on and the p-channel device is off. Therefore, when V_{in} is at L or H voltage, no current flows through the inverter. For these reasons a CMOS circuit has low standby power requirements. However, current flows through the inverter when V_{in} switches because both devices are partially on as they switch from on-to-off and vice versa. This is why the power dissipated increases with switching frequency.

CMOS gates have "zero" input current and large dc fanout.

The input and output characteristics are different from TTL gates. The MOS insulated gate results in no direct input current except for

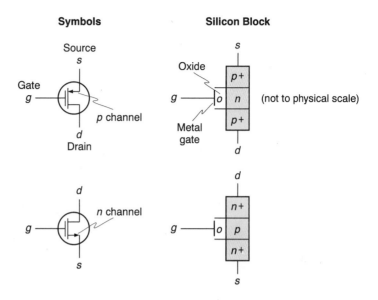

FIGURE 2.21
MOS Transistors

parasitic leakage currents (less than a microampere). Modern CMOS inverters are designed to deliver the same output current levels at high and low output voltages. For example, MOS device specifications provide for 24 milliamperes with V_{ol} = 0.5 volts and V_{oh} = 3.8 volts in the AC/ACT families. This symmetrical capability results in almost equal rise and fall times in the switching waveforms. Transient currents can be many milliamperes as intrinsic circuit capacitance is charged and discharged.

Speed of transitions is limited by circuit capacitance.

The CMOS DC specifications for ACT family are in Figure 2.23. The dc input current is less than a microampere due to the insulated gate. With milliamperes of output current, the fanout is very high when driving only CMOS gates. But TTL gates require the driver to sink 0.4 milliamperes at low voltage. This limits the fanout to 24/0.4 = 60, which is still high and is not recommended as a practical matter.

FIGURE 2.22
CMOS Inverter

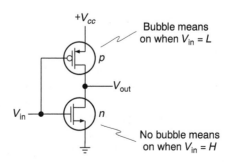

FIGURE 2.23
**CMOS DC
Specifications* for
ACT family**

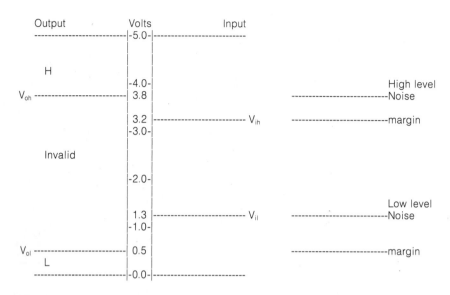

* See data book for exact values.

CMOS behavior is similar to the TTL behavior shown in Figures 2.13 through 2.17. The plots are very similar but the numbers are different because CMOS technology is radically different from TTL technology. For example, the "gain" of the TTL transfer characteristic in Figure 2.14 on page 64 is about 3.2v/0.2v = 16, while the "gain" of a corresponding CMOS transfer characteristic can be over 100.

SUMMARY

Diodes

The diode is a two-terminal device whose impedance depends upon the magnitude and polarity of the voltage across the terminals. In elementary terms a modern forward-biased diode has a low impedance, whereas the impedance is very high when the diode is reversed biased.

Transistors

The silicon npn block representing the transistor (Figure 2.3 on page 52) may be thought of as two superimposed pn junction diodes (Figure 2.1 on page 50) using the same p block which is very thin. One way to define transistor action is the ability of the base current to modulate the collector-emitter current.

TTL Circuit Analysis

The 74LS00 "nand" gate and the 74LS05 inverter illustrate how analog parts are used to construct digital circuits.

TTL Characteristics

Any TTL gate has at least one input and one output. We need to know the rules, or criteria, for connecting outputs to inputs if we are to wire together a multitude of gates. This implies knowing the input and output VI (voltage-current) characteristics as well as the transfer characteristic (output voltage versus input voltage). In addition to these static dc characteristics defining logical operations, a gate has transient characteristics defining computational performance.

Open Collector and Other Gates

Gates with open collector outputs provide means for connection to discrete digital circuits such as the wired AND connection.

The Schmidt trigger circuit uses positive feedback to create a transfer function with high gain and hysteresis. The Schmidt trigger is used to convert a slowly slewing waveform into a waveform with fast H to L and L to H transitions.

CMOS Characteristics

CMOS is the acronym formed from the words "complementary metal oxide semiconductor." The performance of modern MOS transistors qualifies them for the major role in digital logic that they now have. In addition, a CMOS circuit uses essentially zero power when not switching (i.e., while standing by). The four major CMOS logic families are designated HC, HCT, AC, and ACT. HCT and ACT parts are designed with TTL compatibility. The AC/ACT families are significantly faster than the HC/HCT families that appeared first.

REFERENCES

Advanced CMOS Logic Data Book, Dallas: Texas Instruments, 1988

Blakeslee, T. R., *Digital Design with Standard MSI and LSI,* 2nd Ed., New York: Wiley, 1979

Mead, C. and L. Conway, *Introduction to VLSI Systems,* Reading, MA: Addison-Wesley, 1980

Fletcher, W. L., *An Engineering Approach to Digital Design,* Englewood Cliffs, NJ: Prentice-Hall, 1979

Seidensticker, R. B., *The Well-Tempered Digital Design,* Reading, MA: Addison-Wesley, 1986

Shoji, M., *CMOS Digital Circuit Technology,* Englewood Cliffs, NJ: Prentice-Hall, 1988

TTL Logic Data Book, Dallas: Texas Instruments, 1988

Winkler, D. and F. Prosser, *The Art of Digital Design,* Englewood Cliffs, NJ: Prentice-Hall, 1987

Name _____   SID# _____   Section _____
         (last)   *(initials)*

Approved by _____   Date _____

Grade on report _____

LSTTL dc characteristics

1. Input current versus input voltage

   _____ Plot of I_{in} versus V_{in} (0 to 5 volts)

2. Output voltage versus input voltage

   _____ Plot of V_{out} versus V_{in} (0 to 5 volts)

3. Output low voltage versus output low current.

   _____ Plot of V_{ol} versus I_{ol} (0 to 10 milliamperes)

4. Output high voltage versus output high current.

   _____ Plot of V_{oh} versus I_{oh} (0 to 2 milliamperes)

Reference: Chapter 12 Project Information

Parameters are identified in the figures.

1. Plot I_{in} vs V_{in} of a 74LS04 gate (Figure P01.1).

FIGURE P01.1

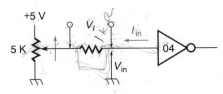

2. Plot V_{out} vs V_{in} of a 74LS04 gate (Figure P01.2).

FIGURE P01.2

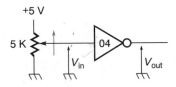

3. Plot V_{ol} vs I_{ol} of a 74LS04 gate (Figure P01.3).

FIGURE P01.3

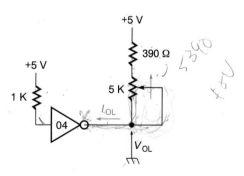

4. Plot V_{oh} vs I_{oh} of a 74LS04 gate (Figure P01.4).

FIGURE P01.4

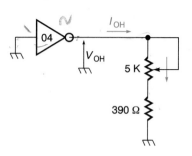

PROBLEMS

2.1 Show that KT/e = 26mv at 300°K.

2.2 Plot $f = 10^{-15} (e^x - 1)$ for x ranging from -10 to 32. Use a linear scale for x and a logarithmic scale for f.

2.3 Calculate V_b, V_{OL}, and I_{OL} for the circuit in Figure P2.1a.

2.4 Calculate the maximum allowable integer value for n when $V_{OL} = 0.5$ volts and $I_{IL} = 0.4$ milliamperes for the circuit in Figure P2.1b.

2.5 Find values for R_1 and R_2 when $V_I = 2.2$v, $I_c = 10$ma and $\beta = 50$ for the circuit in Figure P2.1c. [The LED voltage drop is 1.4 volts at 10ma.]

2.6 Derive the equation for V_{ce} in Figure 2.6.

2.7 Plot V_{b1} as a function of V_{in} where V_{b1} is the voltage at the base of Q_1 in Figure 2.8. V_{in} ranges from 0 to 5 volts.

2.8 Plot I_{in} versus V_{in} (Figure 2.8). V_{in} ranges from 0 to 5 volts.

2.9 Four currents are shown in the circuit of Figure 2.8 with input H_{IN}. Calculate the current values after all resistor values are reduced by half.

FIGURE P2.1

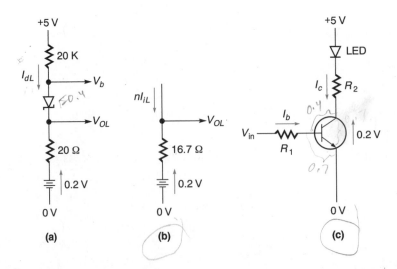

(a) (b) (c)

2.10 Calculate I_{IL} in milliamperes for the circuit of Figure 2.8 with input L_{IN} when the input voltage is 0.5 volts and the 20K R_1 is replaced by a 15K resistor.

2.11 In Figure 2.9 let the input voltage at both inputs be 2.4 volts. Calculate the currents flowing in the resistors and the voltages at the transistors' bases.

2.12 In Figure 2.9 let the input voltage at both inputs be 0.5 volts. Calculate the currents flowing in the resistors and the voltages at the transistors' bases.

2.13 The 74LS05 is essentially a constant current generator I of about 25 milliamperes (Figure 2.12). When driven by I, a capacitor C gains or loses charge q and voltage v according to q = Cv. The rise-and-fall time rate of a digital waveform is the differential dv/dt. Find the equation for dv/dt. Calculate dv/dt when I = 25ma and C = 50pf. Put dv/dt in the form of volts per nanosecond.

2.14 The 74LS high output equivalent circuit is a 3.7-volt battery V_b in series with R = 92 ohms (Figure 2.13). Assume the gate output is connected to a 50 pf capacitor C whose other lead is grounded. When the gate output is low, assume C is charged to zero volts. When a gate input changes the output tries to switch to the H level. It cannot do so until the capacitor charges to some voltage H. In this circuit the output voltage $v = V_b (1 - e^{-t/RC})$. Plot v as a function of normalized time x = t/RC. Let x range from 0 to 10. Put RC in the form of nanoseconds.

2.15 In Figure 2.16 estimate the ringing frequency in megahertz. What is the period in nanoseconds? If the pin inductance L is 8 nanohenry, what is the capacitance C in picofarad?

2.16 Calculate R_{max} and R_{min} for a 74LS05 open collector gate driving six 74LS gate inputs (Figure 2.18).

2.17 The velocity of light in free space is $3 \cdot 10^8$ meters/second. Use dimensional analysis and convert the units into centimeters/nanosecond (cm/ns). Why are these units more relevant to digital designs?

3 COMBINATIONAL LOGIC THEORY

Logic theory partitions naturally into two parts: combinational logic theory for circuits without memory and sequential logic theory for circuits with memory. The major goal of this chapter is to acquire knowledge of the theoretical basis, mathematical tools, and design aids of switching theory. This knowledge coupled with modern sequential logic theory allows the design of an amazing array of digital systems. We start with combinational logic because it is a prerequisite to sequential logic.

Combinational logic theory is a switching algebra that is a special case of Boolean algebra. Like any other algebra, the basis of switching algebra is a set of axioms. Huntington's axioms are the basis of switching algebra. Given these axioms, theorems are proved. Given axioms and theorems, additional theorems are proved in the spirit of Euclid's geometry. During this process switching algebra's constants, variables, and operators are discussed. Next functions that represent logic to be implemented in hardware are built. In most of the above, the truth table is used as an aid. The truth table lists all input variable combinations and specifies the output for each input combination. The special role of DeMorgan's theorems is also discussed.

Given a function can it be simplified? Yes—in principle. Algebraic methods based on the theorems are illustrated, but only briefly, because the Karnaugh map offers a preferable graphic alternative to algebraic simplification.

Two canonical forms of logic equations are helpful in your work. They are helpful because all logic equations are canonical forms or simplifications of canonical forms. Also, truth tables are lists of canonical forms, and the Karnaugh map is a graphical display of canonical forms. Canonical forms facilitate the assembly of switching function elements into specific logic functions.

Because a design process usually begins with words not equations, incomplete word descriptions manifest themselves as missing entries in truth tables or Karnaugh maps. Your understanding of switching theory allows you to fill in the missing entries with reasonable estimates so that the design process can proceed.

In order to proceed you are shown how to build bridges that allow you to move to and from switching equations, truth tables, and Karnaugh maps. The bridge building processes are facilitated by numerical descriptions of functions. Finally you are shown that the very important concepts of table entered and map entered variables allow you to deal with more complex problems in a straightforward manner.

INTRODUCTION   Combinational logic theory and useful design aids are presented in this chapter. Combinational logic analysis results in a formal description of a logic circuit, a description that is in the form of an algebraic expression or equivalent graphical formats known as truth tables and Karnaugh maps respectively. Switching algebra is used to perform the analysis.

Logic *design* typically starts from an informal word description. The design challenge is to convert the informal description to a formal description so that you can proceed to synthesize a solution. Combinational logic synthesis procedures convert a formal description into a logic circuit.

3.1 Axiomatic Basis of Switching Algebra

Huntington proved that the following six axioms provide a complete basis for *Boolean algebra*. We define a switching algebra consisting of a set of elements S and the operators +, *, and ' where the following axioms are true.

AXIOM 1 Commutation
For all elements x, y of S we have commutation:

$$x + y = y + x \qquad \text{(A1)}$$

and

$$x * y = y * x. \qquad \text{(A1D)}$$

AXIOM 2 Distribution
For all elements x, y, z of S we have distribution:

$$x * (y + z) = (x * y) + (x * z) \qquad \text{(A2)}$$

and

$$x + (y * z) = (x + y) * (x + z). \qquad \text{(A2D)}$$

AXIOM 3 Neutral Elements Exist
There are distinct elements 0 and 1 of S such that for any element x of S we have

$$x + 0 = x \qquad \text{(A3)}$$

and

$$x * 1 = x. \qquad \text{(A3D)}$$

AXIOM 4 Inverse Elements Exist
For any element x of S there is an element of S, denoted as x', such that

$$x + x' = 1 \qquad \text{(A4)}$$

and

$$x * x' = 0. \qquad \text{(A4D)}$$

AXIOM 5 There exist at least two elements x,y in S such that x does not equal y.

AXIOM 6 If elements x and y are members of the set S, then elements $x * y$ and $x + y$ are also members of S.

When there are only two elements 0, 1 in S, the Boolean Algebra is called a switching algebra.

$$S = \{0, 1\}$$

Operator Precedence: The conventional operator hierarchy is to execute $'$ first, then $*$, and $+$ last.

Important: switching theory is based on the axioms.

Axioms and Theorems of Switching Algebra: The axioms and theorems of switching algebra are shown in Table 3.1. The theorems are proved in this chapter.

TABLE 3.1 Axioms and Theorems of Switching Algebra

Commutative		
AXIOM 1 $x + y = y + x$		**AXIOM 1D** $x * y = y * x$
Distributive		
AXIOM 2 $x(y + z) = xy + xz$		**AXIOM 2D** $x + yz = (x + y)(x + z)$
Operations with 1 and 0		
AXIOM A3 $x + 0 = x$		**AXIOM 3D** $x * 1 = x$
Complementarity		
AXIOM 4 $x + x' = 1$		**AXIOM 4D** $x * x' = 0$
Idempotent		
THEOREM 2 $x + x = x$		**THEOREM 2D** $x * x = x$

TABLE 3.1   **(continued)**

Involution
THEOREM 3 $(x')' = x$

Complementarity

THEOREM 4 $x + 1 = 1$	**THEOREM 4D** $x * 0 = 0$

Consensus
THEOREM 6 $xy + x'z + yz = xy + x'z$
THEOREM 6D $(x + y)(x' + z)(y + z) = (x + y)(x' + z)$

Simplification

THEOREM 5 $x + xy = x$	**THEOREM 5D** $x(x + y) = x$
THEOREM 7 $xy + xy' = x$	**THEOREM 7D** $(x + y)(x + y') = x$
THEOREM 8 $x + x'y = x + y$	**THEOREM 8D** $x(x' + y) = xy$
LEMMA 1	
$x + (x' + y) = 1$	$x(x'y) = 0$

DeMorgan

THEOREM 9 $(x + y)' = x'y'$	
and for n variables	$(x + y + \cdots + n)' = x'y' \cdots n'$
THEOREM 9D $(xy)' = x' + y'$	
and for n variables	$(xy \cdots n)' = x' + y' + \cdots + n'$

Associative

THEOREM 10 $x + (y + z) = (x + y) + z$	**THEOREM 10D** $x(yz) = (xy)z$

3.2   ## Duality

The first four axioms present their equations in two parts as pairs. Each pair has the following property: one part is changed into the other part when all constants 0 and 1 are exchanged and all operators $*$, $+$ are exchanged. This property defines the *dual* of a Boolean expression. The property implements the *principle of duality*.

> **Principle of Duality**   If a Boolean statement is true then the dual of the statement is true.

Duality may help solve
difficult problems.

Because this principle of duality is true for all axioms, the principle is
applicable to any theorem in switching algebra; the dual of any theorem is true (Example 3.1). The procedures implementing duality (and
not the Boolean algebra) are what is important in the following examples. (Boolean algebra is used in Examples 3.2 and 3.3 without explanation in order to not scatter the examples of duality throughout the
text.)

EXAMPLE 3.1 **Duality is True for All Axioms and Theorems**

The two equations of Axiom 3 are duals.

$$x + 0 = x$$
$$\downarrow \quad \downarrow$$
$$x * 1 = x$$

EXAMPLE 3.2 **Omission of Parentheses Leads to Precedence Rule Violation**

Omitting parentheses erroneously can change operator precedence
when forming the dual. The algebra is not the point here.

$$g = x'y + xy'$$
$$= x' * y + x * y'$$
$$\downarrow \quad \downarrow \quad \downarrow$$
$$h = x' + y * x + y'$$
$$= x' + yx + y'$$
$$= 1 \qquad \text{Wrong!}$$

This is wrong because omission of parentheses leads to violation of
precedence rules. See example 3.3.

EXAMPLE 3.3   **Correctly Forming the Dual**

When forming the dual do not complement or otherwise change any variable. Also take care to add parentheses before forming the dual in order to avoid change of operator precedence. The algebra is not the point here. Compare this example to Example 3.2.

$$g = x'y + xy'$$
$$= (x' * y) + (x * y')$$
$$= (x' * y) + (x * y')$$
$$\downarrow \qquad \downarrow \qquad \downarrow$$
$$h = (x' + y) * (x + y')$$
$$= (x' + y)(x + y') \qquad \text{Correct!}$$

3.3

Proving Theorems

All new theorems can be proved using the axioms and the old theorems by two methods: an algebric process and the method of *perfect induction*. The perfect induction method is presented in Section 3.11.5. You can prove the dual of a theorem is true by invoking duality (for example, see Theorem 6D on page 103). When proving theorems, there is a benefit beyond the derived result. The process itself is instructive because the details of the techniques used in the proofs generally are useful when processing combinational equations. Here are two theorems proved by an algebraic process.

Proofs develop algebraic skills.

THEOREM 1 The Element 0 is Unique

If there are two zero elements 0 and p, then by Axiom 3 for any element x

$$x + 0 = x$$

and

$$x + p = x.$$

Substitute $x = p$ in the first equation and $x = 0$ in the second.

$$p + 0 = p$$

and

$$0 + p = 0$$

Since $0 + p = p + 0$ by Axiom 1, we have

$$p = 0 \quad \text{Q.E.D.}$$

THEOREM 1D The Element 1 is Unique

Suppose there are two elements 1 and q. Then by Axiom 3 for any element x

$$x * 1 = x$$

and

$$x * q = x.$$

Substitute $x = q$ in the first equation and $x = 1$ in the second.

$$q * 1 = q$$

and

$$1 * q = 1$$

Since $q * 1 = 1 * q$ by Axiom 1 we have

$$q = 1 \quad \text{Q.E.D.}$$

3.4 Constants

There are only two constants in switching algebra: *true, T, and false, F,* represented by the symbols 1 and 0 respectively. The symbols 1 and 0 are not numbers here. Different meanings may be given to 1 and 0 (see Table 3.2). Voltage was discussed in Chapter 2.

3.5 Variables

Switching algebra is not an algebra of numbers. Switching algebra is an algebra of *states* represented by the constants 0 and 1. A variable is in stage 0 or it is in state 1.

The algebra of the constants follows from the axioms. In the next section, this list is derived from the axioms.

$$0 + 0 = 0 \qquad 0 * 0 = 0$$
$$0 + 1 = 1 \qquad 0 * 1 = 0$$
$$1 + 0 = 1 \qquad 1 * 0 = 0$$
$$1 + 1 = 1 \qquad 1 * 1 = 1$$

With some exceptions (i.e., $1 + 1 = 1$), the constants and variables used in switching theory axioms and theorems follow rules that are

TABLE 3.2   **Possible Meanings of 0 and 1**

Four possible binary pairs interpreted as the meaning of 0 and 1.

CONSTANT	SWITCH$_1$	SWITCH$_2$	VOLTAGE$_1$	VOLTAGE$_2$
0	open	closed	5 volts	0 volts
1	closed	open	0 volts	5 volts

identical to the rules of elementary arithmetic and algebra. But even one exception forces us to emphasize switching algebra is not an algebra of numbers. The constants 0 and 1 do not represent quantities.

Variables are assigned one of two states: 0 or 1.

Variables are assigned the constants 0 or 1. Again, these constants are not numbers, because 0 does not mean "zero" and 1 does not mean "one" in switching algebra. 0 and 1 represent states. 0 and 1 may mean a switch is on or off, a voltage is high or low, or some other binary pair (Table 3.2). Nevertheless, there is no harm thinking in terms of numbers 0 and 1 as long as the context is not ignored.

Switching algebra has many applications in addition to digital logic circuits. Switching algebra may be applied to states of capacitor charges, magnetic domains, relay contacts, and other mechanical devices.

In switching algebra any letter of the alphabet is used to represent a switching variable, because 0 and 1 are the only constants. Using symbols for constants is rare in this algebra.

A literal is a variable or the complement of a variable, e.g., x, x′, y, y′, z, z′. The term literal is used to simplify discussions. It is easier to say "the term is a product of literals" than to say "the term is a product of variables or the complements of variables."

3.6   Operators

AND, OR, NOT are the three logical operators.

The operators ′, *, and + used in the axioms have the names NOT, AND, and OR respectively. We derive the properties of NOT, AND, and OR directly from the axioms and two new theorems. The properties are summarized in Table 3.3 on page 93.

Operator symbols replace the operator words as follows.

$$\text{NOT } x = x'$$

$$x \text{ AND } y = x * y$$

$$x \text{ OR } \quad y = x + y$$

In decimal arithmetic the addition operator is defined by a table of all possible sums. We need similar tables for the logical operators $'$, $*$, and $+$. How are combinations such as $1 + 0$ evaluated? The axioms are used to evaluate all possible combinations of 0 and 1.

The simplest functions use one operator.

$$g(x,y) = x'$$

$$g(x,y) = xy \qquad \text{(* can be omitted without error)}$$

$$g(x,y) = x + y$$

Using these functions we apply the axioms to build tables defining the three Boolean operators.

AND   The AND operator equation for two input variables x and y and for output g is

$$g = xy$$

The variables x and y can be replaced only by the two constants 0 and 1. In how many ways can x and y be replaced with 0 and 1?

Think of the function xy as having two positions or slots, one for each variable. Clearly each slot can be filled two ways—by a 1 or by a 0. Therefore the number of combinations is $2 * 2 = 2^2 = 4$. The combinations are the x,y pairs 0,0   0,1   1,0   1,1. You can display what you are doing in a table.

x	y	g
0	0	
0	1	
1	0	
1	1	

What is g for all combinations of the constants 0 and 1 when substituted for x and y in the equation $g = xy$?

$$g = x * y$$

$$g = 0 * 0 = ?$$

$$g = 0 * 1 = ?$$

$$g = 1 * 0 = ?$$

$$g = 1 * 1 = ?$$

We reach back to the axioms to find the answers. The axioms are the only source of the calculations needed to evaluate $g = x * y$ for all combinations of x,y.

Axiom 3D tells us that

$$x * 1 = x$$

so that

$$g = 0 * 1 = 0 \quad \text{when } x = 0$$

and

$$g = 1 * 1 = 1 \quad \text{when } x = 1.$$

Axiom 1D tells us that

$$x * y = y * x$$

so that

$$0 * 1 = 1 * 0$$

and thus

$$g = 1 * 0 = 0.$$

The results allow us to fill the table as follows.

x	y	g	
0	0		
0	1	0	Axiom 3D
1	0	0	Axiom 3D and 1D
1	1	1	Axiom 3D

One table entry is still missing because there is no axiom that can help evaluate $g = 0 * 0$. However, Theorem 2D (see below) provides the solution:

$$x * x = x \quad \text{implies} \quad 0 * 0 = 0 \quad \text{when } x = 0.$$

AND function is true when all inputs are true.

This result completes the table defining the AND ($*$) operator.

x	y	xy	
0	0	0	Theorem 2D
0	1	0	Axiom 3
1	0	0	Axioms 3 and 1
1	1	1	Axiom 3

The AND function output is true when all input variables are true.

THEOREM 2D For All x in the Set S: $x * x = x$

Proof:
$$x * x = xx + 0 \qquad \text{A3}$$
$$= xx + xx' \qquad \text{A4D}$$
$$= x(x + x') \qquad \text{A2}$$
$$= x * 1 \qquad \text{A4}$$
$$= x \qquad \text{Q.E.D.} \qquad \text{A3D}$$

OR The OR operator equation for two input variables x and y and for output g is

$$g = x + y$$

Axiom 3:
$$x + 0 = x$$
$$g = 0 + 0 = 0$$
$$g = 1 + 0 = 1$$

Axiom 1:
$$x + y = y + x$$
$$1 + 0 = 0 + 1$$

so that $\quad g = 0 + 1 = 1.$

No axiom can evaluate the $1 + 1$ combination. Theorem 2 (see below) provides the solution.

$$x + x = x \quad \text{implies} \quad 1 + 1 = 1 \quad \text{when } x = 1.$$

OR function is true when any input is true.

This completes the table defining the OR (+) operator.

x	y	$x + y$	
0	0	0	Axiom 3
0	1	1	Axioms 3 and 1
1	0	1	Axiom 3
1	1	1	Theorem 2

The OR function output is true when one or more input variables are true.

THEOREM 2 For All x in the Set S: $x + x = x$

$$
\begin{aligned}
\text{Proof:} \quad x + x &= (x + x)(1) & \text{A3D} \\
&= (x + x)(x + x') & \text{A4} \\
&= x + xx' & \text{A2D} \\
&= x + 0 & \text{A4D} \\
&= x & \text{A3}
\end{aligned}
$$

NOT function comple-
ments a variable.

NOT   The NOT operator equation for one input variable x and for output g is

$$g = x'.$$

Axiom 4, 4D: $x + x' = 1$ $x * x' = 0$.

For $x = 0$ $0 + x' = 1$

Axiom 3D: $0 + z = z$ and $z = 1$ implies $x' = 1$ when $x = 0$.

For $x = 1$ $1 * x' = 0$

Axiom 3: $1 * z = z$ and $z = 0$ implies $x' = 0$ when $x = 1$.

x	x'
0	1
1	0

The NOT function forms the complement of a variable.

The axioms and theorems were used to create tables for each operator. The NOT, AND, and OR *truth tables,* equations, and symbols are shown in Table 3.3. The symbols are discussed in Chapter 4.

Truth tables represent
functions in graphic
form.

These tables are called truth tables because they display when the function defined by the table is true as well as when the function is false for all combinations of inputs.

3.7   Functions

In switching algebra $g(x)$ is taken to mean a function of x as is done in calculus. Any switching function g is represented in three major ways: equation, truth table, and *Karnaugh* map. Bridges from one representation to another are needed. Here the bridges between equations and

TABLE 3.3 NOT, AND, and OR Truth Tables, Equations, and Symbols

Operator truth tables:

NOT		AND			OR		
x	g	x	y	g	x	y	g
0	1	0	0	0	0	0	0
1	0	0	1	0	0	1	1
		1	0	0	1	0	1
		1	1	1	1	1	1

Operator equations:

$$g = x'$$ $$g = xy$$ $$g = x + y$$

Operator symbols:

truth tables are built in four ways. These bridges are needed to move on to the very important Karnaugh map method (Section 3.12 on page 125).

3.7.1 Bridge from Sum of Terms Equation to Truth Table

The equation $g = x + y$ is called a *sum of terms* because + appears to represent a summation operator. In fact the + symbol represents the OR operator, and the sum is a *logical sum of terms* (OR of ANDs). (Equations like $g = xy$ are one-term sums.) In the process of evaluating the elementary NOT, AND, and OR functions truth tables have been introduced.

Substitute all combinations of 0 and 1 to create truth table.

$$g = x' \quad (g \text{ equals NOT } x)$$

$$g = xy \quad (g \text{ equals } x \text{ AND } y)$$

$$g = x + y \quad (g \text{ equals } x \text{ OR } y)$$

The exclusive or, XOR, function is more complex. XOR is an example of a composite operator requiring two NOT, two AND, and one OR operators.

$$g(x,y) = x'y + xy' = x \text{ xor } y = x \oplus y \quad (g \text{ equals } x \text{ XOR } y)$$

Even though, x, x', y, y' appear in the equation, the equation is still treated as $g(x,y)$ using two variables and not four. This follows from the fact that the complement of a variable is not independent of the variable (Theorem 3).

EXAMPLE 3.4   **XOR Sum of Terms Evaluation Using a Truth Table**

The XOR equation is evaluated for all x,y combinations. In this example the process is easier because the derived NOT, AND, and OR tables are used instead of using the axioms. The x',y' columns ease the task.

x	y	x'	y'	$g = x'y + xy' = ?$
0	0	1	1	$g = 1 * 0 + 0 * 1 = 0$
0	1	1	0	$g = 1 * 1 + 0 * 0 = 1$
1	0	0	1	$g = 0 * 0 + 1 * 1 = 1$
1	1	0	0	$g = 0 * 1 + 1 * 0 = 0$

ROW	x	y	$x \oplus y$
0	0	0	0
1	0	1	1
2	1	0	1
3	1	1	0

THEOREM 3 For All x in the Set S: $(x')' = x$
   The complement of the complement of x equals x.
      Proof: if x' is the complement of x then by axioms 4 and 4D

$$x + x' = 1 \quad \text{and} \quad x * x' = 0$$

From these equations we can say the complement of x' is x.

$$(x')' = x \quad \text{Q.E.D.}$$

EXERCISE 3.1   Derive the truth table for $g = x$ XOR y'

Answer:

ROW	x	y	g
0	0	0	1
1	0	1	0
2	1	0	0
3	1	1	1

∎

3.7.2 Bridge from Product of Terms Equation to Truth Table

Equations such as the XOR equation are also written as a product of terms.

$$g(x,y) = (x + y)(x' + y') = x \text{ XOR } y = x \oplus y$$

The $g(x,y)$ equation is called a *product of terms* because the implied $*$ appears to represent a multiplication operator. In fact the implied $*$ symbol represents the AND operator and the product is a *logical product of terms* (AND of ORs). Equations like $g = (x + y)$ are one-term products.

EXAMPLE 3.5   **XOR Product of Terms Evaluation Using a Truth Table**

x	y	x'	y'	$g = (x + y)(x' + y')$
0	0	1	1	$g = (0 + 0)(1 + 1) = 0 * 1 = 0$
0	1	1	0	$g = (0 + 1)(1 + 0) = 1 * 1 = 1$
1	0	0	1	$g = (1 + 0)(0 + 1) = 1 * 1 = 1$
1	1	0	0	$g = (1 + 1)(0 + 0) = 1 * 0 = 0$

ROW	x	y	$x \oplus y$
0	0	0	0
1	0	1	1
2	1	0	1
3	1	1	0

so that $g(x,y) = x \oplus y$ as before

EXERCISE 3.2   Derive the truth table for $g = (x + y')(x' + y)$

Answer:

ROW	x	y	g
0	0	0	1
1	0	1	0
2	1	0	0
3	1	1	1

∎

3.7.3   Bridge from Truth Table to Sum of Terms Equation

Pretend you do not know the equation for XOR. Can you derive it from a truth table definition? The answer is yes. Here is how to do it. Start from the truth table definition of XOR. Add to it the $g(x,y)$ column to emphasize that each row is an evaluation of g.

Each truth table row can generate one term of an equation.

ROW	x	y	g	$g(x,y)$
0	0	0	0	$g(0,0)$
1	0	1	1	$g(0,1)$
2	1	0	1	$g(1,0)$
3	1	1	0	$g(1,1)$

The process that converts a truth table into an equation describing when $g = 1$ is based upon two important facts.

1. Each truth table row represents a term that is a product (AND) of the input variables. (In row$_1$ when $x = 0$ AND $y = 1$ output $g = 1$.)
2. The function g equals the logical sum of product terms (OR of ANDs) representing truth table rows whose value is one. (Output $g = 1$ when the terms representing row$_1$ OR row$_2$ are true.)

Function g equals 1 on two rows. So, two terms are needed.

Sum of terms includes terms from rows where $g = 1$.

Truth table row$_1$ tells you $g = 1$ when $x = 0$ and $y = 1$. What function $g_1(x,y) = 1$ when $x = 0$ and $y = 1$? Observe that $x' = 1$ when $x = 0$. Therefore, in row$_1$

$$x'y = 1 * 1 = 1.$$

Let

$$g_1(x,y) = x'y$$

Similarly for row_2 $g = 1$ when $x = 1$ and $y = 0$. Since $y' = 1$ when $y = 0$ in row_2

$$xy' = 1 * 1 = 1.$$

Let
$$g_2(x,y) = xy'$$

Now use the second fact. Function $g = 1$ in row_1 and row_2. Therefore

$$g = g_1 + g_2 = x'y + xy' = x \oplus y$$

EXAMPLE 3.6   **From Truth Table to Sum of Terms Equation**

Derive $g(x,y)$ from the OR table.

OR	x	y	g
	0	0	0
	0	1	1
	1	0	1
	1	1	1

The OR table has three 1s in the g column implying three terms in the expression for g. The truth table tells us that $g = 1$ when $x'y$ or xy' or xy equal 1. Therefore, the equation for g is

$$
\begin{aligned}
g &= x'y + xy' + xy & \\
&= x'y + x(y' + y) & \text{A2} \\
&= x'y + x(1) & \text{A4} \\
&= x'y + x & \text{A3D} \\
&= (x' + x)(y + x) & \text{A2D} \\
&= (1)(y + x) & \text{A4} \\
&= y + x & \text{A3D}
\end{aligned}
$$

The three-term expression reduces to $x + y$, which is no surprise, because the truth table was originally generated by $g = x + y$.

EXERCISE 3.3   Derive the reduced sum of terms (OR of ANDs) equation for g from the truth table.

ROW	x	y	g
0	0	0	1
1	0	1	1
2	1	0	1
3	1	1	0

Answer: $g = x' + y'$ ■

3.7.4 Bridge from Truth Table to Product of Terms Equation

Once again, pretend you do not know the equation for XOR. Can you derive it in product of terms (AND of ORs) format from a truth table definition? The answer is yes if you focus on when $g = 0$ instead of when $g = 1$.

ROW	x	y	g	$g(x,y)$
0	0	0	0	$g(0,0)$
1	0	1	1	$g(0,1)$
2	1	0	1	$g(1,0)$
3	1	1	0	$g(1,1)$

The process that converts a truth table into an equation describing when $g = 0$ is based upon two important facts.

Product of terms includes terms from rows where $g = 0$.

1. Each truth table row represents a term that is a sum of the input variables. (In row_0 output $g = 0$ when $x + y = 0$.)
2. The function g equals the logical product of sum terms representing truth table rows whose value is zero. (Output $g = 0$ when the terms representing row_0 AND row_3 are false.)

Function g equals zero on two rows. So, two terms are needed.
   Truth table row_0 tells you $g = 0$ when $x = 0$ and $y = 0$. What logical sum of variables function $g_0(x,y) = 0$ when $x = 0$ and $y = 0$? Clearly in row_0 $x + y = 0 + 0 = 0$.
   Let $g_0(x,y) = (x + y)$.

Truth table row_3 tells you $g = 0$ when $x = 1$ and $y = 1$. What logical sum of variables function $g_3(x,y) = 0$ when $x = 1$ and $y = 1$? When x and y equal 1, x' and y' equal 0. Therefore in row_3 $x' + y'$ is 0 because x and y both equal 1.

Let $g_3(x,y) = (x' + y')$.

Now use the second fact. Function $g = 0$ in rows 0 and 3. This implies that

$$g = g_0 * g_3 = (x + y)(x' + y') = x \oplus y$$

EXERCISE 3.4

Derive the reduced sum of terms (AND of ORs) equation for g from the truth table.

ROW	x	y	g
0	0	0	1
1	0	1	1
2	1	0	1
3	1	1	0

Answer: $g = (x' + y')$ ■

Observations on sums of products (OR of ANDs)

1. $g = x + y$ is a sum of two terms. Sum means using the OR operator to build an expression for g including the two terms.

2. When an input variable (x or y) is 1 in the table for a row where $g = 1$, the variable (x or y) appears in the product term of the expression for g corresponding to that row.

3. When an input variable (x or y) is 0 in the table for a row where $g = 1$, the complement of the variable (x' or y') appears in the product term of the expression for g corresponding to that row.

4. $g = 1$ when any product term in the sum is 1. In fact when g is the sum of any number of terms, this observation is true (Theorem 4). Note that x can be any expression, for example, a sum of terms.

THEOREM 4 For All x in the Set S: $x + 1 = 1$

Proof:

$$x + 1 = (x + 1)(1) \qquad \text{A3D}$$
$$= (x + 1)(x + x') \qquad \text{A4}$$
$$= (x + 1 * x') \qquad \text{A2D}$$
$$= x + x' \qquad \text{A3D}$$
$$= 1 \qquad \text{A4}$$

Observations on products of sums (AND of ORs)

1. $g = (x + y)(x' + y')$ is a product of two terms. Product means using the AND operator to build an expression for g including the two terms.

2. When an input variable (x or y) is 0 in the table for a row where $g = 0$, the variable (x or y) appears in the sum term of the expression for g corresponding to that row.

3. When an input variable (x or y) is 1 in the table for a row where $g = 0$, the complement of the variable (x' or y') appears in the term of the expression for g corresponding to that row.

4. $g = 0$ when any sum term in the product is 0. In fact when g is the product of any number of terms, this observation is true (Theorem 4D). Note that x can be any expression, for example, a product of terms.

THEOREM 4D For All x in the Set S: $x * 0 = 0$

$$\begin{array}{lll} \text{Proof:} & x * 0 = (x * 0) + (0) & \text{A3} \\ & = (x * 0) + (x * x') & \text{A4D} \\ & = x(0 + x') & \text{A2} \\ & = x * x' & \text{A3} \\ & = 0 & \text{A4D} \end{array}$$

3.7.5   More about Building a Truth Table

Truth tables are constructed by listing all combinations of the input variables' values with one combination per row and defining the output for each row.

One input variable has two combinations of values.

$$0 \quad 1$$

n inputs have 2^n combinations.

Two input variables have four combinations of values.

$$00 \quad 01 \quad 10 \quad 11$$

Three input variables have eight combinations of values.

$$000 \quad 001 \quad 010 \quad 011 \quad 100 \quad 101 \quad 110 \quad 111$$

And so forth. . . .

These combinations of values may be considered binary numbers. This interpretation allows saying that the combinations are listed in numerical order to build the input side of the truth table independently of the output side. Then the definition of the output function g prescribes the list of 1s and 0s in the g column to complete the table. When more than one output function needs to be defined for the same variable, one output column is added for each output. This point leads to the question: how many output functions can be prescribed for a given number of input variables?

3.7.6 Enumeration of Switching Functions

You have learned that one form for all functions of switching variables is a logical sum of product terms (OR of ANDs).

TABLE 3.4 **All Functions of Two Variables**

$x'y'$	$x'y$	xy'	xy	SUM OF TERMS	g EQUATION	COMMON NAME
0	0	0	0	0	0	
0	0	0	1	xy	xy	AND
0	0	1	0	xy'	xy'	
0	0	1	1	$xy' + xy$	x	
0	1	0	0	$x'y$	$x'y$	
0	1	0	1	$x'y + xy$	y	
0	1	1	0	$x'y + xy'$	$x'y + xy'$	XOR
0	1	1	1	$x'y + xy' + xy$	$x + y$	OR
1	0	0	0	$x'y'$	$x'y'$	#NOR $f = (x + y)'$
1	0	0	1	$x'y' + xy$	$x'y' + xy$	XNOR
1	0	1	0	$x'y' + xy'$	y'	
1	0	1	1	$x'y' + xy' + xy$	$x + y'$	
1	1	0	0	$x'y' + x'y$	x'	
1	1	0	1	$x'y' + x'y + xy$	$x' + y$	
1	1	1	0	$x'y' + x'y + xy'$	$x' + y'$	#NAND $f = (xy)'$
1	1	1	1	$x'y' + x'y + xy' + xy$	1	

See Section 3.10.

Rearranging the list we emphasize the 16 functions of 2 variables.

0	1		
x'	x	y'	y
$x'y'$	$x'y$	xy'	xy
$x' + y'$	$x' + y$	$x + y'$	$x + y$
$x'y' + xy$	$x'y + xy'$		

There are 2^2 or four possible functions of one variable because g is defined by filling two slots each with a 0 or a 1. The four functions are shown below in truth table format.

x	$g_0(x)$	$g_1(x)$	$g_2(x)$	$g_3(x)$
0	0	0	1	1
1	0	1	0	1

There are 2^m functions because m rows have 2^m combinations.

The g equations follow directly from the truth table:

$$g_0 = 0$$

$$g_1 = x$$

$$g_2 = x'$$

$$g_3 = 1$$

The truth table for a function g of two variables has four rows. Now g is 0 or 1 in each row. Therefore, g can have zero to four terms in its logical sum. The method used above shows us that the four terms that may be included in a function of two variables are $x'y'$, $x'y$, xy', and xy.

In how many ways can a group of these terms be selected to form a function? Let 1 mean a term is included and 0 mean a term is excluded from a logical sum. The results are shown in Table 3.4.

For n variables there must be $m = 2^n$ terms. And for m terms there are $f = 2^m$ functions. Therefore $f = 2^{2^n}$. The behavior of this function is astronomical even for small n (Table 3.5).

TABLE 3.5   The Equation $f = 2^{2^n}$

n	f
1	4
2	16
3	256
4	65536
5	4,294,967,296

3.8 Function Simplification

Algebra is used to simplify functions by eliminating terms, combining terms, and eliminating *literals* in terms. A literal is any variable, or its complement, in any term. In other words, terms are made up from literals which represent variables. The object of function simplification is a less complex logic circuit.

Theorems 5, 5D, 6, and 6D help to eliminate terms.

THEOREM 5 For All x, y in the Set S $x + xy = x$

Proof:		
	$x + xy = x * 1 + xy$	A3D
	$= x(1 + y)$	A2
	$= x(1)$	Theorem 4
	$= x$	A3D

THEOREM 5D For All x, y in the Set S $x(x + y) = x$

Proof:		
	$x(x + y) = x * x + x * y$	A2
	$= x + x * y$	Theorem 2D
	$= x * 1 + x * y$	A3D
	$= x(1 + y)$	A2D
	$= x(1)$	Theorem 4
	$= x$	A3D

THEOREM 6 For All x, y, z in the Set S
$xy + x'z + yz = xy + x'z$ (consensus)

Proof:		
	$xy + x'z + yz = xy + x'z + yz(x + x')$	A3D, A4
	$= xy(1 + z) + x'z(1 + y)$	A1D, A2
	$= xy(1) + x'z(1)$	Theorem 4
	$= xy + x'z$ Q.E.D.	A3D

THEOREM 6D For All x, y, z in the Set S
$(x + y)(x' + z)(y + z) = (x + y)(x' + z)$

Proof: This is the dual of Theorem 6.

$$
\begin{aligned}
xy \quad &+ \quad x'z \quad + \quad yz \quad = \quad xy \quad + \quad x'z \\
(x * y) &+ (x' * z) + (y * z) = (x * y) + (x' * z)
\end{aligned}
$$

$$\downarrow \quad \downarrow \quad\quad \downarrow \quad \downarrow \quad \downarrow \quad\quad\quad \downarrow \quad \downarrow \quad\quad \downarrow$$

$$(x + y) * (x' + z) * (y + z) = (x + y) * (x' + z) \quad \text{Q.E.D.}$$

EXAMPLE 3.7 Recognizing Expressions Equivalent to Variables in Theorems

The trick to solving problems is to recognize expressions equivalent to the x and y variables in the theorems.

$$
\begin{aligned}
g &= pq'r' \quad + q'rs \quad + pq's \\
&= r'(pq') + r(q's) + pq's
\end{aligned}
$$

Let $x = r'$, $y = pq'$, $z = q's$, and recognize that $pq's = (pq')(q's)$.

Then $g = xy + x'z + yz = xy + x'z$ (Theorem 6, consensus)

Therefore $g = r'(pq') + r(q's)$

Theorems 2, 7, and 7D help to combine terms.

THEOREM 2 For All x in the Set S $x + x = x$

Proof.		
	$x + x = (x + x)(1)$	A3D
	$= (x + x)(x + x')$	A4
	$= x + xx'$	A2
	$= x + 0$	A4D
	$= x$	A3

THEOREM 7 For All x, y in the Set S $xy + xy' = x$

Proof:		
	$xy + xy' = x(y + y')$	A2D
	$= x(1)$	A4
	$= x$	A3D

THEOREM 7D For All x, y in the Set S $(x + y)(x + y') = x$
Proof: Dual of Theorem 7.

EXAMPLE 3.8 $g = pq'r + pqr$

Let $x = pr$, $y = q$.
Then

$$g = xy' + xy = x \qquad \text{(Theorem 7)}$$

Therefore

$$g = pr$$

Theorems 8 and 8D help us to eliminate literals.

THEOREM 8 For All x, y in the Set S $x + x'y = x + y$

Proof:		
$x + x'y = (x + x')(x + y)$		A2
$= (1)(x + y)$		A4
$= x + y$ Q.E.D.		A3D

THEOREM 8D For All x, y in the Set S $x(x' + y) = xy$

Proof:		
$x(x' + y) = xx' + xy$		A2D
$= 0 + xy$		A4D
$= xy$ Q.E.D.		A3

EXAMPLE 3.9 $pqr's + r = pqs + r$

Let $x = r$, $y = pqs$. Use Theorem 8 $(x + x'y = x + y)$.

EXERCISE 3.5 Use algebra to simplify the function $g = w'x' + wy'z + xy'z$.

Answer: $g = w'x' + y'z$ ∎

EXERCISE 3.6 Use algebra to simplify the function $g = (w + x)(w' + y + z)$.

Answer: $g = w'x + wz + wy$ ∎

3.9 ━━━━ DeMorgan's Theorems

DeMorgan's very important theorems find the complement or inverse of an expression. For two variables DeMorgan provided us with two theorems.

$$(x + y)' = x'y' \qquad \text{Theorem 9}$$

$$(xy)' = x' + y' \qquad \text{Theorem 9D}$$

Extremely important theorems.

An Important Observation Based on Axiom 4 if $x + y = 1$ and $xy = 0$ then x and y are complements.

LEMMA 1 $x + (x' + y) = 1 \qquad x(x'y) = 0$

Note: Lemma 1 is used in the proof that switching variables have the associative property, which is not an axiom. Thus this yet-to-be-proved property is not available for use in the following proof. In other words, you cannot drop parentheses in this proof unless you invoke an axiom or theorem.

$$x + (x' + y) = 1 \qquad x(x'y) = 0$$

Proof:

$$
\begin{aligned}
x + (x' + y) &= (1)[x + (x' + y)] && \text{A3D} \\
&= (x + x')[x + (x' + y)] && \text{A4} \\
&= x + x'(x' + y) && \text{A2D} \\
&= x + x' && \text{T5D} \\
&= 1 && \text{A4}
\end{aligned}
$$

$$
\begin{aligned}
x(x'y) &= 0 + x(x'y) && \text{A3} \\
&= xx' + x(x'y) && \text{A4D} \\
&= x(x' + x'y) && \text{A2} \\
&= xx' && \text{T5} \\
&= 0 && \text{A4D}
\end{aligned}
$$

THEOREM 9 $(x + y)' = x'y'$

and for n variables

$$(x + y + \cdots + n)' = x'y' \cdots n' \qquad \text{(DeMorgan)}$$

Method: Let $p = (x + y)$ and $q = (x'y')$.

If $p + q = 1$ and $pq = 0$ then p and q are complements by Axiom 4.

Then $(x + y)$ and $(x'y')$ are complements.

Proof: $(x + y) + (x'y') = [(x + y) + x'] * [x + y) + y']$ A2D, A1

$$= 1 \qquad\qquad * 1 \qquad\qquad \text{Lemma 1}$$

$$= 1 \qquad\qquad\qquad\qquad\qquad \text{A3D}$$

$$(x + y) * (x'y') = x(x'y') + y(y'x') \qquad\qquad \text{A2, A1D}$$

$$= 0 \qquad\qquad + 0 \qquad\qquad \text{Lemma 1}$$

$$= 0 \quad \text{Q.E.D.} \qquad\qquad\qquad \text{A3D}$$

THEOREM 9D $(xy)' = x' + y'$

and for n variables

$$(xy \cdots n)' = x' + y' + \cdots + n' \qquad \text{(DeMorgan)}$$

Method: Let $p = (xy)$ and $q = (x' + y')$.

If $p + q = 1$ and $pq = 0$ then p and q are complements by Axiom 4.

Then (xy) and $(x' + y')$ are complements.

Proof: $(xy) + (x' + y') = [x + (x' + y')] * [(y + (y' + x')]$

$$\qquad\qquad\qquad\qquad\qquad\qquad \text{A2D, A1}$$

$$= 1 \qquad\qquad * 1 \qquad\qquad \text{Lemma 1}$$

$$= 1 \qquad\qquad\qquad\qquad\qquad \text{A3D}$$

$$(xy) * (x' + y') = (xy)x' + (xy)y' \qquad\qquad \text{A2, A1D}$$

$$= 0 \qquad\qquad + 0 \qquad\qquad \text{Lemma 1}$$

$$= 0 \quad \text{Q.E.D.} \qquad\qquad\qquad \text{A3}$$

DeMorgan's theorems do not explicitly reveal the relations between complementary functions. A generalization of the theorem in the following form has been suggested by Shannon.

$$g(x_1, x_2, \cdots x_n, +, *) = g'(x_1', x_2', \cdots x_n', *, +)$$

DeMorgan's theorem in this form shows that the complement of any function is obtained by making the replacements.

$$x_j \rightarrow x_j'$$

$$+ \rightarrow *$$

$$* \rightarrow +$$

Care is necessary when executing this process because parentheses may be implicit in the function as written. See Section 3.2 on duality.

Product terms become sum of complements.

The complementation concept is of great importance in the design of switching functions. Complementation expresses the idea that for every switching network there exists another network which has exactly complementary properties.

Sum terms become product of complements.

Furthermore, simplification of switching functions is sometimes easier when working with the complementary function. This is neither obvious nor predictable. Sometimes a function's complement has fewer terms or is in a form more amenable to application of theorems and so forth. Working with the function's complement is an option that can be helpful and is worth trying when you are having difficulty with the function itself.

THEOREM 10 $a + (b + c) = (a + b) + c$

10D $a * (b * c) = (a * b) * c$

This is the proof of the associative property.

Proof: Let $y = a + (b + c)$

$$x = (a + b) + c$$

We use a variation on the method of complements. If x and y' are complements then $x = y$.

Part 1.

$$x + y' = x + a'(b'c') \qquad \text{T9D}$$

$$= (x + a') * [x + (b'c')] \qquad \text{A2D}$$

$$= (x + a') * [(x + b') * (x + c')] \qquad \text{A2D}$$

Evaluate each term:

$$x + a' = (a + b) + c + a' \qquad \text{definition}$$

$$= (a + b) + a' + c \qquad \text{A1}$$

$$= 1 + c \qquad \text{Lemma 1}$$

$$= 1 \qquad \text{T4}$$

$$x + b' = 1 \qquad \text{same as } x + a'$$

$$x + c' = (a + b) + c + c'$$

$$= (a + b) + 1 \qquad \text{A4}$$

$$= 1 \qquad \text{T4}$$

Therefore
$$x + y' = 1 * [1 * 1]$$

$$= 1 \qquad \text{A3D, A3D}$$

Part 2.

$$x * y' = [(a + b) + c]y'$$

$$= (a + b)y' + cy' \qquad \text{A2}$$

$$= ay' + by' + cy' \qquad \text{A2}$$

Evaluate each term:

$$ay' = a[a'(b'c')] = 0 \qquad \text{T9, Lemma 1}$$

$$b * y' = b * b' + b * y' \qquad \text{A3, A4D}$$

$$= b * (b' + y') \qquad \text{A2}$$

$$= b * (b' + [a' * (b' * c')]) \qquad \text{T9D}$$

$$= b * \{[b' + (b' * c')] * [b' + a']\} \qquad \text{A2D}$$

$$= b * \{b' * [b' + a']\} \qquad \text{T5}$$

$$= b * \{b'\} \qquad \text{T5D}$$

$$= 0 \qquad \text{A4D}$$

$$c * y' = 0 \qquad \text{see } b * y'$$

Therefore $\quad x * y' = 0 + 0 + 0 = 0$ Q.E.D. $\qquad$ A3

EXERCISE 3.7 Find the complement of $g = y + z'x + xy'z$. Do not simplify.

Answer: $g' = y'(z + x')(x' + y + z')$ ∎

EXERCISE 3.8 Find the complement of $g = (w + x)(w' + y + z)$

Answer: $g' = w'x' + wy'z'$ ∎

SIDELIGHT   **Physical Interpretation of Algebraic Theorems and Axioms**

Logic networks implemented with single-pole-single-throw switches have the advantage of great simplicity. A network of switch contacts exhibits only two conditions of conduction between its terminals: open or closed. The digits 0 and 1 can be assigned to represent these two conditions of operation in either of two ways. The digit 0 can represent the open-circuit condition, and then the digit 1 represents the closed-circuit condition. The other way to make the assignment is to let the digit 1 represent the open-circuit condition so that the digit 0 represents the closed-circuit condition.

In this chapter the theorems of switching algebra have been developed without regard for physical meaning or applications. Here is one possible physical meaning to the theorems by implementing them with networks of switch contacts. The digit 0 represents the open-circuit condition, and the digit 1 represents the closed-circuit condition. This assignment means that any variable x is represented by a normally closed switch contact and the complement x' is represented by a normally open switch contact. Networks for some of the theorems and axioms in Table 3.1 follow.

Theorem or Axiom	Switching Circuit

Theorem or Axiom	Switching Circuit

Canonical Forms

There are two *canonical* forms of logic equations: *standard sum of products* (OR of ANDs) and *standard product of sums* (AND of ORs), which are defined below. A specific equation need not be in canonical form. Why bother with canonical forms? One answer is that all equations are simplifications of the canonical forms. Another answer is that truth tables are lists of terms used in canonical forms.

Reminder: a literal is a variable or the complement of a variable.

3.10.1 Canonical Product Form Definition

A product term (AND) consists of one literal or the logical product of two or more literals. A product term is in canonical product form when there is exactly one occurrence of every function variable in the term. Every variable or its complement is used in the term. Every is the key word. A canonical product term has the special name *minterm*. Lower case m is used for the minterm symbol.

The prefix min is used because each minterm is true for only one combination of variables. This is consistent with the fact each truth table row represents only one combination of variables. Another point of view is that AND'ing a group of literals is analogous to calculating the minimum of their values. For n variables there are 2^n minterms corresponding to 2^n truth table rows. For example, two variables have four minterms and four truth table rows. The four minterm symbols are m_0, m_1, m_2, and m_3.

> Product term is 1 for only one combination of its literals.

	VARIABLES				MINTERMS			
					m_0	m_1	m_2	m_3
Row	x	y	x'	y'	$x'y'$	$x'y$	xy'	xy
0	0	0	1	1	1	0	0	0
1	0	1	1	0	0	1	0	0
2	1	0	0	1	0	0	1	0
3	1	1	0	0	0	0	0	1

An equation, which is the sum of one or more minterms, is in a sum of products (OR of ANDs) canonical form. Any term can be converted into canonical form by the use of Axiom 4: $x + x' = 1$. Other axioms may be involved, but Axiom 4 is the essential one as shown in the following examples.

EXAMPLE 3.10   Truth Table to Sum of Minterms

ROW	x	y	g_1	g_2
0	0	0	1	1
1	0	1	1	1
2	1	0	0	1
3	1	1	0	0

$g_1(x,y) = m_0 + m_1$
$\quad\quad\quad = x'y' + x'y$
$g_2(x,y) = m_0 + m_1 + m_2$
$\quad\quad\quad = x'y' + x'y + xy'$

EXAMPLE 3.11   **Expanding x' into Canonical OR of ANDs Form**

$$g(x,y) = x' \qquad \text{given (not in canonical form)}$$
$$= x' * 1 \qquad\qquad\qquad \text{A3}$$
$$= x' * (y + y') \qquad\qquad \text{A4}$$
$$= x'y' + x'y \qquad\qquad \text{A2, A1}$$
$$= m_0 + m_1$$

EXAMPLE 3.12   **Expanding 1 into Canonical OR of ANDs Form**

$$g(x,y) = 1 \qquad\qquad\qquad\qquad\qquad\qquad \text{given}$$
$$= 1 * 1 \qquad\qquad\qquad\qquad\qquad \text{A3}$$
$$= (x + x')(y + y') \qquad\qquad\qquad \text{A4}$$
$$= x'y' + x'y + xy' + xy \quad \text{A2 repeated, A1}$$
$$= m_0 + m_1 + m_2 + m_3$$

EXERCISE 3.9   Expand $g = x + y'$ into canonical OR of ANDs form.

Answer: $g = x'y' + xy' + xy$ ∎

EXERCISE 3.10   Expand $g = x' + y'$ into canonical OR of ANDs form.

Answer: $g = x'y' + x'y + xy'$ ∎

3.10.2   Canonical Sum Form Definition

A sum term (OR) consists of one literal or the logical sum of two or more literals. A sum term is in canonical sum form when there is exactly one occurrence of every function variable in the term. Every variable or its complement is used in the term. Again, every is the key word. A canonical sum term has the special name *maxterm*. Upper case M is used for the maxterm symbol.

The prefix max is used because each maxterm is true for all combinations of variables except one. This is usually stated in another way: a maxterm is zero for only one combination of variables. Another point of view is that OR'ing a group of literals is analogous to calculating the maximum of their values. When we learn that a maxterm is the complement of some minterm, the single zero is consistent with the fact that each truth table row represents only one combination of variables.

For n variables there are 2^n maxterms corresponding to 2^n truth table rows. For example, two variables have four maxterms and four truth table rows. The four maxterm symbols are M_0, M_1, M_2, and M_3.

Sum term is 0 for only one combination of its literals.

VARIABLES				MAXTERMS			
				M_0	M_1	M_2	M_3
x	y	x'	y'	$x + y$	$x + y'$	$x' + y$	$x' + y'$
0	0	1	1	0	1	1	1
0	1	1	0	1	0	1	1
1	0	0	1	1	1	0	1
1	1	0	0	1	1	1	0

Observe that a maxterm is false for only one combination of variables. Important note: a maxterm is the complement of the minterm in the same row. This is readily proved by using DeMorgan's Theorem.

An equation of one or more maxterms is in a product of sums (AND of ORs) canonical form. Any term can be converted into canonical form by the use of Axiom 3: $x = x + 0$. Other axioms may be involved, but Axiom 3 is the essential one as shown in the following examples.

EXAMPLE 3.13   **Truth Table to Product of Minterms**

ROW	x	y	g_1	g_2
0	0	0	1	1
1	0	1	1	1
2	1	0	0	1
3	1	1	0	0

$g_1 = M_2 * M_3$
$\quad = (x' + y)(x' + y')$
$g_2 = M_3$
$\quad = (x' + y')$

EXAMPLE 3.14   **Expand x' into Canonical AND of ORs Form**

$$g(x,y) = x' \qquad\qquad\qquad \text{Given}$$
$$= x' + 0 \qquad\qquad\qquad \text{A3}$$
$$= x' + yy' \qquad\qquad\qquad \text{A4}$$
$$= (x' + y)(x' + y') \qquad\qquad \text{A2D}$$
$$= M_2 * M_3$$

EXAMPLE 3.15   **Expand 0 into Canonical AND of ORs Form**

$$g(x,y) = 0 \qquad\qquad\qquad\qquad\qquad \text{Given}$$
$$= 0 + 0 \qquad\qquad\qquad\qquad\qquad \text{A3}$$
$$= xx' + yy' \qquad\qquad\qquad\qquad \text{A4D}$$
$$= (xx' + y) * (xx' + y') \qquad\qquad \text{A2D}$$
$$= (x + y)(x' + y)(x + y')(x' + y') \qquad \text{A2D}$$
$$= (x + y)(x + y')(x' + y)(x' + y') \qquad \text{A1}$$
$$= M_0 * M_1 * M_2 * M_3$$

EXERCISE 3.11   Expand $g = xy$ into canonical AND of ORs form.

Answer: $g = (x + y)(x + y')(x' + y)$ ∎

EXERCISE 3.12   Expand $g = y$ into canonical AND of ORs form.

Answer: $g = (x + y)(x' + y)$ ∎

3.10.3   Minterms and Maxterms

Truth table row numbers are the decimal equivalents of the binary number representing the 0, 1 assignments to variables on each row. A function is defined when the function's output column is assigned 0 or 1 on each row. This is why when row numbers become minterm numbers m_j any function can be represented by a list of m_j. If the output function is 1 on row_j then m_j is included in the list (Example 3.10).

Minterm or maxterm number is same as truth table row number.

On the other hand, when row numbers become maxterm numbers M_k any function can be represented by a list of M_k. If the output function is 0 on row_k then M_k is included in the list (Example 3.13).

Examples 3.10 and 3.13 reveal that g_1 (or any g) can be written as a sum of minterms or as a product of maxterms. Look at the subscript numbers; for any g all subscripts are included when m_j and M_k are shown. If you know the m_j list you can readily deduce the M_k list and vice versa.

$$g_1 = m_0 + m_1 = M_2 * M_3 \qquad g_2 = m_0 + m_1 + m_2 = M_3$$

Converting expressions to m_j or M_k   Axiom 2 and Axiom 4 are the keys to the conversion processes.

$$
\begin{aligned}
g(x,y,z) &= x + y \\
&= x + y + 0 \\
&= x + y + zz' \\
&= (x + y + z')(x + y + z) && \text{A2D} \\
&= M_1 * M_0 \\
g(x,y,z) &= x + y \\
&= x(y + y')(z + z') + y(x + x')(z + z') && \text{A4} \\
&= x'yz' + x'yz + xy'z' + xy'z + xyz' + xyz \\
&= m_2 + m_3 + m_4 + m_5 + m_6 + m_7
\end{aligned}
$$

EXAMPLE 3.10 (Repeated)

When $g = 1$ on row_j include minterm m_j.

ROW	x	y	g_1	g_2
0	0	0	1	1
1	0	1	1	1
2	1	0	0	1
3	1	1	0	0

$$
\begin{aligned}
g_1(x,y) &= m_0 + m_1 \\
&= x'y' + x'y \\
g_2(x,y) &= m_0 + m_1 + m_2 \\
&= x'y' + x'y + xy'
\end{aligned}
$$

EXAMPLE 3.13 (Repeated)

When $g = 0$ on row_j include maxterm M_j.

ROW	x	y	g_1	g_2	
0	0	0	1	1	$g_1 = M_2 * M_3$
1	0	1	1	1	$= (x' + y)(x' + y')$
2	1	0	0	1	$g_2 = M_3$
3	1	1	0	0	$= (x' + y')$

EXERCISE 3.13

Use algebra to convert $g = x + yz$ into a sum of minterms.

Answer: $g(x,y,z) = m_3 + m_4 + m_5 + m_6 + m_7$ ■

EXERCISE 3.14

Use algebra to convert $g = x + yz$ into a product of maxterms.

Answer: $g(x,y,z) = M_0 * M_1 * M_2$ ■

3.11 Design Aid: The Truth Table

A design process usually begins with words, not equations. Words are more readily translated into a graphical display, such as a truth table, than into equations. One major reason for this is that most of the time the words incompletely specify the problem. The incompleteness is manifested as missing truth table entries. You cannot know how to write equations when there are "missing entries." However, you can deduce how to fill in missing entries in a truth table in a way that is consistent with the problem statement. How are truth tables used?

First two methods are presented: filling a table from an equation and deriving an equation from a table. Then how to derive four types of equations from a truth table, how to use table entered variables to reduce table size, and how to prove theorems by the table based method of perfect induction are shown.

3.11.1 Numerical Description of Functions

Interpret combinations of states in rows as numbers.

As functions become more complex, term-by-term comparisons become more difficult and the number of errors tends to increase. When *numerical descriptions of functions* are used, the work is facilitated. In fact minterm and maxterm symbols m_j and M_k allow us to assign numbers to the terms. We develop the idea further.

The input side of truth tables is a list of all combinations of input variables. For variables $x\ y\ z$ the binary combination entered into truth table row$_5$ is 1 0 1. One simplification was invoked when the rows were numbered with the decimal equivalents of the binary combinations. Row$_5$ represents minterm m_5 which equals $xy'z$.

If we designate a variable as 1 and a complemented variable as 0, then $xy'z$ becomes 101 and the step to minterm m_5 is natural. The association is reversed for maxterms: 0 for the variable and 1 for the complemented variable so that $x' + y + z'$ becomes 101 and the maxterm for row$_5$ is M_5. [Reminder: $m_5 = M_5'$ and $M_5 = m_5'$.] Carried to its logical conclusion, the shorthand notation for a function is a list of minterms or maxterms. For example, $g = \Sigma\, m(1,2,3,5,6) = \Pi\, M(0,4,7)$.

We suggest the following mechanical procedure for writing out functions with essentially zero errors.

$g(w,x,y) = \Sigma\, m(1,2,3,5,6)$

$g = m_1 \qquad + m_2 \qquad + m_3 \qquad + m_5 \qquad + m_6$ (given)

$g \rightarrow w\ x\ y + w\ x\ y + w\ x\ y + w\ x\ y + w\ x\ y$ (no primes)

$\quad\ 0\ \ 0\ 1\ \ \ 0\ \ 1\ 0\ \ \ 0\ \ 1\ 1\ \ \ 1\ \ 0\ 1\ \ \ 1\ \ 1\ 0$ (binary number)

$g = w'\ x'\ y + w'\ x\ y' + w'\ x\ y + w\ x'\ y + w\ x\ y'$ (minterms)

For $f(w,x,y,z)$ do you prefer to compare literals or numbers in the following list to discover that $m_2 + m_6 = w'yz'$ and $m_4 + m_5 = w'xy'$? Try both ways.

m_2	$w'x'y\ z'$	0010
m_4	$w'x\ y'z'$	0100
m_5	$w'x\ y'z$	0101
m_6	$w'x\ y\ z'$	0110

3.11.2 Equation to Truth Table

The equation $g(w,x,y) = wxy' + w'x + y$ has a truth table. Create the input side of the truth table by observing there are three variables each taking two values 0, 1. Two values can fill three slots in $2 * 2 * 2 = 8$ ways. So write eight rows giving the wxy columns binary numbers 000 through 111. Keeping in mind 0 means false and 1 means true, note that the eight numbers represent the eight possible minterms three variables can have. Now, when does $g = 1$?

The g column is filled by evaluating each term of the equation for g for the eight variable combinations or minterms. When is g true?

Reminder: The first g term wxy' is true when $w = 1$, $x = 1$, and

Do not need equation in canonical form.

$y' = 1$. This means that $y = 0$. The number 110_2 corresponds to $w = 1$, $x = 1$, and $y = 0$. This is why you put a 1 in row_6 of the g output column.

The equation is analyzed as follows.

Key question: when is a term true?

g is true when wxy' is true, so mark a 1 for g on row_6.

g is true when $w'x$ is true. Since $w'x = w'xz' + w'xz$ g is true when $w'xz'$ or $w'xz$ are true (010 or 011, row_2 or row_3).

g is true when y is true. y is true on rows 1, 3, 5, and 7.

g' is the complement of g so evaluating g' is straightforward. Note: always add the g' column. This can be very useful.

ROW	w	x	y	g	g'	g TERM THAT IS TRUE		
0	0	0	0	0	1			
1	0	0	1	1	0	y		
2	0	1	0	1	0		$w'x$	
3	0	1	1	1	0	y	$w'x$	
4	1	0	0	0	1			
5	1	0	1	1	0	y		
6	1	1	0	1	0			wxy'
7	1	1	1	1	0	y		

3.11.3 Truth Table to Equation

Pretend you do not know the equation for g. How do you derive it from the truth table?

Consider row_5. On row_5 the variable values 101 correspond to $w = 1$, $x = 0$, and $y = 1$. One way to write the corresponding term is first to write wxy and then to mark the variables that take the zero state in this term with primes.

$m_5 = \text{minterm } 5 \rightarrow wxy \rightarrow wx'y$ which is true only on row_5.

This technique helps you avoid errors when dealing with a series of terms corresponding to the six rows where g is true. Start with

$$g \rightarrow wxy + wxy + wxy + wxy + wxy + wxy$$

and create the minterms m_1, m_2, m_3, m_5, m_6, m_7 by entering the

Tick (') minterm variables corresponding to zeros.

Tick (') maxterm variables corresponding to ones.

appropriate complements of the variables.

$$g = w'x'y + w'xy' + w'xy + wx'y + wxy' + wxy$$

$$= m_1 \quad + m_2 \quad + m_3 \quad + m_5 \quad + m_6 \quad + m_7$$

This canonical form for g reduces to the original equation.

Four equations can be derived from any truth table   These equations are the answers to the following questions.

When does $g = 1$?   (Example 3.16)
When does $g = 0$?   (Example 3.17)
When does $g' = 1$?   (Example 3.18)
When does $g' = 0$?   (Example 3.19)

The questions have a purpose: to find the equation with the minimum number of terms. Asking when g or $g' = 1$ results in a sum of minterms equation. And, asking when g or $g' = 0$ results in a product of maxterms equation. Furthermore and importantly, the maxterms may be replaced by complements of minterms. The above g function column has two 0s and six 1s. The g function is represented by a sum of six minterms or the product of two maxterms. The g' function is represented by a sum of two minterms or the product of six maxterms.

(Reminder: A minterm m_j is true only for (one) row_j of a truth table and a maxterm M_k is false for (one) row_k of a truth table.)

Here is an easy way to check each maxterm. Convert each term to a number by assuming the primed variables are 1s:

$$w + x' + y' \rightarrow 011 \rightarrow M_3$$

Recapitulating you have:

$$g = m_1 + m_2 + m_3 + m_5 + m_6 + m_7 = M_0 * M_4$$

$$g' = M_1 * M_2 * M_3 * M_5 * M_6 * M_7 = m_0 + m_4$$

EXAMPLE 3.16   When Does g = 1?

$$g = 1 \text{ on rows } 1,2,3,5,6,7.$$

Write out six minterms without primes.

$$g \rightarrow wxy + wxy + wxy + wxy + wxy + wxy$$

Create the minterms 1 2 3 5 6 7 by entering the complements.

$$g = w'x'y + w'xy' + w'xy + wx'y + wxy' + wxy$$

$$= m_1 \quad + m_2 \quad + m_3 \quad + m_5 \quad + m_6 \quad + m_7$$

EXAMPLE 3.17   **When Does $g = 0$?**

$$g = 0 \text{ on rows 0 and 4.}$$

Write out two maxterms without primes in the form of complements of minterms. We believe this action minimizes the potential for error.

$$g \rightarrow (wxy)'(wxy)'$$

Create minterms 0 and 4 by marking the complements.

$$
\begin{aligned}
g &= (w'x'y')'(wx'y')' \\
&= m_0' \quad * \quad m_4' \\
&= M_0 \quad * \quad M_4' \\
&= (w + x + y)(w' + x + y) \quad \text{By DeMorgan's Theorem}
\end{aligned}
$$

EXAMPLE 3.18   **When Does $g' = 1$?**

$$g' = 1 \text{ on rows 0 and 4.}$$

Write out two minterms without primes.

$$g' \rightarrow wxy + wxy$$

Create minterms 0 and 4 by marking the complements.

$$
\begin{aligned}
g' &= w'x'y' + wx'y' \\
&= m_0 \quad + m_4
\end{aligned}
$$

EXAMPLE 3.19   **When Does $g' = 0$?**

$$g' = 0 \text{ on rows 1,2,3,5,6,7}$$

Write out six maxterms without primes in the form of complements of minterms.

$$g' \rightarrow (wxy)' \ (wxy)' \ (wxy)' \ (wxy)' \ (wxy)' \ (wxy)'$$

Create minterms 1,2,3,5,6,7 by marking the complements.

$$
\begin{aligned}
g' &= (w'x'y)' * (w'xy')' * (w'xy)' * (wx'y)' * (wxy')' * (wxy)' \\
&= m_1' \quad\quad * m_2' \quad\quad * m_3' \quad\quad * m_5' \quad\quad * m_6' \quad\quad * m_7'
\end{aligned}
$$

$$= M_1 \quad * M_2 \quad * M_3 \quad * M_5 \quad * M_6 \quad * M_7$$

$$= (w + x + y')(w + x' + y)(w + x' + y')(w' + x + y')$$

$$(w' + x' + y)(w' + x' + y')$$

3.11.4 Table-entered Variables

Coefficients of terms are entered into output columns.

The size of a table is halved when an input variable is treated as a *table entered variable* instead of as an input. Furthermore, the table display is significantly improved in most, if not all, cases.

Normally the coefficient of a product term is the implied "1." Table-entered variables are based on the idea that one or more of the literals in a term may be defined as the coefficient of the term instead of the implied "1." This means the term is reduced to the literals not included in the coefficient. For example $1(xyz)$ becomes $x(yz)$. The new coefficient is x and the term reduces to yz. This idea is developed in two ways.

1. The equation $f = x$ xor y, for example, is usually written as the sum of two minterms with implied coefficient 1. However, a function of two variables has four minterms with four coefficients a_j ($j = 0$ to 3) where each a_j may be any Boolean expression.

$$f = xy' + x'y$$

$$f = 0 * xy + 1 * xy' + 1 * x'y + 0 * x'y'$$

$$f = 0 * m_3 + 1 * m_2 + 1 * m_1 + 0 * m_0$$

$$f = a_3 * m_3 + a_2 * m_2 + a_1 * m_1 + a_0 * m_0 \quad \text{(general case)}$$

Literals may be interpreted as coefficients of terms.

For example, let $a_3 = p$, $a_2 = 0$, $a_1 = pr + p'r'$, $a_0 = 1$. Then the corresponding truth table with table-entered variables p,r is as follows.

ROW	j	k	f	f
0	0	0	a_0	1
1	0	1	a_1	$pr + p'r'$
2	1	0	a_2	0
3	1	1	a_3	p

2. An eight row table is constructed from a three-variable equation. Then a four-row table is constructed from the same equation considered as a two-variable equation with one table entered variable. The meaning of this real equation is not relevant here. The equation is $q^+ = jq' + k'q$.

The three input variables are j, k, and q. Construct the table by evaluating g for all input variable combinations. Each evaluation result is 0 or 1.

ROW	j	k	q	q^+	NOTES
0	0	0	0	0	$q^+ = q$
1	0	0	1	1	
2	0	1	0	0	$q^+ = 0$
3	0	1	1	0	
4	1	0	0	1	$q^+ = 1$
5	1	0	1	1	
6	1	1	0	1	$q^+ = q'$
7	1	1	1	0	

Now let the input variables be j and k, while treating q as a table entered variable. Construct the table by evaluating q^+ for all j, k input variable combinations. Each evaluation result is 0, 1, q', or q. In effect each pair of rows from the above table merge into one row. The notes in the table above anticipate the new results.

ROW	j	k	q^+
0	0	0	q
1	0	1	0
2	1	0	1
3	1	1	q'

EXERCISE 3.15   Construct the table for $q^+ = t \oplus q$ with q as a table entered variable.

Answer:

ROW	t	q^+
0	0	q
1	1	q'

■

EXERCISE 3.16 Construct the table for $q^+ = s + r'q$ with q as a table entered variable.

Answer:

ROW	r	s	q^+
0	0	0	q
1	0	1	1
2	1	0	0
3	1	1	1

■

3.11.5 Method of Perfect Induction

The method of perfect induction is practical in a switching algebra because switching algebra has only the two constants 0 and 1. This method is not practical in algebras using real numbers where there are an infinity of numbers.

Only two constants make this practical.

Proof by perfect induction is implemented by substituting all combinations of values for the variables in the theorem's equation or any Boolean identity. The truth table provides an orderly means for implementing any proof. Create a truth table for each side of the equation and compare the tables row by row. The theorem or identity is proved if all rows are equal.

EXAMPLE 3.20 **Proof by Perfect Induction**

In practice you build tables to prove theorems such as

$$x + 1 = 1 \qquad\qquad x * 0 = 0$$

The relevant tables are

x	1	$x + 1$
0	1	1
1	1	1

x	0	$x * 0$
0	0	0
1	0	0

proving that $x + 1 = 1$ $\qquad\qquad x * 0 = 0.$

EXAMPLE 3.21 **Proof by Perfect Induction**

Here is another theorem: $x + x'y = x + y$

x	y	x'	$x'y$	$x + x'y$	$x + y$
0	0	1	0	0	0
0	1	1	1	1	1
1	0	0	0	1	1
1	1	0	0	1	1

proving by perfect induction that $x + x'y = x + y$.

3.12 Design Aid: The Karnaugh Map

The *Karnaugh map* (K map) allows you to simplify logic functions in a straightforward manner by providing a more useful graphic display of logic functions than truth tables. A map has one square corresponding to each row of a truth table. Since a minterm corresponds to a row, then each minterm corresponds to one square. The *squares* have the same numbers as truth table row numbers. A one is entered into a square if the function is true for the minterm the square represents. A zero is entered if the function is false. In a design process values can be directly entered into a K map. This is an alternative to using a truth table.

K map squares correspond to truth table rows.

3.12.1 Karnaugh Map Formats

K maps for functions of one, two, three, and four variables are in Figures 3.1, 3.2, 3.3, and 3.4. (See Section 3.13 for one way to deal with more than four variables. For tens of input variables tabular methods such as *Quine-McClusky* are used in industry in the context of CAD tools. See the references.) Each square corresponds to a minterm or, in other words, to a truth table row. The numbers in the squares correspond to truth table row numbers. A function is defined by placing a set of 1s and 0s in the squares just as you do with the output column in a truth table. Truth tables with multiple output columns generate one K map per output column.

Horizontal form is consistent with K maps of more variables.

FIGURE 3.1
K Map for functions of one variable

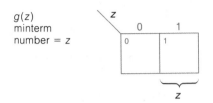

K map coordinates are written in several formats (Figures 3.3, 3.5, and 3.6) or in any redundant combination of these formats (Figure 3.7). We prefer using variables as map coordinates (Figure 3.3) when drawing maps by hand.

K maps with variables as map coordinates are easier to process.

Figure 3.7 emphasizes the relationships of K map coordinates, square numbers, minterms, and minterm numbers. The minterm number, the minterm, and the minterm symbol m_j are entered in each square.

3.12.2   Karnaugh Map Concept

The Karnaugh map is based on the concept of *adjacent states*. Two states are adjacent if they differ in the value of only one variable (variable x is in one state and x' is in the other state). This concept is important because the algebraic combination of two adjacent states results in the elimination of one variable. The Karnaugh map displays all the adjacencies present in a standard sum (function). Thus you can focus on *clusters* of adjacent terms and simplify the function on the map. (Note: *Subcube* is an alias for cluster.)

Key idea: adjacent states.

FIGURE 3.2
K Map for functions of two variables

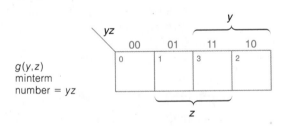

or

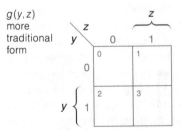

FIGURE 3.3
**K Map for functions
of three variables**

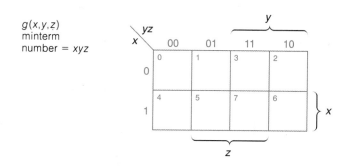

The concept of adjacent states forces the numbers in the K map squares to be out of sequence. In order to facilitate the explanation of why the numbers are out of sequence the numbers in the squares need to be changed to binary format (Figure 3.8). The 0 1 3 2 sequence is selected because the K map concept requires that any horizontal or vertical move from any square to any adjacent square changes only one bit in the number.

$$00 \rightarrow 01 \rightarrow 11 \rightarrow 10$$

$$00 \leftarrow 01 \leftarrow 11 \leftarrow 10$$

A map with this property allows you to use the axiom $x + x' = 1$ so you can implement the concept of adjacent states.

FIGURE 3.4
**K Map for functions
of four variables**

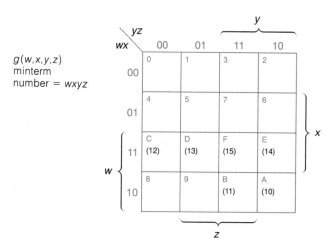

Numbering the squares with hexadecimal numbers is preferable in order to have one digit numbers in each square.

FIGURE 3.5
K Map coordinates using binary numbers

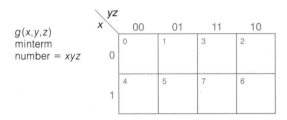

$g(x,y,z)$
minterm
number $= xyz$

EXAMPLE 3.22   **Adjacent States Illustrated by Horizontal and Vertical Moves**

In this map note that any horizontal or vertical move changes only one bit in the original number.

$g(x,y,z)$
minterm
number $= xyz$

	y		
000	001	011	010
100	101	111	110

z x

Horizontal and vertical moves change one variable   Next note that any horizontal or vertical move changes only one bit in the minterm number in a four-variable map (Figure 3.9). The bit change represents a variable changing from r to r' or r' to r.

$$1111 \rightarrow 1101 \text{ implies } wxyz \rightarrow wxy'z$$

$$1111 \rightarrow 1011 \text{ implies } wxyz \rightarrow wx'yz$$

For example, pretend you are sitting in square 0000 of Figure 3.9. The four moves out of this square are the following.

$$0000 \rightarrow 0001 \text{ implies } w'x'y'z' \rightarrow w'x'y'z$$

FIGURE 3.6
K Map coordinates using minterms

$g(x,y,z)$
minterm
number $= xyz$

x \ yz	$y'z'$	$y'z$	yz	yz'
x'	0	1	3	2
x	4	5	7	6

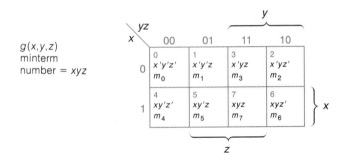

FIGURE 3.7

K Map coordinates using variables and binary numbers

$$0000 \rightarrow 0010 \text{ implies } w'x'y'z' \rightarrow w'x'y\ z'$$

$$0000 \rightarrow 0100 \text{ implies } w'x'y'z' \rightarrow w'x\ y'z'$$

$$0000 \rightarrow 1000 \text{ implies } w'x'y'z' \rightarrow w\ x'y'z'$$

Diagonal moves change two variables.

Diagonal moves are not useful Suppose you are sitting in square 1111 of Figure 3.9. The diagonal move from square 1111 to 1001 changes two variables.

$$1111 \rightarrow 1001 \quad \Rightarrow \quad wxyz \rightarrow wx'y'z$$

If $g = 1$ in those two squares then $g = wxyz + wx'y'z = wz(xy + x'y')$, which cannot be simplified using $x + x' = 1$ or any other method. This is why diagonal moves make no sense in K maps. Therefore, the states in diagonally related squares are not adjacent in the sense that $x + x' = 1$ cannot be used. Also, observe that they have no common boundary except a point.

3.12.3 Clusters

Clusters represent groups of terms. No function simplification results when the cluster is a group of 1 term. A cluster of two squares elimi-

FIGURE 3.8

K Map squares with binary numbers

$g(z)$
minterm
number = z

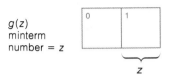

$g(y,z)$
minterm
number = yz

FIGURE 3.9
Four variable K map with binary numbers in the squares

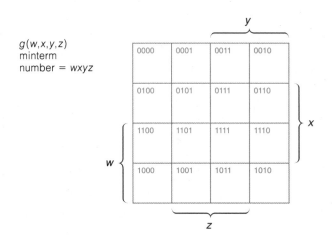

Clusters of adjacent states eliminate variables.

nates one variable from the term the cluster represents, a cluster of four squares eliminates two variables, a cluster of eight squares eliminates three variables, and a cluster including all squares eliminates all variables ($g = 1$ or 0). Cluster a rectangle of 2^n squares to eliminate n variables. Clusters must be rectangles because all pairs of squares in the cluster must be adjacent. [Gerrymandering is not allowed.] The number of ones (zeros) in a cluster must equal a power of 2.

Clusters of two squares eliminate one variable   Suppose $g = 1$ in squares 0 and 1 (Figure 3.10a). Then z is eliminated or drops out of the equation for g.

$$g = y'z' + y'z = y'(z' + z) = y'$$

When $g = 1$ in squares 3 and 1 (Figure 3.10b), y drops out of the equation for g.

$$g = yz + y'z = (y + y')z = z$$

Suppose $g = 1$ in squares 3 and 2 (Figure 3.10c). Then z drops out of the equation for g.

$$g = yz + yz' = y(z + z') = y$$

Think of the map as a cylinder when $g = 1$ in squares 0 and 2 (Figure 3.10d). This time y drops out.

$$g = y'z' + yz' = (y' + y)z' = z'$$

Observe that single squares represent minterms of two variables and that, in this two-variable K map, pairs of squares represent terms of one literal z or z' or y or y'.

FIGURE 3.10

K Map with clusters of two squares

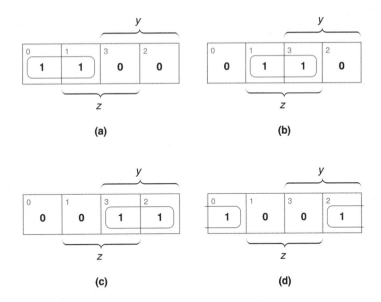

Clusters of four squares eliminate two variables   Typical clusters of four squares are shown in Figure 3.11. The four 1s in a row in the first map could be in any row or column. The square cluster can be located anywhere in the map; you can move it horizontally and/or vertically. The cluster in the third map is also a square cluster if you think of the map as a cylinder. The 1s at the four corners also form a square cluster. The equation above each map is the simplified term represented by the cluster. A square cluster of four eliminates two variables ($n = 2$ since $2^n = 4$). Here is how the equations are taken from the map.

The variables w and x are constant in the four squares of the row of 1s in Figure 3.11a. Therefore, w and x are factors of the four squares' minterms. The variables y and z are represented by their four two-variable minterms in the four squares. Now you can write the equation for g.

$$g = wx(y'z' + y'z + yz + yz') = wx$$

The literals x' and z' are constant in the four corner squares of Figure 3.11d. Therefore x' and z' are factors of the four squares' minterms. The variables w and y are represented by their four two variable minterms in the four squares. The equation for g follows.

$$g = x'z'(w'y' + w'y + wy + wy') = x'z'$$

The $w'y'$ and $x'z$ equations are taken from their K maps in a similar manner.

FIGURE 3.11
K Maps with typical clusters of four squares

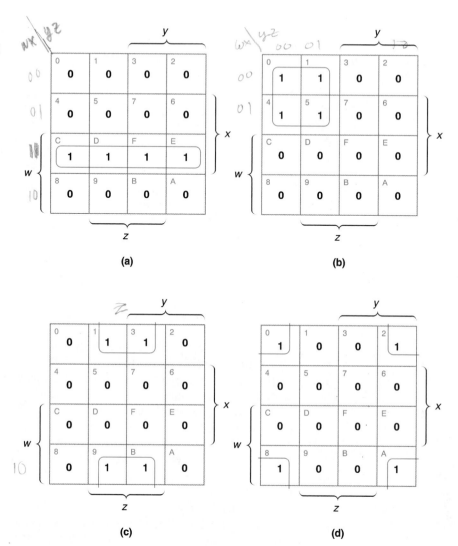

Clusters of eight squares eliminate three variables All of the clusters of eight possible in a four-variable map are shown in Figure 3.12. These always have two adjacent rows or two adjacent columns.

We build upon what we learned from deriving the equations for clusters of four squares. The equations for each of the two columns in Figure 3.12d are two-variable functions. Their sum equals z'. And, note that only z' is constant in the eight squares.

$$\text{column}_0 = y'z' \qquad \text{column}_2 = yz' \qquad \Rightarrow g = z'$$

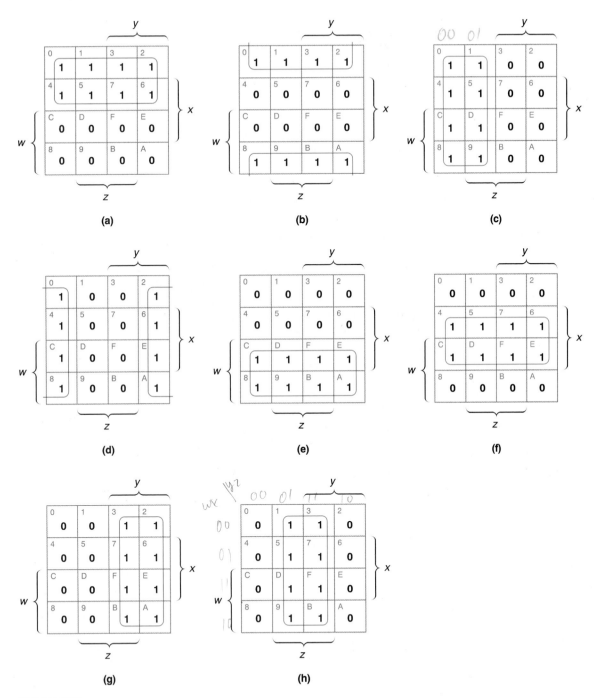

FIGURE 3.12
K Maps with clusters of eight squares

Guiding principle   A guiding principle for selecting clusters of minterms starts the process by selecting those 1 entries in the map which can be included in only one cluster. These unique 1s are included in the largest possible clusters. Select the clusters that include the greatest number of squares with 1s. The usefulness of the K map depends upon your ability to recognize clusters of entries.

Do not violate the guiding principle   In Figure 3.13 one might leap to include squares 0,1,4,5 in a cluster thereby violating the guiding principle. Including in a cluster the squares that can only belong to one cluster is the first task. Having done so, you can see that each of the four 1s are in some cluster making the square cluster of four redundant.

Squares that appear unrelated may be adjacent   The six four-square clusters possible in a three-variable map are shown in Figure 3.14. The $g = z'$ map (Figure 3.14c) shows that squares that appear unrelated are in fact adjacent. Think of the map as a cylinder.

Apply the theorem $x + x' = 1$ to eliminate variables that change value within the cluster. A square cluster of four squares means we can eliminate two variables; one variable for the horizontal adjacencies and the second variable for the vertical adjacencies defining the cluster. For example in the $g = z$ map (Figure 3.14d) horizontal adjacencies eliminate y, and vertical adjacencies eliminate x.

We can eliminate two variables in a rectangular cluster of four squares. In the $g = x$ map (Figure 3.14f) the y and z variables are eliminated by the two pairs of horizontal adjacencies.

Cluster isolated 1's (0's) first.

FIGURE 3.13
Writing and reading a four-variable K map

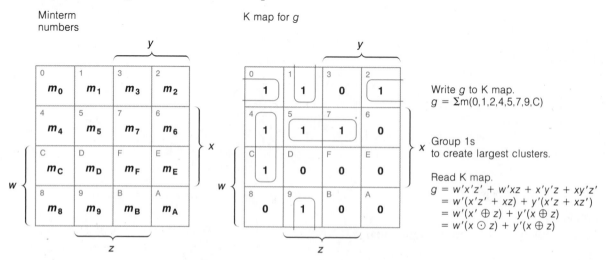

Write g to K map.
$g = \Sigma m(0,1,2,4,5,7,9,C)$

Group 1s
to create largest clusters.

Read K map.
$g = w'x'z' + w'xz + x'y'z + xy'z'$
$\ \ = w'(x'z' + xz) + y'(x'z + xz')$
$\ \ = w'(x' \oplus z) + y'(x \oplus z)$
$\ \ = w'(x \odot z) + y'(x \oplus z)$

FIGURE 3.14
Four-variable K maps with clusters of four squares

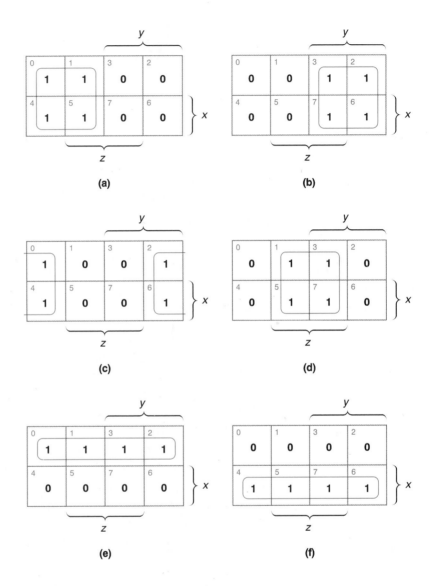

(a)

(b)

(c)

(d)

(e)

(f)

Clusters are not obvious in some maps In this K map (Figure 3.15) the fact that all clusters are clusters of four squares is not readily apparent. Note that minterms 5, 8, and B can only belong to one cluster each. The terms formed by these three clusters are $w'x$, wz', and $x'y$. This minimum sum is also unique.

There may be alternative ways to form clusters A clear example illustrating that a function can be written in more than one minimum sum

Minterm
numbers

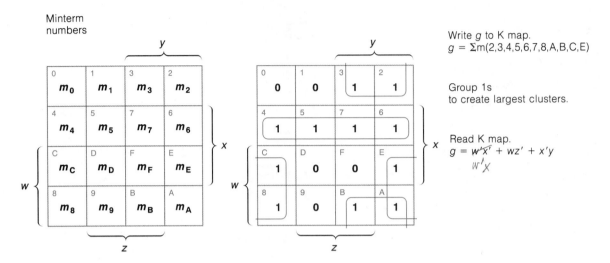

Write g to K map.
$g = \Sigma m(2,3,4,5,6,7,8,A,B,C,E)$

Group 1s
to create largest clusters.

Read K map.
$g = w'x' + wz' + x'y$

FIGURE 3.15
Writing and reading a four-variable K map

form is shown in Figure 3.16. Two forms are possible because there are two ways every 1 can be included in a cluster.

EXAMPLE 3.23   Consensus Term Cluster Overlaps other Clusters

The function g with minterms m_1, m_3, m_4, m_5 has a map whose 1s indicate when $g = 1$ and whose 0s indicate when $g = 0$. The lines around the adjacent pairs of 1s account for all the 1s. The clusters show that $g = x'z + xy'$. The redundant consensus term $y'z$ clearly stands out as a cluster overlapping the other clusters.

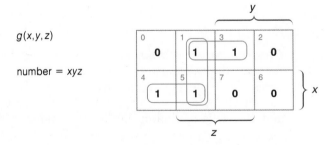

$g(x,y,z)$

number $= xyz$

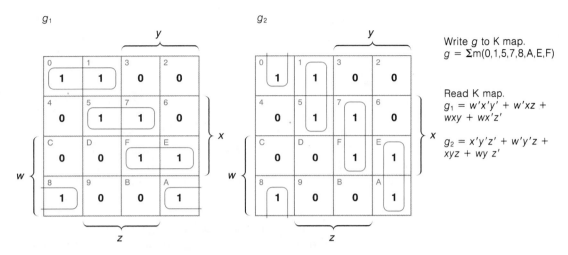

Write g to K map.
$g = \Sigma m(0,1,5,7,8,A,E,F)$

Read K map.
$g_1 = w'x'y' + w'xz + wxy + wx'z'$

$g_2 = x'y'z' + w'y'z + xyz + wy\,z'$

FIGURE 3.16
Writing and reading a four-variable K map

EXERCISE 3.17

Write $g = w'x'z + w'x' + wy'z + xy'z$ to a K map.
Answer:

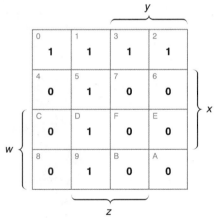

■

EXERCISE 3.18

Read the K map of Exercise 3.17

Answer: $g = w'x' + y'z$

■

3.12.4   Writing the Map

One way to use the map is to map or write a function g directly onto the map, term by term, so that you can read from the map a simplified

Write "small" clusters.
Read larger clusters to
simplify.

equation for the function. The sequence of operations writing and then reading the map in Example 3.24 is:

Given $g = y'z' + yz'$ write→   K map   read→ $g = z'$

The equation read from the map will be a simplified form of the original equation used to write the map if there are clusters greater than those written to the map.

Any function is a sum of all its minterms when the minterms whose coefficients are 0 are included. This point of view allows for a one-to-one correspondence between an equation and its K map.

Recapitulation: A function in canonical form is a sum of minterms. A map has one square for each minterm in a function. When one of those minterms in the sum is true, the function is true ($= 1$). When a minterm not in the sum is true, the function is false ($= 0$). Creating or writing a map is a process whereby for each minterm in the sum a 1 is entered in the map. And, for each minterm not in the sum a 0 is entered in the map. However, most functions are not in canonical form.

The function $g = xyz' + x'y + z$ is mapped onto a K map term by term. The process is illustrated in Example 3.25. The process shows that there is no need to start with an equation in canonical form. In effect the process calculates the value of each minterm as 1 or 0 and enters the values into the K map. The minterm equations are not explicitly determined.

EXAMPLE 3.24   **Writing and Reading a K-map**

Given $g = g(x,y) = 1 * y'z' + 1 * yz'$

Map this function onto a two-variable K map by placing a 1 in each square corresponding to each minterm with coefficient 1 and a 0 in each square corresponding to each minterm with coefficient 0.

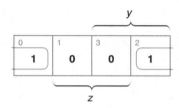

Form a two-square cluster from squares 0 and 2 and read $g = z'$.

EXAMPLE 3.25   **Writing a Non-canonical Form to a K-map**

Now we map a function in a series of steps that with experience can be consolidated into one step. First map each term of the equation $g = xyz' + x'y + z$ onto its own K map. Then OR the individual term maps onto one map.

(a) Map for the xyz' term

(b) Map for the $x'y$ term

(c) Map for the z term

(d) Function map, which is the OR of the individual term maps

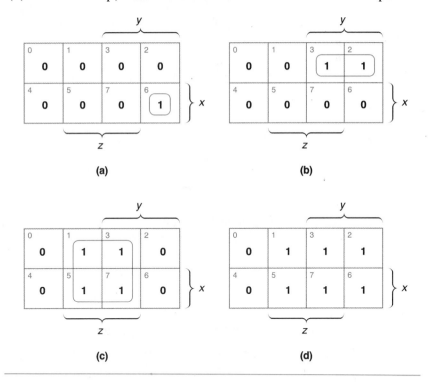

(a)   (b)

(c)   (d)

3.12.5   Reading the Map

The reading process provides an opportunity to simplify the function. The reading process is essentially a process of including adjacent 1s (minterms) or adjacent 0s (maxterms) in clusters with a rectangular shape and with a size that is some power of 2. The smallest number of clusters that include all the 1s (0s) yields a function with the minimum number of terms.

The usefulness of the K map depends upon your ability to recognize clusters of 1 (0) entries. Sometimes you may be particular about the form you want the circuit to take. Then usefulness also depends on your ability to select a set of clusters including all 1s (0s) in the map consistent with the form you have in mind. Read the K map by deciding what clusters of 1s include all the 1s in the map. Keep in mind that clusters contain 1, 2, 4 or 2^n squares. The guiding principle applies (page 134).

Recognize clusters of 2^n squares.

> **Important observation:** Terms can be used in more than one expression because Theorem 2 says $x + x = x$. Since squares represent terms, squares can be included in more than one cluster.

EXAMPLE 3.26   **Reading Clusters Simplifies Terms**

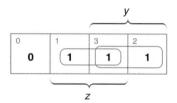

The cluster 1,3 reduces to a z term and the cluster 3,2 reduces to a y term. The three minterms represented by the three 1s reduce to $g = y + z$.

EXAMPLE 3.27   **A K-map for a Function that is Always True**

If there is a 1 in every square the function is alwas true. One cluster includes all terms, so $g = 1$ (the sum of all the minterms equals 1).

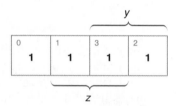

EXAMPLE 3.28 **A Cluster of One Square**

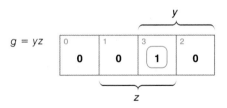

$g = yz$

EXAMPLE 3.29 **Eliminate Variables with Clusters**

Read the map for the function $g = xyz' + x'y + z$.

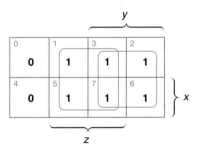

The largest clusters we can use are two clusters of four squares. The four 1s in the cluster above z condense to z and the four 1s in the cluster below y condense to y, so $f = y + z$. The algebraic proof follows.

Proof: $g_1 = x'y'z + xy'z + x'yz + xyz$ (four 1s above z)

 1   5   3   7

$ = (x' + x)y'z + (x' + x)yz = y'z + yz = z$

$g_2 = x'yz' + xyz' + x'yz + xyz$ (four 1s below y)

 2   6   3   7

$ = (x' + x)yz' + (x' + x)yz = yz' + yz = y$

$g = g_1 + g_2 = z + y$ Q.E.D.

Is it obvious that

$$g = xyz' + x'y + z = y + z?$$

Not really. Contrast simplification via K maps to algebraic simplification.

$$g = xyz' + x'y + z$$
$$= y(xz' + x') + z \qquad \text{A2}$$
$$= y(z' + x') + z \qquad \text{Theorem 8}$$
$$= (yz' + z) + yx' \qquad \text{A2, A1}$$
$$= (y + z) + yx' \qquad \text{Theorem 8}$$
$$= y(1 + x') + z \qquad \text{A1, A2}$$
$$= y + z \qquad \text{Theorem 4, A3D}$$

From minterms to equation via a K map   Many functions are specified by truth tables that are lists of minterms or maxterms such as the function g in Figure 3.17. Writing minterms into a K map is simple, however, reading that map is another matter.

The guiding principle in selecting clusters of minterms for a simplified equation is first to select the 1 (0) entries in the map that can be included in only one cluster. Minterm m_7 can only be clustered with m_5 forming the term $w'xz$. Minterm m_1 can only be clustered in the column of four squares forming $y'z$. The remaining two 1s, m_4 and m_0, are part of a square cluster forming the term xy'. Observe that each cluster contains at least one minterm that is in no other cluster. And, since the largest possible clusters are used, this is a fully simplified and unique sum of terms for the function g.

FIGURE 3.17
Writing and reading a four-variable K map

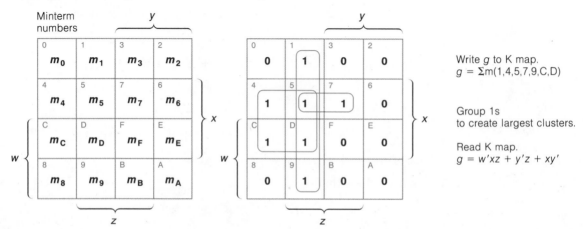

Write g to K map.
$g = \Sigma m(1,4,5,7,9,C,D)$

Group 1s
to create largest clusters.

Read K map.
$g = w'xz + y'z + xy'$

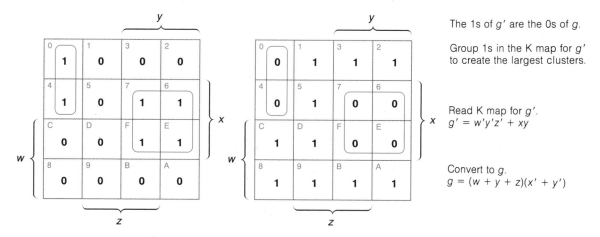

FIGURE 3.18
Reading maxterms from a four-variable K map

The 1s of g' are the 0s of g.

Group 1s in the K map for g' to create the largest clusters.

Read K map for g'.
$g' = w'y'z' + xy$

Convert to g.
$g = (w + y + z)(x' + y')$

Use g' when there are fewer 0s than 1s   Ones or zeros may be read from the map to form g or g'. One criteria for selecting g or g' is the relative number of 1s and 0s. When there are fewer 0s, form g' and then apply DeMorgan's Theorem to form g.

We calculate g' when the number of 0s is less than the number of 1s (Figure 3.18). The K map is a map of g, therefore when $g = 0$ the complementary function g' is 1; g' is the sum of minterms where $g = 0$. And $g = (g')'$ or g is taken directly from the map as the product of

Cluster zeros to read g'.

FIGURE 3.19
Reading maxterms from a four-variable K map

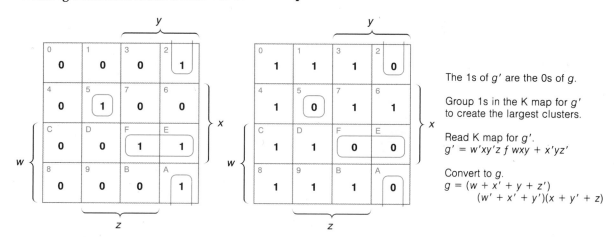

The 1s of g' are the 0s of g.

Group 1s in the K map for g' to create the largest clusters.

Read K map for g'.
$g' = w'xy'z$ f $wxy + x'yz'$

Convert to g.
$g = (w + x' + y + z')$
$(w' + x' + y')(x + y' + z)$

maxterms where $g = 0$. The maxterms are first written as the complements of minterms to minimize error potential. Another example is shown in Figure 3.19.

3.12.6 Don't Care Conditions

A function is defined by a sum of one or more minterms. A minterm is included (the coefficient is 1) or excluded (the coefficient is 0). The possibility arises that the coefficient of a minterm is not specified: i.e., it can take the value 0 or 1. Then we say we don't care if the minterm is included or excluded in the function. The corresponding truth table or K map entry may be 0 or 1. However, the don't care status is marked with a dash.

Don't care conditions are imposed by external sources when certain combinations of external source outputs never occur because of the logic controlling the external sources.

Clusters define don't cares as 1 or 0.

When reading a K map with dashes entered in one or more squares, you are free to assign each dash a 0 or a 1. The assignments seek to attain the goal of maximizing each cluster's size. In this way you completely specify the function. (Once the assignments are made, the function will generate a definite output if the corresponding input condition does occur even though it is not supposed to occur.)

Clusters must include at least one 1 or 0.

Caution: Do not use clusters of dashes. All clusters must include at least one 1 or one 0.

EXAMPLE 3.30   **Four Variable K Map with Don't Care Terms**

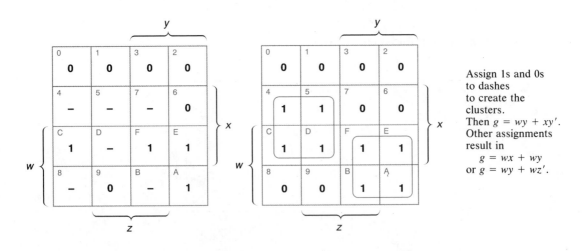

Assign 1s and 0s to dashes to create the clusters.
Then $g = wy + xy'$.
Other assignments result in
$g = wx + wy$
or $g = wy + wz'$.

EXAMPLE 3.31   Four Variable K Maps with Don't Care Terms

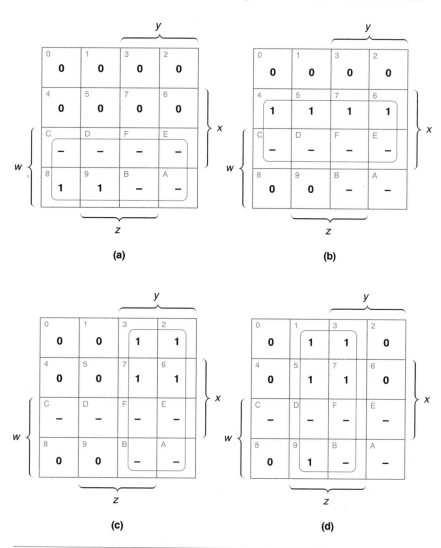

(a)   (b)

(c)   (d)

3.12.7   OR and AND Operations with K Maps

The effect of OR and AND operators is shown in Figure 3.20 and Figure 3.21. The applicable theorems for common terms are $1 + 1 = 1$ and $1 * 1 = 1$. And for other terms, applicable theorems are $1 + 0 = 1$

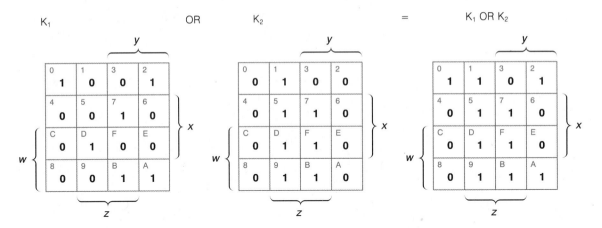

FIGURE 3.20
OR Operations on K maps

and $1 * 0 = 0$. The intuitive proof that these operations yield correct results follows.

Apply $m_j + m_j = m_j$, $m_j * m_j = m_j$, and $m_j * m_k = 0$ to the equations for the K maps in Figures 3.20 and 3.21.

Let $\quad K_1 = m_0 + \qquad m_2 + \qquad m_7 \qquad + m_A + m_B + m_D$

Let $\quad K_2 = \qquad m_1 + \qquad m_5 + m_7 + m_9 \qquad + m_B + m_D + m_F$

$K_1 + K_2 = m_0 + m_1 + m_2 + m_5 + m_7 + m_9 + m_A + m_B + m_D + m_F$

Let $\quad K_3 = m_0 + m_1 + m_2 + m_3 + m_8 + m_9 + m_A + m_B$

Let $\quad K_4 = m_1 + m_2 + m_5 + m_6 + m_9 + m_A + m_D + m_E$

FIGURE 3.21
AND Operations on K maps

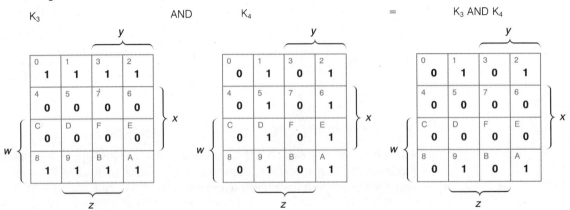

$$K_3 * K_4 = m_0 * K_4 + m_1 * K_4 + m_2 * K_4 + m_3 * K_4 + m_8 * K_4 + m_9 * K_4$$
$$+ m_A * K_4 + m_B * K_4$$
$$= 0 \qquad + m_1 * m_1 + m_2 * m_2 + 0 \qquad + 0 \qquad + m_9 * m_9$$
$$+ m_A * m_A + 0$$
$$= 0 \qquad + m_1 \qquad + m_2 \qquad + 0 \qquad + 0 \qquad + m_9$$
$$+ m_A \qquad + 0$$
$$= m_1 + m_2 + m_9 + m_A$$

3.13 ▬▬▬ Map-entered Variables

The K map becomes unwieldy even with five variables; with more than five variables the K map rapidly becomes unmanageable. The map-entered variable reduces the required map size, thereby extending the K map's practical usefulness. The number of function variables equals the number of K map dimensions plus the number of map-entered variables. How are variables entered into a map?

Up to this point the only K map entries have been 1, 0, and – (for don't care). This set of entries implies all variables in a function's equation are assigned a dimension of a K map.

K map entries are coefficients of product terms in the equation. Consider the simplest case. When one literal in a product term of n literals is treated as a coefficient, does the literal replace the 1 normally entered into the K map? The answer is yes. This idea is the basis for map-entered variables. In effect this idea is the same as the table-entered variable idea (Section 3.11.4).

> Coefficients of terms are entered into squares.

> Literals may be interpreted as coefficients of terms.

3.13.1 Writing the Map

The equation $g = xy + yz'$ has three variables: x, y, z. The number of K map variables reduces to two if x is treated as a coefficient. When x or x' is not in a term, the coefficient is 1.

$$g = xy + yz' = x(y) + 1(yz')$$

We know how to map the yz' term because the coefficient is 1. It maps into square 2. The coefficient of the y term is x. If the coefficient is one, the y term would map into squares 3 and 2. So, instead of 1 enter the coefficient x into squares 3 and 2 (Figure 3.22). The final step is to logically OR multiple entries in a square. In this case $1 + x$ becomes 1 (Figure 3.23). Next, we show why the method is a general solution.

FIGURE 3.22
K Map for $g = xy + yz'$ with map-entered variable x

$g = xy + yz'$

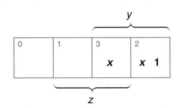

FIGURE 3.23
K Map for $g = xy + yz'$ after logical OR of entries

$g = xy + yz'$

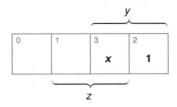

FIGURE 3.24
K Map for $g = xy + yz'$ with all minterms mapped

$g(x, y, z)$

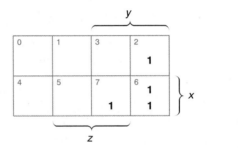

FIGURE 3.25
K Map for $g = xy + yz'$ with map entered variable x

$g = xy + yz' = xyz + xyz' + xyz' + x'yz'$

$g(x, y, z)$

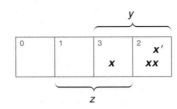

FIGURE 3.26
K Map for $g = xy + yz'$ with map entered variable x

$g = xy + yz'$

$g(x, y, z)$

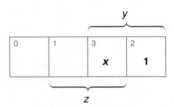

The equation $g = xy + yz'$ has three variables: x, y, z. Designate x as a map-entered variable and then reduce the K map to the two dimensions y, z.

$$g = xy + yz'$$

·(form all minterms)

$$g = xy(z + z') + (x + x')yz'$$

$$g = xyz + xyz' + xyz' + x'yz'$$

(make x, x' coefficients)

$$g = x(yz) + x(yz') + x(yz') + x'(yz')$$

Map the four minterms into a 3-variable K map (Figure 3.24).

Collapse the x dimension. The four 1s from Figure 3.24 convert to three x and one x' because x or x' is a coefficient of each two dimensional minterm (Figure 3.25).

Since $(x + x' + x) = 1$ the two x's and one x' in square 2 of Figure 2.25 are replaced by a 1 to obtain the final map (Figure 3.26) with map entered variable x.

3.13.2 Reading the Map

Here is one algorithm for reading K maps with map entered variables.

1's are treated as don't cares because $a + a' = 1$.

A. Set all map entered variables to 0. Read all 1s.
B. Restore one map entered variable and read the map while treating all 1s as don't cares. Set the map entered variable back to 0.
C. If there is another map-entered variable, repeat step two or else quit.

EXAMPLE 3.32

$g(x, y, z, a, b)$

1. Step A. Set $a = b = 0$. Read the column of 1s as yz.
2. Step B. Restore a. With $b = 0$ and the 1s as don't cares read the four square cluster above z as az because a is the coefficient. Set $a = 0$.↘
3. Step C. b is another map-entered variable. Go to Step B.
4. Step B. Restore b. With $a = 0$ and the 1s as don't cares read the two square cluster below y as $bx'y$ because b is the coefficient. Set $b = 0$.
5. Step C. There are no more map entered variables. Quit.

Therefore $g = yz + az + bx'y$

EXAMPLE 3.33

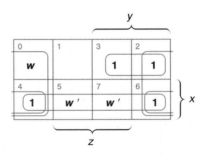

1. Step A. Set $w = w' = 0$ (strange but correct). Read the 1s as $x'y + xz'$.
2. Step B. Restore w. With $w' = 0$ and the 1s as don't cares, read the four-square cluster above z' as wz' because w is the coefficient. Set $w = 0$.
3. Step C. w' is another map-entered variable. Go to Step B.
4. Step B. Restore w'. With $w = 0$ and the 1s as don't cares, read the four-square cluster in row x as $w'x$ because w' is the coefficient. Set w' to 0.
5. Step C. There are no more map-entered variables. Quit.

Therefore $g = wz' + w'x + xz' + x'y$

$= wz' + w'x + x'y$

Note: xz' is a consensus term.

EXERCISE 3.19   Read the K map.

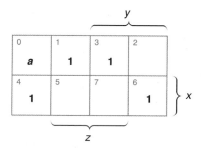

One Answer: $g = ax'y' + (x \text{ xor } z)$ ∎

EXERCISE 3.20   Read the K map.

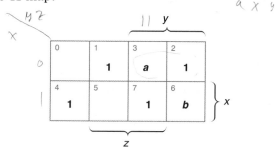

One Answer: $g = ax'y + bxy + (x \text{ xor } y \text{ xor } z)$ ∎

EXAMPLE 3.34 **Reading a K-map with Map-Entered Variables**

$g(w,x,y,z,a,b,c)$

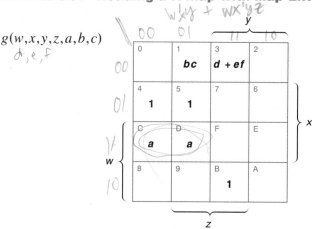

1. Step A. Set $a = bc = d + ef = 0$. Read the ones as $w'xy' + wx'yz$.

2: Step B. Restore a. With $bc = d + ef = 0$ and the 1s as don't cares, read the four-square cluster to the left of x as axy' because a is the coefficient. Set $a = 0$.

3. Step C. bc is another map-entered variable. Go to step B.

4. Step B. Restore bc. With $a = d + ef = 0$ and the 1s as don't cares read the two square cluster in the $y'z$ column as $bcw'y'z$ because bc is the coefficient. Set $bc = 0$.

5. Step C. $(d + ef)$ is another map-entered variable. Go to step B.

6. Step B. Restore $(d + ef)$. With $a = bc = 0$ and the 1s as don't cares, read the two square cluster in the yz column as $(d + ef)x'yz$ because $(d + ef)$ is the coefficient. Set $(d + ef) = 0$.

7. Step C. There are no more map-entered variables. Quit.

Therefore $g = w'xy' + wx'yz + axy' + bcw'y'z + (d + ef)x'yz$

SUMMARY

Axiomatic Basis of Switching Algebra
Huntington proved that six axioms provide a complete basis for *Boolean algebra*.

Principle of Duality
If a Boolean statement is true then the dual of the statement is true.

Constants
There are only two constants in switching algebra: *true T and false F* represented by the symbols 1 and 0 respectively. The symbols 1 and 0 are not numbers here. Different meanings may be given to 1 and 0.

Variables
Switching algebra is not an algebra of numbers. Switching algebra is an algebra of *states* represented by the constants 0 and 1. A variable is in state 0 or it is in state 1.

Variables are assigned the constants 0 or 1. Again, we do not say numbers because 0 does not mean ''zero'' and 1 does not mean ''one'' in this algebra. 1 and 0 may mean a switch is on or off, a voltage is high or low, or some other binary pair (Table 3.2). Nevertheless there is no harm thinking in terms of numbers 0 and 1 so long as the context is not ignored.

Logical Operators
NOT, AND, and OR truth tables, equations, and symbols

Operator truth tables:

NOT	x	g	AND	x	y	g	OR	x	y	g
	0	1		0	0	0		0	0	0
	1	0		0	1	0		0	1	1
				1	0	0		1	0	1
				1	1	1		1	1	1

Operator equations:

$$g = x' \qquad g = xy \qquad g = x + 7$$

Operator symbols:

Functions
In switching algebra $g(x)$ is taken to mean a function of x as is done in the calculus. Any switching function g is represented in three major ways: equation, truth table, and Karnaugh map.

Methods are given for bridging equations and truth tables:
1. From sum of terms equation to truth table,
2. From product of terms equation to truth table,
3. From truth table to sum of terms equation; and
4. From truth table to product of terms equation.

Function Simplification
The object of function simplification is a less complex logic circuit. Simplification of complex functions by algebraic means is possible but difficult. However, a method more expedient than algebra uses the Karnaugh map. Nevertheless, using algebra skillfully is very important because there are many "small" circuits worth simplifying. Table 3.1 on pages 83–84 lists the axioms and theorems of switching algebra.

DeMorgan's Theorem

DeMorgan's very important theorems find the complement or inverse of an expression. For two variables DeMorgan provided us with two theorems:

$$(x + y)' = x'y' \qquad (xy)' = x' + y'$$

Related to DeMorgan's Theorem is an important observation based on Axiom 4:

if $x + y = 1$ and $xy = 0$ then x and y are complements.

Canonical Forms

There are two canonical forms of logic equations: standard sum of products (OR of ANDs) and standard product of sums (AND of ORs). A specific equation need not be in canonical form. Why bother with canonical forms? One answer is: all equations are simplifications of the canonical forms. Other answers are: truth tables are lists of terms used in canonical forms, and K maps are displays of functions in canonical form.

Minterms and Maxterms

Truth table row numbers are decimal equivalents of the binary number representing the 0, 1 assignments to variables on each row. And the function's output column is assigned 0 or 1 on each row. When row numbers become minterm numbers m_j any function can be represented by a list of m_j. If the output function is 1 on row_j then m_j is included in the list.

When row numbers become maxterm numbers M_j any function can be represented by a list of M_j. If the output function is 0 on row_j then M_j is included in the list.

Design Aid: The Truth Table

A design process usually begins with words not equations. Words are more readily translated into a graphical display such as a truth table than into equations. One major reason for this is that most of the time the words incompletely specify the problem. The incompleteness is manifested as missing truth table entries. We do not know how to write equations when there are "missing entries." However, we are usually able to fill in missing entries in a truth table in a way that is consistent with the problem statement.

Numerical Description of Functions

As functions become more complex, writing down the terms becomes more difficult and the number of errors tends to increase. The following mechanical procedure allows for writing down functions with essentially zero errors.

$$g = m_1 \quad + m_2 \quad + m_3 \quad + m_5 \quad + m_6 \quad \text{(given)}$$

$$g \rightarrow w\,x\,y \ + w\,x\,y \ + w\,x\,y \ + w\,x\,y \ + w\,x\,y \quad \text{(no primes)}$$

$$\quad 0\ 0\ 1 \quad\ 0\ 1\ 0 \quad\ 0\ 1\ 1 \quad\ 1\ 0\ 1 \quad\ 1\ 1\ 0 \quad \text{(binary number)}$$

$$g = w'x'y \ + w'x\,y' \ + w'x\,y \ + w\,x'y \ + w\,x\,y' \quad \text{(minterms)}$$

Equation to Truth Table

Every function g has a truth table. The input side of the truth table is created by observing that each variable takes two values 0, 1. Two values can fill n slots in 2^n ways. Write out 2^n rows of binary numbers 0 to $2^n - 1$. Keeping in mind 0 means false and 1 means true, note that the 2^n numbers represent the 2^n minterms of n variables. Now, when does $g = 1$? The g column is filled with the 1 or 0 resulting from evaluation of each term of the equation for g for the 2^n variable combinations or minterms.

Truth Table to Equation

The equation is derived from the truth table by including minterms for which $g = 1$ in the table. For a three-variable table, the variable values 101 in row$_5$ correspond to $w = 1$, $x = 0$, and $y = 1$. A simple way to write the corresponding minterm uses the numerical description of each term.

$$m_5 = \text{minterm } 5 \rightarrow wxy \rightarrow wx'y \quad \text{which is true only on row}_5.$$

Four equations can be derived from any truth table. The equations answer the following questions. When does $g = 1$? When does $g = 0$? When does $g' = 1$? When does $g' = 0$? The purpose of the questions is to find the minimal equation.

Table-entered Variables

The size of a table is halved when an input variable is treated as a *table-entered variable* instead of as an input. Furthermore, the table display is significantly improved in most, if not all, cases.

Method of Perfect Induction
Proof by perfect induction is implemented by substituting all combinations of values for the variables in the theorem's equations. The truth table provides an orderly means for implementing any proof. The theorem is proved if all combinations yield the same result on both sides of the equation.

Design Aid: The Karnaugh Map
The Karnaugh map (K map) allows you to simplify logic functions in a straightforward manner by providing a more useful graphic display of logic functions than truth tables. A map has one square corresponding to each row of a truth table. Since a minterm corresponds to a row, then each minterm corresponds to one square. The squares have the same numbers as truth table row numbers. A 1 is entered into a square if the function is true for the minterm a square represents. A 0 is entered if the function is false.

Karnaugh Map Concept
The Karnaugh map is based on the concept of adjacent states. Two states are adjacent if they differ in the value of only one variable. This concept is important because the algebraic combination of two adjacent states results in the elimination of one variable. The Karnaugh map displays all the adjacencies present in a standard sum (function). Thus, you can focus on clusters of adjacent terms and simplify the function on the map.

Clusters
Clusters represent groups of terms. No function simplification results when the cluster is a group of 1 term. Clustering two squares eliminates one variable, clustering four squares eliminates two variables, and clustering eight squares eliminates three variables, and clustering all squares eliminates all variables ($g = 1$ or 0). In other words cluster 2^n rectangular squares to eliminate n variables.

Writing the K Map
One way to use the map is to map or write a function g onto the map so that you can read a simplified equation for the function from the map. For example, the sequence of operations writing and then reading the map is:

Given $g = y'z' + yz'$ write $\rightarrow$ K map   read$\rightarrow$ $g = z'$

Reading the Map

Read the K map by deciding what rectangular clusters of 1s include all the 1s in the map. Keep in mind that clusters contain 1, 2, 4 or 2^n squares.

Don't Care Conditions

A function is defined by a sum of one or more minterms. A minterm is included (the coefficient is 1) or excluded (the coefficient is 0). The possibility arises that the coefficient of a minterm is not specified, i.e., it can take the value 0 or 1. Then we say we don't care if the minterm is included or excluded in the function. The corresponding truth table or K map entry may be 0 or 1. However, the don't care status is marked with a dash.

Map-entered Variables

The K map becomes unwieldy even with five variables; with more than five variables the K map rapidly becomes unmanageable. The map-entered variable reduces the required map size, thereby extending the K map's practical usefulness. The number of function variables equals the number of K map dimensions plus the number of map-entered variables.

REFERENCES

Boole, G. *An Investigation of the Laws of Thought*. Macmillan, 1854. Reprint. Dover, 1954.

Caldwell, S. H. *Switching Circuits and Logic Design*. John Wiley & Sons, 1958.

Huntington, E.V. "Sets of Independent Postulates for the Algebra of Logic." *Transactions of the American Mathematical Society* 5 (July 1904): 288–309.

Karnaugh, M. "The Map Method for Synthesis of Combinational Logic circuits." *Transactions of the AIEE, Communications and Electronics* 72, Part 1 (November 1953): 593–599.

McCluskey, E. J. "Minimization of Boolean Functions." *Bell System Technical Journal* 35 (November 1956): 1417–1444.

McCluskey, E. J. *Logic Design Principles*. Prentice-Hall, 1986.

Quine, W. V. "The Problem of Simplifying Functions." *American Mathematical Monthly* 59 (October 1952): 521–531.

Quine, W. V. "A Way to Simplify Truth Functions." *American Mathematical Monthly* 62 (November 1955): 627–631.

Roth, C. H. *Fundamentals of Logic Design* 4th ed. St. Paul: West Publishing, 1992.

Shannon, C. E. "A Symbolic Analysis of Relay and Switching Circuits" *Transactions of the AIEE* 57 (1938): 713–723.

Testing of combinational circuits requires evaluation of truth tables. The signal generator you will assemble and test in this project will be used as a generator of truth tables with one to four input variables. The basic advantage of this method lies in the fact an oscilloscope can be used to view the results. However, the knowledge necessary for understanding how and why truth tables are generated by the signal generator is not presented until much later in this text. This problem could be avoided by specifying switching to generate truth tables. The choice is to not burden you with this laborious method. Instead, you are offered a modern method, and you are asked to use your intuition as you proceed. (Relevant knowledge is available in Sections 4.1, 7.2, and a 163 data sheet.)

A *timing diagram* is a graphical illustration of a circuit's logical behavior as a function of time. A timing diagram's horizontal axis has the time dimension. The vertical axis is used for H and L voltage levels. A *digital waveform* is a plot of the H and L levels a variable takes as a function of time. The *clock period* is the time between successive transitions in the same direction of the clock waveform (Figure 6.13). A useful abstraction of the clock waveform is a series of up arrows representing corresponding clock transitions (Figure 7.22 replicated below).

The four-bit counter (family member 163) has four outputs q_3, q_2, q_1, q_0—one output per bit. The digital waveforms the outputs produce while the 163 is counting are shown in Figure 7.22. If logical 1 is assigned to the H level and logical 0 to the L level you can interpret the waveforms as generators of a sequence of 1s and 0s as shown above the waveforms in Figure 7.22. If you rotate the figure clockwise by 90°, you will find the 1s and 0s represent the sixteen rows of a four-input truth table.

FIGURE 7.22
Four-bit up counter truth table rows & q_j waveforms

State	0	1	2	3	4	5	6	7	8	9	A	B	C	D	E	F
q_0	0	1	0	1	0	1	0	1	0	1	0	1	0	1	0	1
q_1	0	0	1	1	0	0	1	1	0	0	1	1	0	0	1	1
q_2	0	0	0	0	1	1	1	1	0	0	0	0	1	1	1	1
q_3	0	0	0	0	0	0	0	0	1	1	1	1	1	1	1	1

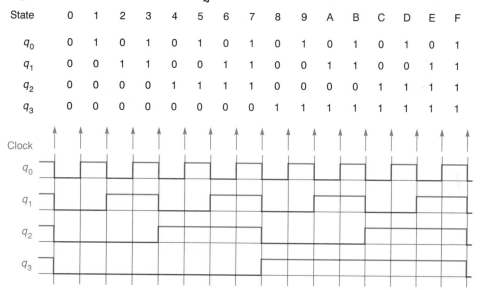

Name _____   SID# _____   Section _____
    (last)   *(initials)*

Approved by _____   Date _____

Grade on report _____

Signal Generator tests AND, OR, and XOR

1. Clock-counter circuit function.

Synchronize scope to R_{co}.
Display one 16-clock counter cycle over 8cm.

_____ Clock
_____ q_0
_____ q_1
_____ q_2
_____ q_3
_____ plot of all q_j, clock, and r_{co} on one timing diagram

2. Verify knowledge of state values.

Point out state   0 _____
                  5 _____
                 14 _____

3. Enter state numbers on the timing diagram.

_____ entered

4. Explain the gate output waveforms.

_____ 00
_____ 02
_____ 86

Reference: Chapter 12 Project Information

1. Build the clock-counter circuit (Figure P02.1) in the upper right hand side of the bread-board. This will serve as a test signal generator in every project.

2. On graph paper plot the waveforms of the outputs clock, R_{co}, q_0, q_1, q_2, q_3. Plot the six output signals in the order listed. Plot a reasonable sketch of what you see. Do not plot ideal waveforms. Identify the reference waveform and mark the triggering point on the waveform. Hint: plot waveforms in a 16-clock period cycle.

   What signal do you use as a reference waveform to trigger the oscilloscope in order to observe the signal generator outputs in correct relation to each other? What principle guides you to a choice?

   Do the plots make sense? Why?

3. If q_0, q_1, q_2, q_3 are Boolean variables, what are the state values? A state value is the binary number $q_3 q_2 q_1 q_0$ (Chapter 1). Show the values as a last line of the timing diagram.

4. Connect signal generator outputs q_0, q_1 to the two inputs (x, y) of the following circuits. Use the reference waveform to trigger the oscilloscope. Plot the output g of each circuit over at least four clock periods. Why four? Make truth tables from the plotted data. What is the equation for the function of each circuit?

   4.1 74LS00 gate
   4.2 74LS02 gate
   4.3 74LS86 gate

FIGURE P02.1

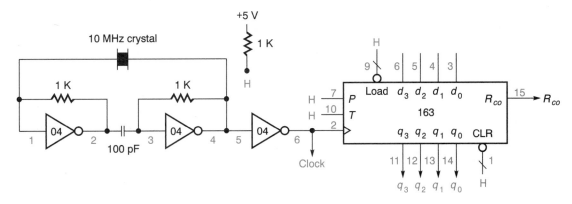

Name _____ SID# _____ Section _____

      (last)   *(initials)*

Approved by _____ Date _____

Grade on report _____

Verifying theorems with logic circuits

1. _____ $g_1 = x + x'$ $q_0 = x$
    _____ $g_2 = 1$

2. _____ $g_1 = x * x'$ $q_0 = x$
    _____ $g_2 = 0$

3. _____ $g_1 = x + 1$ $q_0 = x$
    _____ $g_2 = 1$

4. _____ $g_1 = x + 0$ $q_0 = x$
    _____ $g_2 = x$

5. _____ $g_1 = x + x'y$ $q_0 = x, q_1 = y$
    _____ $g_2 = x + y$

6. _____ $g_1 = x + ab$ $q_0 = a, q_1 = b, q_2 = x$
    _____ $g_2 = (x + a)(x + b)$

7. _____ $g_1 = (xy)'$ $q_0 = x, q_1 = y$
    _____ $g_2 = x' + y'$

8. _____ $g_1 = (x + y)'$ $q_0 = x, q_1 = y$
    _____ $g_2 = x'y'$

9. _____ Plots of outputs *vs* inputs for all cases

10. _____ Proofs of al equations using Boolean algebra

Use logic circuits to verify the theorems assigned for this project from the set of eight. Assignment of circuit pairs to do for this project:

Build the g_1 and g_2 pairs of circuits one pair at a time. Connect the signal generator outputs q_2, q_1, q_0 to the specified inputs. Plot the g_1, g_2 outputs including hazards.

PROVE THAT $g_1 = g_2$ INPUTS

1. $g_1 = x + x'$ $q_0 = x$
   $g_2 = 1$

2. $g_1 = x * x'$ $q_0 = x$
   $g_2 = 0$

3. $g_1 = x + 1$ $q_0 = x$
   $g_2 = 1$

4. $g_1 = x + 0$ $q_0 = x$
   $g_2 = x$

5. $g_1 = x + x'y$ $q_0 = x, q_1 = y$
   $g_2 = x + y$

6. $g_1 = x + ab$ $q_0 = a, q_1 = b, q_2 = x$
   $g_2 = (x + a)(x + b)$

7. $g_1 = (xy)'$ $q_0 = x, q_1 = y$
   $g_2 = x' + y'$

8. $g_1 = (x + y)'$ $q_0 = x, q_1 = y$
   $g_2 = x'y'$

PROBLEMS

3.1 Make truth tables for the following.

(a) $g = xyz$
(b) $g = x + y + z$
(c) $g = x \oplus y \oplus z$

3.2 Use algebra to prove theorems. Justify each step with an axiom or theorem from Table 3.1 on pages 83–84.

(a) $x + x = x$
(b) $xx = x$
(c) $x + 1 = 1 + x = 1$
(d) $x * 0 = 0 * x = 0$
(e) $x + xy = x$
(f) $x(x + y) = x$
(g) $x + x'y = x + y$
(h) $x(x' + y) = xy$
(i) $xy + x'z + yz = xy + x'z$
(j) $(x + y)(x' + z)(y + z) = (x + y)(x' + z)$

3.3 Use algebra to prove the following are true.

(a) $x' \oplus y' = x \oplus y$
(b) $(x \oplus y)' = x' \oplus y = x \oplus y'$
(c) $ax \oplus ay = a(x \oplus y)$
(d) $x \oplus 1 = x'$
(e) $x \oplus 0 = x$
(f) $x'y \oplus xy' = x \oplus y$

3.4 Use algebra. Prove the following f, g pairs are complements without using DeMorgan's theorems.

(a) $f = xy$　　　　　　　$g = x' + y'$
(b) $f = x + y$　　　　　　$g = x'y'$
(c) $f = y'z + xz'$　　　　$g = x'z' + yz$

3.5 Use algebra. Prove the following are true.

(a) $[(x \oplus y)' \oplus (y \oplus z)']' = (x \oplus z)'$
(b) $[(x \oplus y)' \oplus z]' = [x \oplus (y \oplus z)']'$
(c) $x \oplus y \oplus xy = x + y$
(d) $(x \oplus y)' = x \oplus y \oplus 1$
(e) $[(x \oplus y) + x]' = x'y'$

3.6 Use algebra to simplify the functions.

(a) $g = (xy)'z + xz + yz$
(b) $g = (x'y')' + xz + yz$
(c) $g = w'x' + wy'z + xy'z$
(d) $g = x'y + yz + xz + xyz$
(e) $g = (x + p)(x + q)(x + r)(x + s)$
(f) $g = y + z'x + xy'z$
(g) $g = xyz + wyz + x'wz$
(h) $g = x + (x'y')'(x' + yz' + xyz')$
(i) $g = (x + p)(x' + q)(r + p + q)$
(j) $g = y + x'y' + xy + xy'$
(k) $g = (x + x'p)(x' + xq)(r + r'p + p'q)$
(m) $g = wx + wz + x'z' + wy'z'$
(n) $g = (w + x' + y')(w + w'x' + w'z)$
(p) $g = (w + x)(w' + y + z)$
(q) $g = vz + vwx'y + wx'z'$
(r) $g = w'xy' + w'z + xy'z'$
(s) $g = w'xy' + w'xz' + yz$

3.7 Make truth tables for these functions. What simplified functions do they equal?

(a) $f = xy + x'z + yz$
(b) $f = (x + y)(x' + z)(y + z)$
(c) $f = x + x'y$

3.8 Write the equations for f in sum of products (OR of ANDs) format (do not simplify). Use DeMorgan's theorem on f and write the equations for f' in product of sums (AND of ORS) format (do not simplify).

(a)				(b)					(c)				
x	y	f	f'	x	y	z	f	f'	x	y	z	f	f'
0	0	1	0	0	0	0	0	1	0	0	0	0	1
0	1	1	0	0	0	1	1	0	0	0	1	1	0
1	0	1	0	0	1	0	0	1	0	1	0	1	0
1	1	0	1	0	1	1	0	1	0	1	1	0	1
				1	0	0	1	0	1	0	0	1	0
				1	0	1	1	0	1	0	1	0	1
				1	1	0	1	0	1	1	0	0	1
				1	1	1	0	1	1	1	1	1	0

3.9 Write the equations for f' in sum of products (OR of ANDs) format (do not simplify). Use DeMorgan's theorem and write the equations for f in product of sums (AND of ORS) format (do not simplify).

(a) see 3.8(a)
(b) see 3.8(b)
(c) see 3.8(c)

3.10 Convert each f to canonical minterm and maxterm forms.

(a) $f(x,y,z) = z$
(b) $f(y,z) \quad = y + z$
(c) $f(y,z) \quad = 0$
(d) $f(y,z) \quad = 1$

3.11 Which are true?

(a) $\qquad x'y'z + x'yz' = ?\ x'$
(b) $\qquad (xy' + xz')' = ?\ x' + yz$
(c) $(a \oplus b) \oplus (a \oplus c) = ?\ a \oplus (b \oplus c)$
(d) $\qquad ab \oplus ac = ?\ a(b \oplus c)$
(e) $\quad (a \oplus b)(a \oplus c) = ?\ a \oplus bc$

3.12 Write out in AND, OR, and NOT form.

(a) $f = a \oplus b$
(b) $f = ab \oplus b$
(c) $f = a'b \oplus ab'$

3.13 Expand $f(a,b,c)$ to canonical product of sums (AND of ORs).

(a) $f = ac' + ab'$
(b) $f = a' + bc$
(c) $f = (a + c)(b' + c)(a' + b)$
(d) $f = a'b + abc + bc + a$

3.14 Expand $f(a,b,c)$ to canonical sum of products (OR of ANDs).

(a) $f = a(b + c)$
(b) $f = bc' + ab' + a'c$
(c) $f = (a' + c)(a + b')$
(d) $f = (ab + bc)a + b'c$

3.15 Find dual and complement.

(a) $f = [(ab' + c')d + e]b$
(b) $f = abc' + (a + b' + d')(a'bd + e')$

3.16 Read a map (page 168). Write equations for f and f' as sums of products. Complement f and f' to convert to product of sums form.

3.17 Simplify using K maps (and algebra if necessary).

(a) $f = xz + y'z + y(xz' + x'z)$
(b) $f = (xy' + z)(x + y')$
(c) $f = (x + y' + x'y)z'$
(d) $f = wx'y' + xy + y(z + x'z')$
(e) $f = xy + z(w + w'x' + y')$
(f) $f = wxy + z'(wy + wx) + xy'z$
(g) $f = xy'z' + w'(y + z)$
(h) $f = w'xy'z' + w'x'y'z' + wy'z' + wxyz$

3.18 Plot the f, g functions on a K map. Use f and g symbols in lieu of 1s. Are the f, g pairs complements? Why?

(a) $f = yz$ $\qquad\qquad g = y' + z'$
(b) $f = y + z$ $\qquad\quad g = y'z'$
(c) $f = y'z + xz'$ $\qquad g = x'z' + yz$

3.19 Make a truth table for $f = xy + x'z + yz$ with table-entered variable y.

3.20 Make a truth table for $f = (x + y)(x' + z)(y + z)$ with table-entered variable y.

3.21 Make a truth table for $f = vz + vwx'y + wx'z'$ with table-entered variable y.

3.22 Make a K map for $f = xy + x'z + yz$ with table-entered variable y.

3.23 Make a K map for $f = (x + y)(x' + z)(y + z)$ with table-entered variable y.

3.24 Make a K map for $f = vz + vwx'y + wx'z'$ with table-entered variable y.

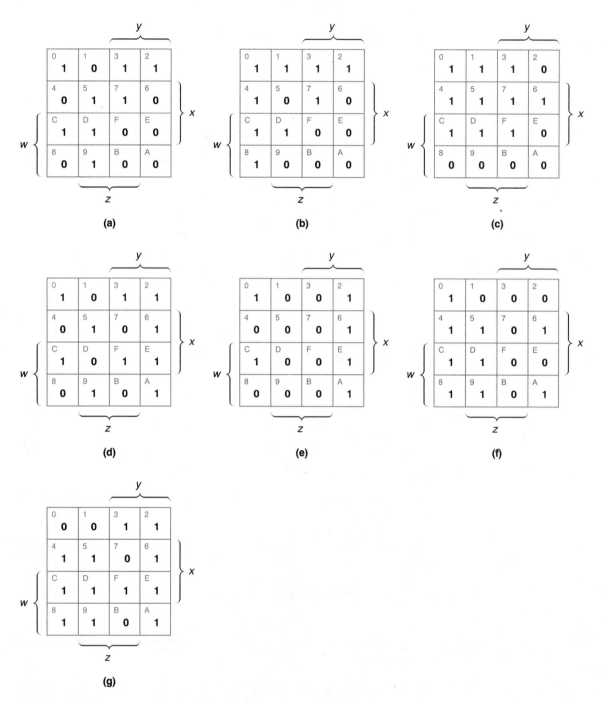

(a)   (b)   (c)

(d)   (e)   (f)

(g)

PROBLEM 3.16

3.25 A marker prints "pass" on relays if the pull-in time is less than 20ms and prints "pass" on circuits if the response time is greater than or equal to 20ms. (The pull-in time is the time it takes the relay contacts to transfer.) Create a truth table, K map, and equation for the "pass" function.

3.26 A room light is controlled independently by a wall switch at each entrance. The room has two entrances. Flipping any switch changes the light from off-to-on or on-to-off. Create a truth table, K map, and equation for the light-is-on function.

3.27 A red-green-yellow traffic light malfunctions when two or more colors are on. Create a truth table, K map, and equation for the malfunction report.

3.28 A switching circuit has three inputs x, y, z and one output g. The output g is true only when any two inputs are true at the same time. Create a truth table, K map, and equation for the output. Design the circuit.

4 COMBINATIONAL MIXED-LOGIC CIRCUITS

The major goal of this chapter is to connect logical states, 0 and 1, to physical voltage states, H (high) and L (low). The connection allows us to employ physical gates as logical operators so we can use switching theory to design assemblies of gates that implement logical functions.

We show that physical devices called gates can physically realize a two-state switching algebra. The assignment of logical 0 and 1 to the physical states of any gate terminal can be made in two ways when mixed logic is used. At some terminals voltage H = True = 1; at other terminals voltage L = True = 1. Having two ways allows for logical variables to be active high or active low. For example, a push button is best wired so as to select ground when activated. On the other hand, an npn transistor driving a lamp requires a high input to turn on the lamp.

Then we demonstrate that the logical operator a gate implements depends on the logical assignments of 1 as H or L voltage at various gate terminals. We employ standard logic family parts in order to start building your experience with integrated circuits used in practice.

We move on to the process for translating a physical device voltage table into a logical truth table by using mixed logic. This allows for what seems to be a natural set of assignments where the resulting logical equations for the device have no complemented variables. We also discover that logical NOT is implemented by the mixing of logical assignments at the two ends of a wire in a circuit.

Then we are ready to analyze logic circuits: a practical method for mixed-logic analysis is presented, followed by a practical method for mixed-logic circuit synthesis.

INTRODUCTION

A *logic circuit* is an assembly of one or more gates. A *gate* is a physical electronic device that implements a logic operator. These days the assembly is most likely to take the form of an integrated circuit. Having begun with a single gate, the electronics industry is now at the point where a modern integrated circuit may have 500,000 or more gates. In any event, the industry has evolved so-called logic families. In turn, any specific assembly of parts from the logic families is supplemented, if not replaced, by a very-large-scale integrated (*VLSI*) circuit known as an application-specific integrated circuit (*ASIC*). The ASIC is outside the scope of our discussion.

Logic families include gates and *circuit building blocks* that ease the design process. We start with the gates and work our way up the ladder of complexity. Our first goal is to learn how to translate our design equations into logic circuits we can use to perform tasks.

4.1 ▰▰▰▰ Gates Are Logical Operators

AND gate executes * operator.

OR gate executes + operator.

The * logical operator is implemented by an AND gate; the + logical operator is implemented by an OR gate. One way to implement the ' logical operator is with an inverter gate; the other way does not make use of a gate, as is demonstrated later. Single-gate circuits using these gates and the equations they represent are shown in **Figure 4.1**. The bubbles shown at some of the gate terminals and circuit nodes in these circuits will be fully discussed later. The point here is to show that a variety of circuits can represent the same equation: that synthesis is not unique. The digital logic symbols reveal the physical as well as the logical behavior of the circuit. Whereas the logical point of view deals with switching variables as the 1's and 0's of gate inputs and outputs, the physical point of view deals with the voltages at gate inputs and outputs that represent switching variables. So a physical gate has a physical voltage table as well as a logical truth table. The two points of view merge when voltage levels represent 1 and 0.

4.2 ▰▰▰▰ True/False Physical Representations

True and False are represented by voltages in digital electronic circuits known as the transistor–transistor logic (TTL) and complementary metal oxide semiconductor (CMOS) classes of families. This digital logic is powered by a 5-volt power supply. (Some CMOS families can be powered by 12- and 15-volt power supplies.) For the emitter-coupled logic (ECL) family one could argue whether True and False are represented by currents or by voltages in the circuits. This family is an important special case we do not discuss here.

Although modern power supply voltages are 0 volts and 5 volts, circuit outputs do not reach these limits. For example, actual low

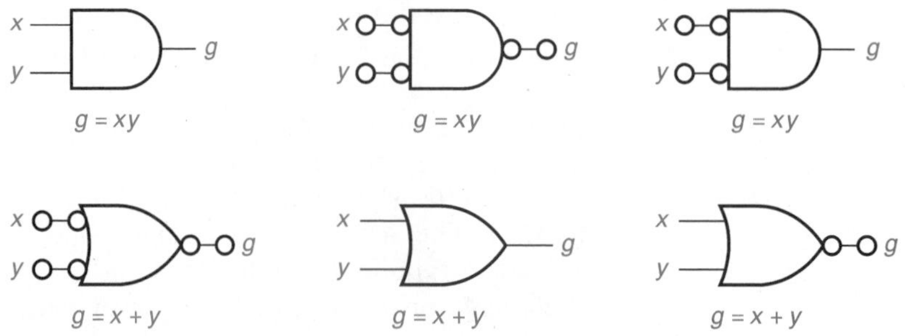

FIGURE 4.1
Typical single-gate circuits

$g = xy$ $g = xy$ $g = xy$

$g = x + y$ $g = x + y$ $g = x + y$

H and L represent a two-state voltage system.

power Shottky transistor transistor logic (LSTTL) logic circuit low voltage level (L) is some value from 0 to 0.8 volts (Figure 4.2). And the logic circuit high voltage level (H) is some value from 2.0 to 5 volts. The band between 0.8 and 2.0 volts is a forbidden, no-logic-state band. When a constant circuit voltage is found in this band, something is defective. The H and L voltages are not single values because every parameter in a manufacturing process has production tolerances associated with it. Thus the physical point of view deals with voltages at the H and L levels. This two-state system of H and L voltage levels physically represents the two-state logic system (0, 1).

The actual range of values for the H and L bands depends on technology and on logic family. Logic families use *bipolar* technology (e.g., 74LS, 74S, 74ALS, 74AS), *CMOS* technology (e.g., 74AC, 74ACT), and the bipolar–CMOS combination (*BiCMOS*) technology (e.g., 74ABT, 74BCT).

4.3 Mixed Logic: True and False May Be at Low or High Voltage

A logic design where True is represented by H and L voltage (in different parts of the circuit) is called a *mixed-logic design*. True (T or 1) and False (F or 0) can be represented in two ways: active high and active low.

1. The $++$ assignment is active high.

$$\text{High voltage H} = 1 = \text{True}$$

$$\text{Low voltage L} = 0 = \text{False}$$

2. The $--$ assignment is active low.

$$\text{High voltage H} = 0 = \text{False}$$

$$\text{Low voltage L} = 1 = \text{True}$$

Active   The assignment of voltage level H to logical 1 means switching variables are active when high; that is, they perform their action when high. Such switching variables are said to be *active high*. The

FIGURE 4.2

Typical two-state voltage system

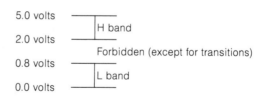

assignment of voltage level L to logical 1 means switching variables perform their action when low; they are active when low.

Positive-logic systems use only active high assignments $(++)$.

Negative-logic systems use only active low assignments $(--)$.

Mixed logic uses $++$ and $--$ in a circuit.

Mixed-logic systems use $++$ and $--$.

Asserted   Asserted, which means a variable is active (True), is independent of whether the active condition is represented by H or by L. Asserted emphasizes action. And not asserted means inactive (False).

Two kinds of bubbles: gate and circuit node.

Bubbles   There are two kinds of bubbles. A bubble at a gate terminal implies that the signal at the terminal is active low. A different type of bubble at a source or destination circuit node means the variable at that node has the $--$ assignment (Figure 4.3a). This is *not* inversion. However, you may prefer a special symbol attached to a variable at a source or a destination in lieu of a bubble (see the highlighted discussion of Notation for Active High and Low Variables). We recommend the two-kinds-of-bubble notation because it usually results in fewer errors, less writing, and greater schematic clarity and readability.

The absence of a bubble at a circuit node indicates that the variable at that circuit node is true (logical 1) when the voltage is high (H). This means that the signal source or destination end of a wire without a bubble identifies a switching variable that is active high at that circuit node (Figure 4.4).

If we were to remove wires from the circuits in Figures 4.3 and 4.4, then we could say that on a wire such as ———, truth is represented by an H voltage; and on a wire such as o———o, truth is represented by an L voltage.

FIGURE 4.3
Active low source and destination—two ways

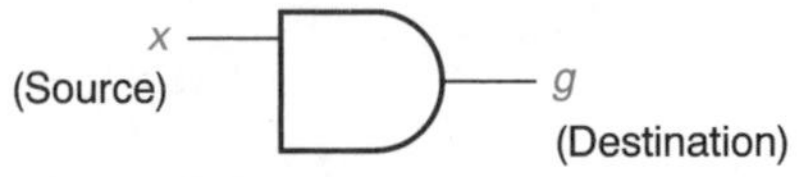

x O—◁ (Source) ▷o—O g (Destination)

(a)

x_L —◁ ▷o— g_L

(b)

FIGURE 4.4
Active high source and destination

x — (Source) ▷— g (Destination)

Notation for Active High and Low Variables   Making
$++$ and $--$ active high and low assignments to variables
implies a need for a notation to represent the assignments in
schematics and text.

Data Book Notation: We call a bubble at gate terminal a
"bubble of the first kind." These are the bubbles found in
data books. The bubble of the first kind at a gate terminal
means merely that the gate input or output is asserted when
the voltage is low (L). (And that is all it means. Interpreting
the presence of a bubble as inversion of the input or output
may be incorrect and is unnecessary. We prefer to make no
interpretation.) The absence of a bubble at a gate terminal
means the gate input or output is asserted when the voltage
is high (H).

Furthermore, in data books variables associated with
variables are marked with an overbar ($\bar{x}$). The bar does not
mean the complement of the variable; it means what the
bubble means—the gate input or output is asserted when
the voltage is low (L). That is all there is to it.

Notation for Text and Schematics: No special notation is
needed in positive- or negative-logic systems. In positive-
logic systems all variables x are understood to be active
high, and, in effect, active low variables are represented by
the complements x'. In negative-logic systems all variables
y are understood to be active low, and, in effect, active high
variables are represented by the complements y'.

In mixed-logic systems, where variables are active high
or low in different parts of the circuit, a special notation is
required in order to know by inspection the active level at
any circuit node.

One special notation is the use of the subscript "H" for
active high variables and the subscript "L" for active low
variables. For instance, the variable g is written as g_L at
active low nodes and as g_H at active high nodes.

Another special notation is to prefix active low variables
with a forward slash / and to leave active high variables
unmarked. Thus, the variable g is written as $/g$ at active low
nodes and as g at active high nodes.

Our preferred notation for mixed-logic systems is to
mark the circuit nodes of active low variables with a new
bubble of the second kind.

Clearly one can make up many other suitable notations.
Since no official standard exists, the notation you use is a
matter of personal choice.

FIGURE 4.5
**Sequence of input
assignments**

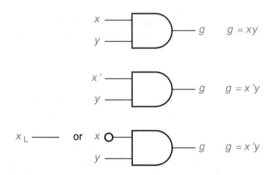

$g \quad g = xy$

$g \quad g = x'y$

$x_L \quad —— \quad$ or $\quad x \ \circ—$

$g \quad g = x'y$

Bubbles replace the
operator.

Bubbles represent complemented variables in mixed-logic schematics:
a mixed-logic schematic clearly indicates the active level on all wires
by using bubbles in lieu of complemented variables. The sequence of
input assignments for x shown in Figure 4.5 followed by the sequence
of output assignments for g shown in Figure 4.6 should help to clarify
this point.

   This implies that sources x' ——— and $x \circ$——— (or x_L ———)
are equivalent. This equivalence is demonstrated by the following
assignment table and source symbols:

$++$	$++$	$--$
x	x'	x
0	1	1
1	0	0
x ———	x' ———	$x \ \circ$——— Source symbols
		x_L ———

This sequence implies that destinations ——— g' and ———$\circ$ g (or
——— g_L) are equivalent. This equivalence is demonstrated by the

FIGURE 4.6
**Sequence of output
assignments**

$g \qquad g = xy$

$g' \qquad g' = xy \rightarrow g = (xy)'$

$\circ g$ or ——— $g_L \qquad g = (xy)'$

following assignment table and destination symbols:

$++$ g	$++$ g'	$--$ g
0	1	1
1	0	0

————— g ————— g' ————$\circ$ g Destination symbols
————— g_L

4.4 Logic Family Part Numbers

The number specifies the circuit.

In logic families of parts, the full number is 74LS04 or 74ALS138 or 74ACT11800 or 74BCT240, and so forth. The 74xx or 74yyy represents the prefix identifying the family. The number identifies the logic circuit. The circuit numbers are constant with time and change of family. We want to know the circuit numbers and what they represent.

We drop the prefix and only use the number because the number specifies the circuit function. We may speak of the 00 gate, for example, but we know that the 74LS00 package contains four 00 gates. Do not let our convenient use of "00" confound you. Furthermore, IC schematics are not given in this text simply because data books are so readily available.

4.5 Logic True Conventions and the 04 Gate

The 04 (e.g., 74LS04) is called an inverter. The 04 voltage table reveals that the output Y is the inverse, or opposite, of the input A. The voltage table is physical. We emphasize this reality by saying the voltage table has nothing to do with logic. That may seem strange, since the 04 exists to implement logic. Nevertheless, voltage is not connected to logic values until we make assignments. Before we add wires, put a gate in a circuit, and make assignments, we need to understand that action at any one gate terminal influences action at other gate terminals. Physical gate terminals are not independent; they are interdependent.

Designating input A to be active high (Figure 4.7) forces output Y to be designated active low. Or designating A to be active low forces Y to be designated active high. (Designations illustrated in Figure 4.7 are *not* assignments.) On the other hand, if we designate both A and Y to be active low (or high) we have a situation where no input action

FIGURE 4.7
04 voltage tables

		Active high A		Active low A	
A	Y	A	Y	A	Y
L	H	L	H	**L**	**H**
H	L	**H**	**L**	H	L

Gate Symbol

causes an output action. In addition, the dual active low (or high) assignment raises the question: what is the gate symbol? Therefore we have this important conclusion:

> Active designations of gate terminals are constrained by the gate's voltage table.

Connecting the 04 to logic One application of the 04 is to effect reassignment of active level from H to L voltage, and vice versa. For example, if a switching variable is active high at the 04 input, then that variable is active low at the 04 output, and vice versa. The two 04 symbols (Figure 4.7) reflect these choices.

Assignments connect voltage to logic.

In a positive-logic system all assignments to variables at various circuit nodes are active high (+ +). Each assignment is independent of all other assignments. (Contrast this independence with the interdependent 04 gate terminals.) The independence of assignments to variables is emphasized by adding wires to create new circuit nodes for variables. Then variables do not appear at gate terminals (**Figure 4.8**). The symbols in Figure 4.8 are composed from the symbol elements shown in Table 4.1. The equation representing the 04 in a positive-logic system is $g = x'$, which is the NOT function (Figure 4.8).

Add wires to create circuit nodes.

TABLE 4.1   Symbol Elements

Only a few elements are used to construct logical function symbols:

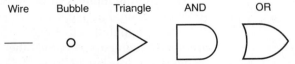

| Wire | Bubble | Triangle | AND | OR |

FIGURE 4.8

04 in a positive-true-logic system

	Variable	Assignment	Assignment Symbol
	x	$1 = T := H$	$++$
	g	$1 = T := H$	$++$

Voltage		Logic		
04		$++$	$++$	← To remind us of the assignments
A	Y	x	g	
L	H	0	1	*Equation:* $g = x'$
H	L	1	0	$g = \text{NOT } x$

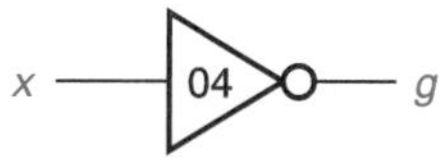

$$x \longrightarrow \boxed{04} \!\!\!\!\!\triangleright\!\!\circ\!\!\!- g$$

Can assign $++$ or $--$ to any circuit node.

In a mixed-logic system, assignments to some variables at various circuit nodes are active high $(++)$ and assignments to variables at other circuit nodes are active low $(--)$. A mix of assignments is used. As before, each assignment is independent of all other assignments, and wires are added to create new circuit nodes for variables. For example, the 04, with wires attached, has two circuit nodes to which we can make independent assignments in four ways (Figure 4.9).

Voltage table defines a physical gate.

The first assignment combination in Figure 4.9 (all $++$) is what is used in positive-logic systems. The last assignment combination (all $--$) is what is used in negative-logic systems. Observe that $g = x'$ when the assignments in a system are all the same. In these cases the 04 implements the NOT operator. Figure 4.9 also shows that $g = x$ when the assignments are mixed. Now the 04 implements the logical identity operator. A physical interpretation of this 04 function is reassignment of the voltage level. That is, the active level is changed from L to H voltage, and vice versa, when the x and g assignments are a mixed pair: $++ \; --$ or $-- \; ++$.

In mixed-logic systems the basic function of inverters is reassigning logical 1 to the complement of the input voltage level. Then H or L voltage can represent True in different parts of a circuit.

In Figure 4.9 two new situations have appeared: The wires

$$\circ\!\!-\!\!\!-\!\!\!-\!\!\!- g \qquad \text{and} \qquad -\!\!\!-\!\!\!-\!\!\!-\!\!\circ g$$

have different active levels at their two ends. When this happens, the equation for the 04 is $g = x'$. Can we conclude that the logical NOT

A	Y	++ x	++ g	++ x	-- g	-- x	++ g	++ x	-- g
L	H	0	1	0	0	1	1	1	0
H	L	1	0	1	1	0	0	0	1

Equation: $g = x'$ $g = x$ $g = x$ $g = x'$

Function: Inverter   Identity   Identity   Inverter

Circuit:

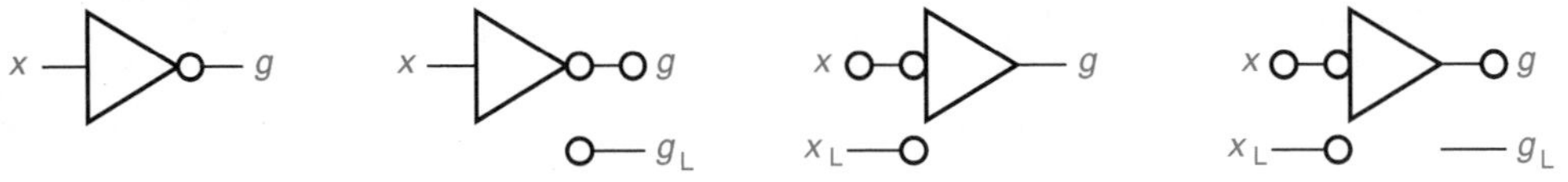

FIGURE 4.9
04 in a mixed-logic system

operator is implemented when the assertion levels are different at the two ends of a wire? The answer is yes.

Consider the two-inverter schematic shown in Figure 4.10.

The first inverter drives output g_2 and output g_3. Output g_2 has a $--$ assignment. Output g_3 has a $++$ assignment.

When input y is H, both output g_2 and output g_3 are L. Input y and output g_2 are asserted. Output g_3 is not asserted.

The second inverter drives outputs g_4 and g_1. Output g_4 has a $--$ assignment. Output g_1 has a $++$ assignment.

When input y is H, both output g_4 and output g_1 are H. Input y and output g_1 are asserted. Output g_4 is not asserted.

Since output $g_3 = y'$, the first inverter can be perceived to be a NOT operator. From the mixed-logic point of view, however, the first inverter reassigns the active level from H to L, producing output $g_2 = y$. And the assignment mismatch produces output $g_3 = y'$.

Since output $g_1 = y$, the two inverters can be perceived as two cascaded NOT operators. From the mixed-logic point of view, the first inverter reassigns the active level from H to L and the second inverter reassigns the active level from L back to H, producing output $g_1 = y$. And the assignment mismatch produces output $g_4 = y'$. There-

INPUT	OUTPUTS			
$++$	$++$	$++$	$--$	$--$
y	g_1	g_3	g_2	g_4
0	0	1	0	1
1	1	0	1	0

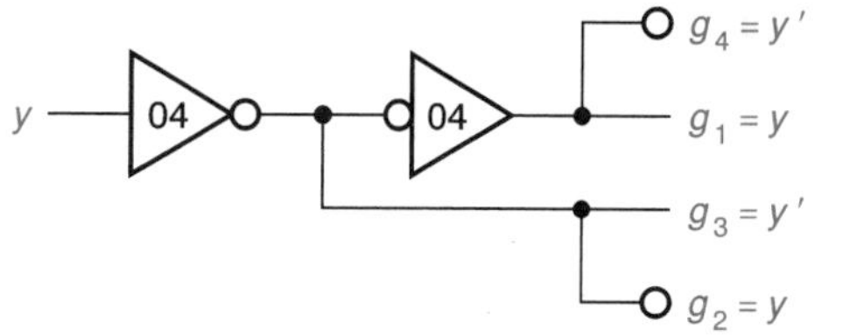

FIGURE 4.10
04 generates all assignments

fore we draw the following conclusion:

An inverter is not necessary to implement the NOT function.

4.6 Logic True Conventions and the 00 Gate

The 00 gate, such as the 74LS00, implements the voltage table shown in Figure 4.11a. The voltage table portrays the relations between the physical gate terminals. We deduce two statements from the table.

Simultaneously driving A AND B high drives Y low.
Individually driving A OR B low drives Y high.

Designating inputs A and B to be active high (Figure 4.11b) forces output Y to be designated active low. Designating A and B to be active low (Figure 4.11c) forces Y to be designated active high. However, designating A active high and B active low, or vice versa, forces no consistent output action. Therefore the 00 is represented by only two out of four possible symbols. The 00 symbols do not have any wires attached to them. The symbols without wires represent the physical gate standing alone. The statements motivate us to give one symbol the AND shape and the other the OR shape. The choice of shapes anticipates using the 00 for logical operations.

Input active levels imply gate symbol.

The active high table (Figure 4.11b) shows there is action at the output only when both inputs are active. This is why the AND shape is used. The bubble at the Y output illustrates the fact that Y is driven low when both A and B are driven high. This is why bubbles are omitted at the A and B inputs of the AND shape.

The active low table (Figure 4.11c) shows there is action at the output when either input is active. The omission of a bubble at the Y

FIGURE 4.11
00 voltage table and symbols

(a)			(b) ACTIVE HIGH A B			(c) ACTIVE LOW A B		
A	B	Y	A	B	Y	A	B	Y
L	L	H	L	L	H	**L**	**L**	**H**
L	H	H	L	**H**	H	**L**	H	**H**
H	L	H	**H**	L	H	H	**L**	**H**
H	H	L	**H**	**H**	L	H	H	L

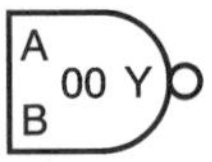

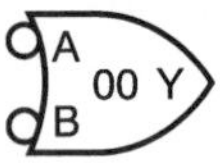

output and the two bubbles at the A and B inputs of the OR shape illustrate the fact that Y is driven high when either A or B is driven low.

Next, after a short discussion concerning gate names, we connect the physical gate to logic.

Gate names Unfortunately, the 00 circuit is called a NAND gate in data books, unfortunate because they assign logical 1 to H voltage and because the 00 can be used as AND or OR or NAND or NOR or four other operators, as you will see. Also, it is regrettable that names given to physical gates are the same or closely related to the logic operators NOT, AND, and OR. Don't be misled into thinking a gate called NAND performs only the NAND logic function. This is why our names for gates are the logic family member's number—the neutral numbering system is not misleading.

Forget gate names. Use gate numbers.

The logic connection The *assignment* is the bridge from voltage tables to logic tables, and vice versa. Assignments are made to switching variables at circuit nodes, which are created by adding wires to the 00 physical gates (Figure 4.12). Variables are sources and destinations at the new circuit nodes generated by the free wire ends. When we draw a wire omitting a bubble at the free end, the de facto assignment is $++$, or active high. When we add a bubble, the de facto assignment is $--$, or active low. The point is that we cannot make a drawing without making assignments. This is why we have to choose. In Figure 4.12 we choose to match active level assignments at circuit nodes to the active levels at the gate terminals. In tabular form this is what was done in Figure 4.12 for the symbol with the AND shape.

Assignment connects voltage to logic.

n terminals imply 2^n assignments.

A gate has only two symbols.

FIGURE 4.12

00 connected to sources and destinations

Terminal A becomes	++ Node x	Terminal B becomes	++ Node y	Terminal Y becomes	−− Node g
L	0	L	0	H	0
L	0	H	1	H	0
H	1	L	0	H	0
H	1	H	1	L	1

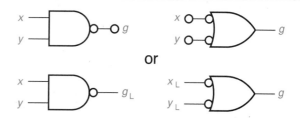

A 3-terminal gate has 8 logic functions.

There are 2^3, or eight, possible sets of assignments for the 00 because each of the three circuit nodes connected by wires to the terminals can be assigned ++ or −−. Thus, the 00 can implement eight different functions, with NAND being only one of the eight.

The assignment set ++ ++ −− converts the 00 voltage table into the truth table for the AND operator. In Figure 4.13 the logical and physical circuit representations merge into one composite symbol. The AND shape is augmented by wires with and without bubbles at the circuit nodes representing the logical assignments. The logic cir-

FIGURE 4.13

00 and the ++ ++ −− assignment set: AND

VOLTAGE TABLE			++	++	−−	← Assignments
A	B	Y	x	y	g	LOGIC CIRCUIT
L	L	H	0	0	0	
L	H	H	0	1	0	
H	L	H	1	0	0	
H	H	L	1	1	1	

$$g = xy$$

$$g_L = xy$$

FIGURE 4.14
00 and the $----$.
$++$ assignment set:
OR

VOLTAGE
TABLE $--$ $--$ $++$ ← Assignments

A	B	Y	x	y	g	LOGIC CIRCUIT
L	L	H	1	1	1	$g = x + y$
L	H	H	1	0	1	
H	L	H	0	1	1	
H	H	L	0	0	0	or

$g_H = x + y$

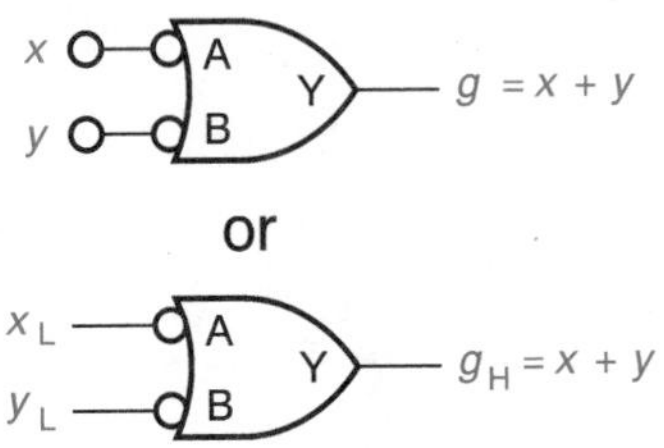

cuit's function $g = xy$ is derived from the truth table with no concern for the origins of the truth table or the composite symbol.

The assignment set $----++$ converts the 00 voltage table into the truth table for the OR operator. That is why the 00 symbol with OR shape is shown in Figure 4.14. The logic circuit function derived from the truth table is $g = x + y$.

Now we can illustrate the important point that physical gates do not process 1's and 0's, but literally process voltages. In both the

FIGURE 4.15
00 and the $++++$
$++$ assignment set:
NAND

VOLTAGE
TABLE $++$ $++$ $++$ $++$ ← Assignments

A	B	Y	x	y	g	g'	
L	L	H	0	0	1	0	From the truth table
L	H	H	0	1	1	0	$g = x'y' + x'y + xy' = x' + y'$
H	L	H	1	0	1	0	
H	H	L	1	1	0	1	$g' = xy$

$g' = xy$ $[g = (xy)']$ $g = x' + y'$ $[g = (xy)']$

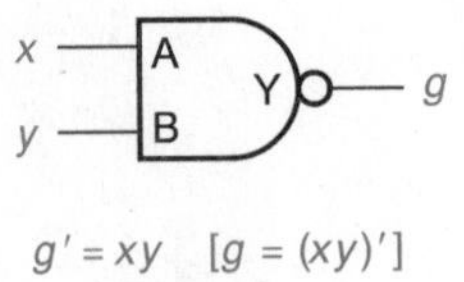

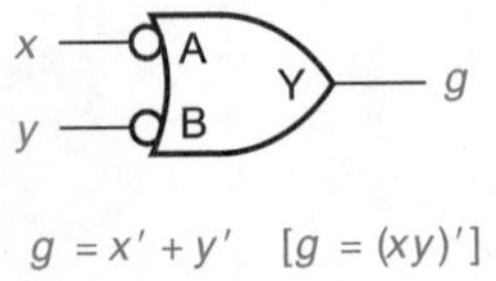

FIGURE 4.16

00 and the $----$ $--$ assignment set: NOR

VOLTAGE TABLE			$--$	$--$	$--$	$--$ ← Assignments	
A	B	Y	x	y	g	g'	
L	L	H	1	1	0	1	From the truth table
L	H	H	1	0	0	1	$g = x'y' = (x + y)'$
H	L	H	0	1	0	1	
H	H	L	0	0	1	0	$g' = xy + xy' + x'y$
							$= x + y$

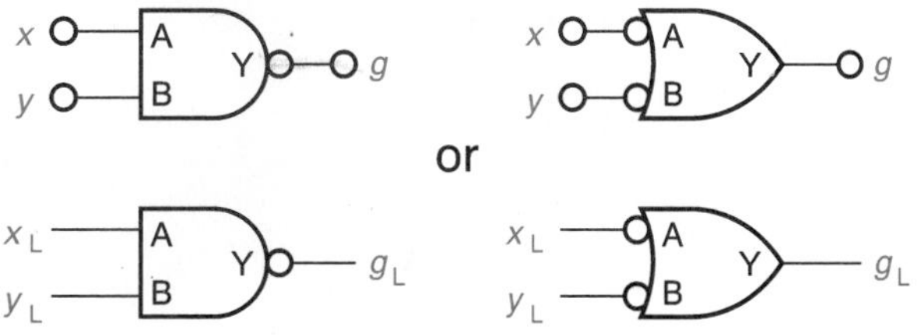

or

$g = x'y'$ $[g = (x + y)']$ $g' = x + y$ $[g = (x + y)']$

AND and OR logic circuits shown in Figures 4.13 and 4.14:

> If x is at the H level and y is at the H level, then g is at the L level.

The logical assignments do not change the physical circuit.

Assignment sets implementing the NAND and NOR operators are displayed in Figures 4.15 and 4.16. In Figure 4.15 DeMorgan's theorem is used to convert the equations for both circuits to the NAND format. In Figure 4.16 the equations convert to the NOR format. As

FIGURE 4.17

00 assignment sets and logical operations

VARIABLE ASSIGNMENT			LOGICAL OPERATOR USING THE 00	LOGIC EQUATION
x	y	g		
$--$	$--$	$--$	NOR	$g = x'y' = (x + y)'$
$--$	$--$	$++$	OR	$g = x + y$
$--$	$++$	$--$		$g = x'y$
$--$	$++$	$++$		$g = x + y'$
$++$	$--$	$--$		$g = xy'$
$++$	$--$	$++$		$g = x' + y$
$++$	$++$	$--$	AND	$g = xy$
$++$	$++$	$++$	NAND	$g = x' + y' = (xy)'$

inputs don't match up

was ascertained earlier, the logical complement is implemented when the assertion levels are different at the two ends of a wire.

The 00 circuit provides the four named logic functions listed in Figure 4.17. Furthermore, Figure 4.17 displays four additional logic functions that do not have names that are available by assignment. The eight assignments generate the four * and four + functions of two variables (also see Table 3.4 on page 000).

EXERCISE 4.1   Derive the 00 logical truth table logic equation for the $--++$ $--$ assignment.

Answer:

VOLTAGE TABLE			$--$	$++$	$--$	$--$	← Assignments
A	B	Y	x	y	g	g'	
L	L	H	1	0	0	1	From the truth table
L	H	H	1	1	0	1	$g = x'y$
H	L	H	0	0	0	1	
H	H	L	0	1	1	0	$g' = x + y'$

∎

EXERCISE 4.2   Derive the 00 logical truth table logic equation for the $++--$ $--$ assignment.

Answer:

VOLTAGE TABLE			$++$	$--$	$--$	$--$	← Assignments
A	B	Y	x	y	g	g'	
L	L	H	0	1	0	1	From the truth table
L	H	H	0	0	0	1	$g = xy'$
H	L	H	1	1	0	1	
H	H	L	1	0	1	0	$g' = x' + y$

∎

4.7   Logic True Conventions and the 02 Gate

The 02 gate, such as the 74LS02, implements the voltage table shown in Figure 4.18a. The voltage table portrays the relations between the physical gate terminals. We deduce two statements from the table.

Individually driving A OR B high drives Y low.
Simultaneously driving A AND B low drives Y high.

Designating inputs A and B to be active high (Figure 4.18b) forces output Y to be designated active low. Designating A and B to be active low (Figure 4.18c) forces Y to be designated active high. However, designating A active high and B active low, or vice versa, forces no consistent output action. Therefore the 02 is represented by only two out of four possible symbols. The 02 symbols do not have any wires attached to them. The symbols without wires represent the physical gate standing alone. The highlighted statements above motivate us to give one symbol the OR shape and the other the AND shape. The choice of shapes anticipates using the 02 for logical operations.

The active high table (Figure 4.18b) shows there is action at the output when either input is active. This is why the OR shape is used. The bubble at the Y output illustrates the fact that Y is driven low when either A or B is driven high. This is why bubbles are omitted at the A and B inputs of the OR shape.

The active low table shows there is no action at the output unless both inputs are active. The omission of a bubble at the Y output and the two bubbles at the A and B inputs of the AND shape illustrate the fact that Y is driven high when both A and B are driven low.

Unfortunately, the 02 is called a NOR gate in data books, unfortunate because, like the 00, the 02 can be used as AND or OR or NAND or NOR or four other operators that do not have names.

The logic connection We repeat the process used with the 00. The assignment is the bridge from voltage tables to logic tables, and vice versa. Assignments are made to switching variables, so we create circuit nodes for variables by adding wires to the 02 physical gates (Figure 4.19). As before, we choose to match active level assignments at circuit nodes to the active levels at the gate terminals.

FIGURE 4.18
02 voltage table and symbols

(a)

A	B	Y
L	L	H
L	H	L
H	L	L
H	H	L

(b) ACTIVE HIGH A B

A	B	Y
L	L	H
L	**H**	L
H	L	L
H	**H**	L

(c) ACTIVE LOW A B

A	B	Y
L	**L**	**H**
L	H	L
H	**L**	L
H	H	L

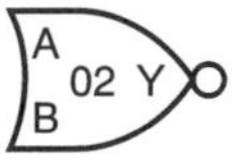

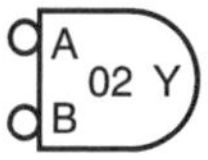

FIGURE 4.19
**02 connected to
sources and
destinations**

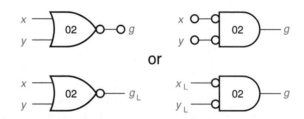

The same four sets of mixed logic assignments that were made for
the 00 are made for the 02 in Figure 4.20 to show once again that a
physical circuit can perform many logic functions. The logic functions
shown below the truth tables in the figure are derived by writing and
simplifying the standard sum equations taken from each truth table:

$$g = xy' + x'y + xy = x + y \qquad\qquad \text{OR}$$

$$g = x'y' = (x + y)' \qquad\qquad \text{NOR}$$

$$g = xy \qquad\qquad \text{AND}$$

$$g = xy' + x'y + x'y' = x' + y' = (xy)' \qquad \text{NAND}$$

As we found with the 00, changing assignments changes the 02
gate's logical function; however, once again keep in mind that the
physical circuit does not change.

The 02 circuit provides the four named logic functions listed in
Figure 4.21, which also displays four other logic functions that do not
have names that are available by assignment. The eight assignments
generate the four * and four + functions of two variables (also see
Table 3.4 on page 101).

FIGURE 4.20
**02 with four
assignment sets:
OR, NOR, AND,
NAND**

Assignments →			+ +	+ +	− −	+ +	+ +	+ +	− −	− −	+ +	− −	− −	− −
x	y	g	x	y	g	x	y	g	x	y	g	x	y	g
L	L	H	0	0	0	0	0	1	1	1	1	1	1	0
L	H	L	0	1	1	0	1	0	1	0	0	1	0	1
H	L	L	1	0	1	1	0	0	0	1	0	0	1	1
H	H	L	1	1	1	1	1	0	0	0	0	0	0	1

Logic function: $g = x + y$ $g = (x + y)'$ $g = xy$ $g = (xy)'$

Name: OR   NOR   AND   NAND

FIGURE 4.21

02 assignment sets and logical operations

VARIABLE ASSIGNMENT			LOGICAL OPERATOR USING THE 02	LOGIC EQUATION
x	y	g		
$--$	$--$	$--$	NAND	$g = x' + y' = (xy)'$
$--$	$--$	$++$	AND	$g = xy$
$--$	$++$	$--$		$g = x' + y$
$--$	$++$	$++$		$g = xy'$
$++$	$--$	$--$		$g = x + y'$
$++$	$--$	$++$		$g = x'y$
$++$	$++$	$--$	OR	$g = x + y$
$++$	$++$	$++$	NOR	$g = x'y' = (x + y)'$

EXERCISE 4.3   Derive the 02 logical truth table logic equation for the $-- ++$ $++$ assignment.

Answer:

VOLTAGE TABLE			$--$	$++$	$++$	$++$	← Assignments
A	B	Y	x	y	g	g'	
L	L	H	1	0	1	0	From the truth table
L	H	L	1	1	0	1	$g = xy'$
H	L	L	0	0	0	1	
H	H	L	0	1	0	1	$g' = x' + y$

■

EXERCISE 4.4   Derive the 02 logical truth table logic equation for the $++ --$ $++$ assignment.

Answer:

VOLTAGE TABLE			$++$	$--$	$++$	$++$	← Assignments
A	B	Y	x	y	g	g'	
L	L	H	0	1	1	0	From the truth table
L	H	L	0	0	0	1	$g = x'y$
H	L	L	1	1	0	1	
H	H	L	1	0	0	1	$g' = x + y'$

■

4.8 ═══════ ## Logic True Conventions and the 86 Gate

The 86 gate, such as the 74LS86, implements the voltage table shown in Figure 4.22a, which portrays the relations between the physical gate terminals. We deduce two statements from the table.

> Simultaneously driving A high AND B low, or vice versa, drives Y high.
> Simultaneously driving A AND B high, or A AND B low, drives Y low.

Designating A to be active high and B to be active low (or vice versa) forces Y to be designated active high (Figure 4.22b). In this case the assignments to A and B are different. Designating both A and B to be active low, or both to be active high, forces Y to be designated active low (Figure 4.22c). In this case the assignments to A and B are the same. Since no other designations force consistent output action, the 86 is represented by only two symbols.

The logic connection   We repeat the process used with the 00. The assignment is the bridge from voltage tables to logic tables, and vice versa. Assignments are made to switching variables, so we create circuit nodes for variables by adding wires to the 86 physical gate (Figure 4.23).

In Figure 4.23 we make four sets of mixed-logic assignments for the 86; this time the physical circuit performs the same logic function in each case. In Figure 4.24 we make the four other sets of mixed-logic assignments for the 86. Figures 4.23 and 4.24 show that the eight assignment sets create only two functions, XOR and XNOR.

XOR has 0 or 2 $--$ assignments.

XNOR has 1 or 3 $--$ assignments.

FIGURE 4.22
86 voltage table and symbols

(a)			(b) ACTIVE A, B DIFFERENT			(c) ACTIVE A, B SAME		
A	B	Y	A	B	Y	A	B	Y
L	L	L	L	L	L	**L**	**L**	**L**
L	H	H	**L**	**H**	**H**	L	H	H
H	L	H	**H**	**L**	**H**	H	L	H
H	H	L	H	H	L	**H**	**H**	**L**

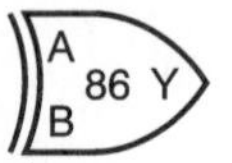

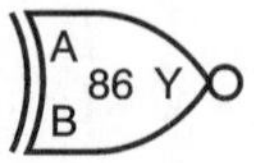

Assignments→			+ +	+ +	+ +	+ +	− −	− −	− −	+ +	− −	− −	− −	+ +
x	y	g	x	y	g_0	x	y	g_1	x	y	g_2	x	y	g_3
L	L	L	0	0	0	0	1	1	1	0	1	1	1	0
L	H	H	0	1	1	0	0	0	1	1	0	1	0	1
H	L	H	1	0	1	1	1	0	0	0	0	0	1	1
H	H	L	1	1	0	1	0	1	0	1	1	0	0	0

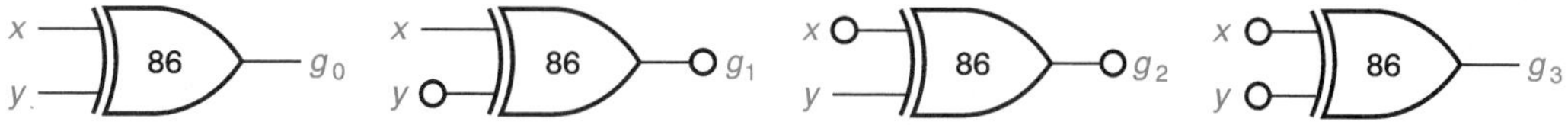

x —[86]— g_0 (y)   x —[86]O g_1 (y O)   x O —[86]O g_2 (y)   x O —[86]— g_3 (y O)

The rows are not in numerical row order in each truth table. For each truth table, $g = x'y + xy' = x$ xor $y = x \oplus y$.

FIGURE 4.23
86 with four assignment sets: XOR

Observe that g is the XOR operator when the number of active low assignments is *even* (0 or 2). When there is an *odd* number of active low assignments, g is the XNOR operator. The eight assignments are listed in Figure 4.25. Next we derive the equations for the eight symbols in Figures 4.23 and 4.24 as well as explain why the XOR symbol is used in Figure 4.24 instead of the XNOR symbol.

FIGURE 4.24
86 with four assignment sets: XNOR

Assignments→			+ +	+ +	− −	+ +	− −	+ +	− −	+ +	+ +	− −	− −	− −
x	y	g	x	y	g_4	x	y	g_5	x	y	g_6	x	y	g_7
L	L	L	0	0	1	0	1	0	1	0	0	1	1	1
L	H	H	0	1	0	0	0	1	1	1	1	1	0	0
H	L	H	1	0	0	1	1	1	0	0	1	0	1	0
H	H	L	1	1	1	1	0	0	0·	1	0	0	0	1

The rows are not in numerical row order in each truth table. For each truth table, $g = x'y' + xy = x$ xnor $y = x \odot y$.

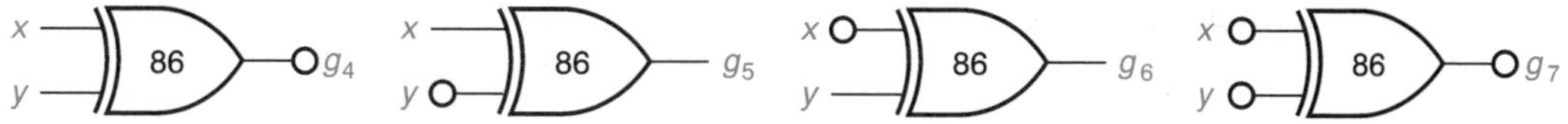

x —[86]O g_4 (y)   x —[86]— g_5 (y O)   x O —[86]— g_6 (y)   x O —[86]O g_7 (y O)

Figure 4.25

86 assignment sets and logical operations

VARIABLE ASSIGNMENT			LOGICAL OPERATOR USING THE 86 LOGIC EQUATION	
x	y	g		
$--$	$--$	$--$	XNOR	$g = x'y' + xy$
$--$	$--$	$++$	XOR	
$--$	$++$	$--$	XOR	
$--$	$++$	$++$	XNOR	
$++$	$--$	$--$	XOR	
$++$	$--$	$++$	XNOR	
$++$	$++$	$--$	XNOR	
$++$	$++$	$++$	XOR	$g = x'y + xy'$

The following analysis shows that four out of eight possible assignments result in the XOR operator and that the other four result in the XNOR function. X_j refers to theorem j in Table 4.2.

ASSIGNMENT			BUBBLES	XOR EQUATION (XOR SYMBOL)—FIGURE 4.23	
x	y	g			
$++$	$++$	$++$	none	$g = x \oplus y$	
$++$	$--$	$--$	at y, g	$g = (x \oplus y')' = (x \odot y)' = x \oplus y$	(X_{11}, X_9)
$--$	$++$	$--$	at x, g	$g = (x' \oplus y)' = (x \odot y)' = x \oplus y$	(X_{12}, X_9)
$--$	$--$	$++$	at x, y	$g = x' \oplus y' = \qquad\qquad x \oplus y$	(X_{10})

ASSIGNMENT			BUBBLES	XNOR EQUATION (XOR SYMBOL)—FIGURE 4.24	
x	y	g			
$++$	$++$	$--$	at g	$g = (x \oplus y)' = x \odot y$	(X_9)
$++$	$--$	$++$	at y	$g = x \oplus y' = x \odot y$	(X_{11})
$--$	$++$	$++$	at x	$g = x' \oplus y = x \odot y$	(X_{12})
$--$	$--$	$--$	at x, y, g	$g = (x' \oplus y')' = x \odot y$	(X_{13})

Important to know.

TABLE 4.2 **XOR—Theorems and Algebra of Constants**

The XOR symbol is $\oplus$, and the XNOR symbol is $\odot$.

THEOREMS

X1	$x \oplus 0 = x$	
X1D	$x \oplus 1 = x'$	
X2	$x \oplus x = 0$	
X2D	$x \oplus x' = 1$	
X3	$x \oplus x \oplus \cdots \oplus x = 0$	Even number of terms
X3D	$x \oplus x \oplus \cdots \oplus x = x$	Odd number of terms
X4	$x \oplus y = y \oplus x$	
X5	$(x \oplus y) \oplus z = x \oplus (y \oplus z) = x \oplus y \oplus z$	
X6	$x \oplus y = z$ implies $\quad x \oplus z = y$	
	$\qquad\qquad\qquad\qquad\quad y \oplus z = x$	
	$\qquad\qquad\qquad\qquad\quad x \oplus y \oplus z = 0$	
X7	$xy \oplus xz = x(y \oplus z)$	

However: $\qquad (x + y) \oplus (x + z) \neq x \oplus (y + z)$

$\qquad\qquad\quad (x \oplus y) + (x \oplus z) \neq x \oplus (y + z)$

$\qquad\qquad\quad (x \oplus y) \cdot (x \oplus z) \neq x \cdot (y \oplus z)$

X8	$x \oplus y = xy' + x'y = (x + y)(x' + y')$
X9	$(x \oplus y)' = xy + x'y' = x \odot y$
X10	$x' \oplus y' = x \oplus y$
X11	$x \oplus y' = x \odot y$
X12	$x' \oplus y = x \odot y$
X13	$(x' \oplus y')' = x \odot y$

ALGEBRA OF CONSTANTS

1. $0 \oplus 0 = 0$
2. $1 \oplus 1 = 0$
3. $1 \oplus 0 = 0 \oplus 1 = 1$

Bubble replaces ' operator.

Now we explain why the XOR symbol was used in Figure 4.24 instead of the XNOR symbol. When the XNOR symbol in Figure 4.26 is used, the analysis starts with XNOR operator in initial equation. Observe that a double complement occurs in two cases.

ASSIGNMENT			BUBBLES	XNOR EQUATION (XNOR SYMBOL)—FIGURE 4.26
x	y	g		
+ +	+ +	– –	at g	$g = x \odot y$
+ +	– –	+ +	at y	$g = (x \odot y')' = (x \oplus y)' = x \odot y$
– –	+ +	+ +	at x	$g = (x' \odot y)' = (x \oplus y)' = x \odot y$
– –	– –	– –	at x, y, g	$g = (x' \odot y') = x \odot y$

Table 4.2 lists theorems involving XOR and XNOR operations and results from the XOR algebra of constants. The theorems are readily proved from the axioms and from the theorems in Table 3.1 on pages 83–84. Four points are worthy of emphasis.

1. The important equation

$$xy + x'y' = (xy' + x'y)'$$

$$m_3 + m_0 = (m_2 + m_1)'$$

and its inverse

$$(xy + x'y')' = xy' + x'y$$

$$(m_3 + m_0)' = m_2 + m_1$$

FIGURE 4.26
**86 with four
assignment sets and
two symbol sets**

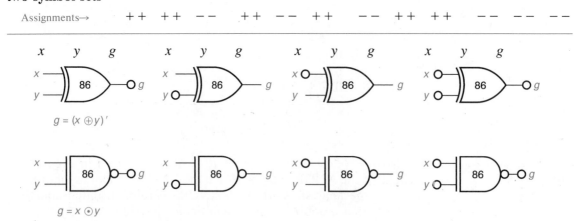

Assignments→   + + + + – –   + + – – + +   – – + + + +   – – – – – –

$g = (x \oplus y)'$

$g = x \odot y$

occur often enough to justify being able to recognize them without hesitation.

2. The XNOR function is the complement of the XOR function.

3. COIN (for "coincidence") is an alias for XNOR.

4. The XNOR function is implemented by the 266 and 810 logic family members. The 266 has open collector outputs (Section 2.5).

4.9

Mixed-Logic Analysis: Equations from Circuits

There are many ways to analyze a circuit. Furthermore, the method used depends on the desired form of the result. Perhaps only experience makes this clear. Nevertheless, shortly we will explore a method that always produces a result. (A different kind of method is illustrated by example at the end of this section.) First we need to solidify some ideas about equivalents.

Process converts physical to logical.

A logic function is not changed when:

a. An active low wire ○────○ replaces an active high wire ──── in a circuit:

b. The NOT operator ′ is implemented by wires such as ○──── and ────○ because their active levels are different at the two ends of the wire. This is why, at any circuit node, the complement of a variable at one active level may be replaced by the variable itself when the active level is complemented. There are four cases.

y ○──── replaces y' ──── $(y_\text{L}$ ──── replaces y_H' ────$)$

y ○────○ replaces y' ────○ $(y_\text{L}$ ────○ replaces y_H' ────○$)$

────○ y replaces ──── y' $($──── y_L replaces ──── $y_\text{H}')$

○────○ y replaces ○──── y' $($○──── y_L replaces ○──── $y_\text{H}')$

c. A Gate's AND (OR) symbol is replaced by the corresponding OR (AND) symbol. This requires terminals with bubbles on the AND (OR) to be replaced by terminals without bubbles on the OR (AND), and vice versa.

Analysis procedure   There are three basic steps. The number of times any step is used depends on the circuit. Upcoming Examples 4.1 through 4.5 illustrate the following analysis procedure.

STEP 1   Replace gates by the physical duals to minimize mismatches of active levels at the two ends of any wire. (Start at the circuit's output(s) and work towards the inputs.) For example:

STEP 2   Start at the circuit's input(s) and work towards the output(s). Ignore the bubbles and execute the ∗ operator for AND gates. Ignore the bubbles and execute the + operator for OR gates. For the circuits in Step 1:

$$\text{AND} \qquad\qquad \text{OR}$$
$$g \rightarrow xy \qquad\qquad g \rightarrow x + y$$

STEP 3   Now take the bubbles into account: Execute the ' operator for active level mismatches at OR gate inputs and outputs. Execute the ' operator for active level mismatches at AND gate inputs and outputs. For the circuits in Step 1:

$$\text{AND} \qquad\qquad \text{OR}$$
$$g' = x'y \qquad\qquad g = x + y'$$

Mark the gate output pins with the output equation.

EXAMPLE 4.1   **Mixed-Logic Circuit Analysis: Minterm**

Consider the schematic of a function of three variables $g(x, y, z)$. Let all variables be active high. The positive-logic-system active high circuit uses complemented variables in the schematic.

$$x' \,—\!\!\Big[\!\!\!\!\supset\!\!— \; g = x'yz'$$

A mixed-logic schematic eliminates complemented variables by complementing the active level, thereby replacing the complemented variables with variables that are not complemented. In this case x' and z' are replaced by x and z at active low circuit nodes. For example, x $\circ\!\!——$ is equivalent to an active high x', because active low variable x is at the H level when False. The bubble replaces the ' operator.

$$g = x'yz' = m_2$$

If we convert to the OR shape, the number of mismatches is not changed in this example. All terminals without bubbles on the AND shape are replaced by terminals with bubbles on the OR shape.

$$g' = (x + y' + z) \text{ so that } g = x'yz' = m_2$$

EXAMPLE 4.2   Mixed-Logic Circuit Analysis

Consider the schematic of a function of three variables $g(x, y, z)$. Let all variables be active high. The positive-logic-system active high circuit uses complemented variables in the schematic.

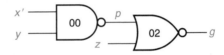

First replace the output OR shape with an AND shape, bubble with no-bubble, and no-bubble with bubble. Then eliminate x' by replacing x' with x-bubble.

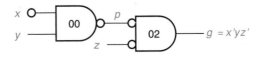

Start at the input. Execute the $*$ operator: $p \rightarrow xy$. Execute the $'$ operator for the $\circ$——— wire so that $p = x'y$. Execute the $*$ operator at the output gate: $g \rightarrow pz$. Execute the $'$ operator at the output gate for the ———$\circ$ wire: $g = pz' = x'yz'$.

EXAMPLE 4.3   Mixed-Logic Circuit Analysis: Maxterm

Consider the schematic of a function of three variables $g(x, y, z)$. Let all variables be active high. The positive-logic-system active high circuit uses complemented variables in the schematic.

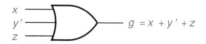

A mixed-logic schematic eliminates complemented variables by complementing the active level, thereby replacing the complemented variables with variables that are not complemented. In this case y' is replaced by y at an active low circuit node. The bubble replaces the $'$ operator.

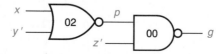

If we convert to the AND shape, the number of mismatches is increased from one to three. All terminals without bubbles on the OR shape are replaced by terminals with bubbles on the AND shape.

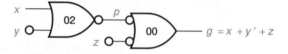

EXAMPLE 4.4   Mixed-Logic Circuit Analysis

Consider the schematic of a function of three variables $g(x, y, z)$. Let all variables be active high. The positive-logic-system active high circuit uses complemented variables in the schematic.

First replace the output AND shape with an OR shape, bubble with no-bubble, and no-bubble with bubble. Then eliminate y' and z' by replacing y' and z' with y-bubble and z-bubble.

Start at the input. Execute the $+$ operator: $p \rightarrow x + y$. Execute the $'$ operator for the o————— wire so that $p = x + y'$. Execute the $+$ operator at the output gate: $g \rightarrow p + z$. There are no $'$ operators at the output gate, and so $g = p + z = x + y' + z$.

EXAMPLE 4.5 **Mixed-Logic Circuit Analysis: XOR**

1. Start the analysis at the output of the circuit and work backwards toward the inputs.

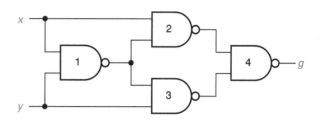

The wires driving output gate 4 have mismatched active levels. Replace the 00 AND symbol with the 00 OR symbol (Step 1). The replacement changes two "NOT-function" wires (○————) into two active low wires (○————○). This application of a mixed-logic idea immensely simplifies the analysis of this or any other circuit.

If 00 gates 2 and 3 driving the output gate are changed to OR symbols, two NOT-functions are created. Do not change 00 gates 2 and 3.

Either the AND or the OR symbol for 00 gate 1 at the input can be used, because there are two wires with mismatched active levels in either case. Retain the AND symbol for gate 1.

2. Starting at the circuit inputs, execute the ∗ operator for AND gate 1 and mark the gate 1 output as xy. This is active low.

Execute the ∗ operator for AND gate 2 and the ′ operator for the xy input, and mark the gate 2 output as $x(xy)'$.

Execute the ∗ operator for AND gate 3 and the ′ operator for the xy input, and mark the gate 3 output as $y(xy)'$.

Execute the + operator for OR gate 4, and mark the gate 4 output as $g = x(xy)' + y(xy)'$, which is the desired result. In this case algebraic manipulation is required to simplify the equation for g:

$$g = (x + y)(xy)' = (x + y)(x' + y') = xy' + x'y = x \oplus y$$

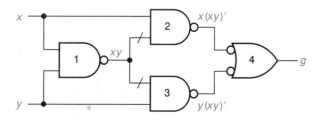

Another way to analyze a logic circuit is illustrated in Example 4.6, which ignores logic to focus on voltage processing. The method traces paths through the circuit from input to output by evaluating gate processing of H and L levels. The first goal is the creation of a voltage table of inputs and outputs. The second goal is the selection of an appropriate assignment set making a connection to logic.

The analysis that builds the second truth table row in Example 4.6 starts by tracing voltage as follows:

$$x = L \Rightarrow q = H, p = H \quad \text{then} \quad y = H, p = H \Rightarrow r = L$$
$$\text{then} \quad r = L \Rightarrow g = H$$

Similar analysis produces the other three rows of the voltage table. Now focus on voltage table columns x, y, and g in order to select appropriate assignments, which are $+ + + + + + (g_1)$ or $+ + + + - -$ (g_2).

EXAMPLE 4.6   Circuit Analysis by Path Tracing of Voltage

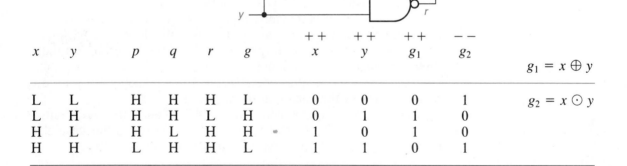

x	y	p	q	r	g	$\begin{array}{c}++\\x\end{array}$	$\begin{array}{c}++\\y\end{array}$	$\begin{array}{c}++\\g_1\end{array}$	$\begin{array}{c}--\\g_2\end{array}$	
										$g_1 = x \oplus y$
L	L	H	H	H	L	0	0	0	1	$g_2 = x \odot y$
L	H	H	H	L	H	0	1	1	0	
H	L	H	L	H	H	1	0	1	0	
H	H	L	H	H	L	1	1	0	1	

Example 4.6 illustrates the following statement:

> Physical gates do not process 1's and 0's; they literally process voltages.

Moreover, as Example 4.6 makes clear:

> We can ignore the logic symbols and the bubbles at the circuit nodes when dealing with the physical levels. We simply

trace paths through circuits verifying behavior by using voltage tables and checking L and H levels.

4.10 Mixed-Logic Synthesis: Circuits from Equations

Process converts logical to physical.

Any Boolean equation can be written as a sum of product terms (OR of ANDs) or as a product of sum terms (AND of ORs) where literals make up either type of term. Both forms can be implemented by two levels of logic: AND gates driving OR gates, and vice versa. The important point is that complements of variables do *not* have to be implemented between the AND and the OR levels. (This is not obvious at this point.) Therefore they need only be implemented at the inputs or at the outputs, which means the complements can be omitted in any equation for the purposes of finding a logic circuit structure. Then the complements are restored by adjusting the structure's input and output active levels. Here is one way to synthesize a circuit of gates.

STEP 1   Given a truth table, a K map, an equation, or some combination thereof, obtain the equation to be synthesized. Put the equation into two-level format (sum of products or product of sums).

STEP 2   Rewrite the equation deleting all ′ operators. Do not consolidate terms. Synthesize a two-level AND, OR circuit directly from the rewritten equation.

STEP 3   Show the active levels of gate terminals. Convert each AND and OR to the specified, or desired, gate symbol by adding bubbles. Use the AND equivalent gate symbol for AND's and the OR equivalent gate symbol for OR's (Figure 4.27). Show active levels of all input variables. Add a bubble at each input terminal corresponding to active low variables.

FIGURE 4.27
Mixed-logic gate symbols

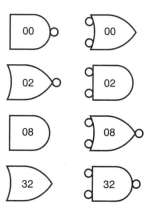

STEP 4   Obtain the equation of the newly converted circuit by analysis.

STEP 5   Compare the original equation to the new equation to ascertain which variables, if any, require active low sources and destinations. The active low sources and destinations are implemented by adding inverters or bubbles to input and output circuit nodes. This process introduces the correct terminal conditions for each input and output variable.

EXAMPLE 4.7   Mixed-Logic Circuit Synthesis

A circuit is defined when the active levels of all variables are specified and the equation is given. Sometimes the circuit is further constrained by specifying the gate types to be used. For example, variables g, w, x are active high, variables y, z are active low, only two-input 00 type gates may be used, and the equation is $g = w'x + yz'$.

STEP 1   Put the equation into two-level format.

$$g = w'x + yz'$$

STEP 2   Delete all ' operators. Do not consolidate terms. Synthesize an OR of ANDs circuit.

$$g = wx + yz$$

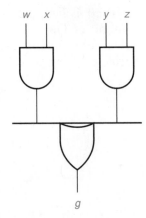

STEP 3   Convert each AND and OR to the specified gate symbol by adding bubbles. Mark active low input variables with a bubble.

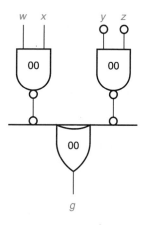

STEP 4   The equation for the new circuit is

$$g = wx + y'z'$$

STEP 5   The original equation is $g = w'x + yz'$. This requires an active low
             source for w and an active high source for y. The sources are used to
             implement the required $'$ operators: w_H is replaced by $w \circ\!\!-\!\!-\!\!-\!\!-\!\!-$ and
             y_L is replaced by $y -\!\!-\!\!-\!\!-\!\!-$.

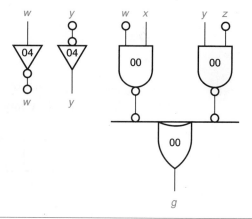

EXAMPLE 4.8   **Mixed-Logic Circuit Synthesis: NAND
               NAND logic**

All variables are active high, only NAND gates like 00, 10, and 20 may
be used, and the equation is

$$g = z'[x + (y \oplus w)] + zyx'$$

STEP 1   Put the equation into two-level format.

a. Convert $\oplus$ operator to eliminate two levels of logic:

$$g = z'(x + yw' + y'w) + zyx'$$

b. Multiply out to eliminate one more level of logic:

$$g = z'x + z'yw' + z'y'w + zyx'$$

STEP 2   Delete all ' operators. Do not consolidate terms. Synthesize an OR of ANDs circuit.

$$g = zx + zyw + zyw + zyx$$

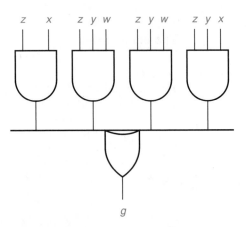

STEP 3   Convert each AND and OR to the specified gate symbol by adding bubbles. Mark active low input variables with a bubble.

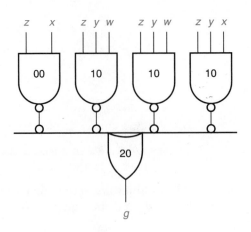

STEP 4   The equation for the new circuit is

$$g = zx + zyw + zyw + zyx$$

STEP 5   The original equation is

$$g = z'x + z'yw' + z'y'w + zyx'$$

This requires active low sources for w, x, y, and z. The sources are implemented by adding inverters. Bubbles implement active low circuit inputs.

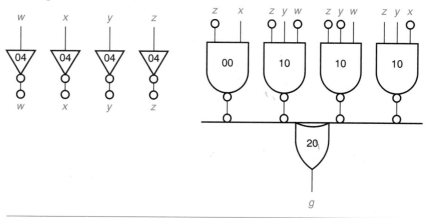

EXAMPLE 4.9   **Mixed-Logic Circuit Synthesis: NOR NOR Logic**

All variables are active high, only NOR gates like 02, 25, and 27 may be used, and the equation is

$$g = z'[x + (y \oplus w)] + zyx'$$

STEP 1   Put the equation into two-level format.

a. Convert $\oplus$ operator to eliminate two levels of logic:

$$g = z'[x + yw' + y'w] + zyx'$$

b. Multiply out to eliminate one more level of logic:

$$g = z'x + z'yw' + z'y'w + zyx'$$

STEP 2   Delete all $'$ operators. Do not consolidate terms. Synthesize an OR of ANDs circuit.

$$g = zx + zyw + zyw + zyx$$

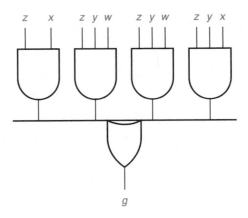

STEP 3   Convert each AND and OR to the specified gate symbol by adding bubbles. Mark active low input variables with a bubble.

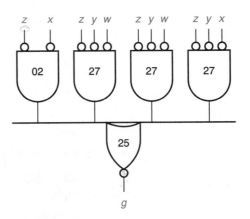

STEP 4   The equation for the new circuit is

$$g' = z'x' + z'y'w' + z'y'w' + z'y'x'$$

STEP 5   The original equation is

$$g = z'x + z'yw' + z'y'w + zyx'$$

This requires active low sources for w, x, y, and z. The output requires an active high destination. The sources and destination are implemented by adding inverters. Bubbles implement active low circuit inputs. The active low output is inverted to reassign it to active high as specified.

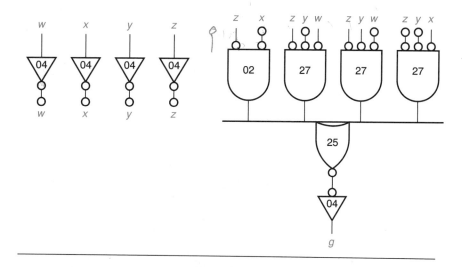

SUMMARY

Gates Are Logical Operators

The ∗ logical operator is implemented by an AND gate; the + logical operator is implemented by an OR gate; and one way to implement the ′ logical operator is with an inverter gate.

True/False Physical Representations

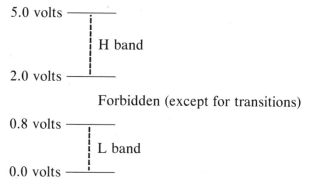

Note: The values 2.0 and 0.8 are logic-family dependent.

Mixed Logic: True and False May Be at Low or High Voltage

A logic design where True is represented by H and L voltage (in different parts of the circuit) is called a mixed-logic design. True (T or 1) and False (F or 0) can be represented in two ways: active high and active low.

1. The + + assignment is active high.

$$\text{High voltage } H = 1 = \text{True}$$

$$\text{Low voltage } L = 0 = \text{False}$$

2. The − − assignment is active low.

$$\text{High voltage } H = 0 = \text{False}$$

$$\text{Low voltage } L = 1 = \text{True}$$

Active: The assignment of voltage level H to logical 1 means switching variables are active when high; that is, they perform their action when high. The assignment of voltage level L to logical 1 means switching variables perform their action when low; they are active when low.

Positive-logic systems use only active high assignments + +. Negative-logic systems use only active low assignments − −. Mixed-logic systems use + + and − − assignments.

Asserted: Asserted, which means a variable is active (True), is independent of whether the active condition is represented by H or by L. Asserted emphasizes action.

Bubbles: There are two kinds of bubbles. A bubble at a gate terminal implies that the signal at the terminal is active low. A different type of bubble at a source or destination circuit node means the variable at that node has the − − assignment (Figure 4.3a). This is not inversion. However, you may prefer a special symbol attached to a variable, such as x_L, at a source or a destination in lieu of a bubble (see Notation for Active High and Low Variables). The two-kinds-of-bubble notation usually results in fewer errors, less writing, and greater schematic clarity and readability.

On a wire such as ———, truth is represented by an H voltage. On a wire such as ○———○, truth is represented by an L voltage.

Bubbles represent complemented variables in mixed logic schematics: A mixed-logic schematic clearly indicates the active level on all wires by using bubbles in lieu of complemented variables.

Logic Family Part Numbers

In logic families of parts, the full number is 74LS04 or 74ALS138 or 74ACT11800 or 74BCT240, and so forth. The 74xx or 74yyy represents the prefix identifying the family. The number identifies the logic circuit. The circuit numbers are constant with time and change of family. We want to know the circuit numbers and what they represent. We drop the prefix and only use the number because the number specifies the circuit function.

Logic True Conventions and Gates

Voltage tables represent physical gates. The voltage table portrays the relations between the physical gate terminals. Symbols for any gate are deduced from the voltage table. In fact, active designations of gate terminals are constrained by a gate's voltage table. Active high designations at a gate's inputs results in one symbol for the gate. Active low designations at a gate's inputs results in a second symbol for the gate. One symbol has the AND shape and the other has the OR shape. The inverter and XOR gates have unique shapes.

The voltage table has nothing to do with logic until wires are added to the gate terminals, the wired gate is embedded into a circuit, and assignments are made. For example, this is why there are eight 00 logic functions (see Figure 4.17), one of which is the NAND function.

The wires o———— and ————o have different active levels at their two ends. When this is the case, a logical NOT function is implemented. This is why an inverter is unnecessary to implement the NOT function. Nevertheless, one application of inverters is implementation of the NOT function.

Gate names: Unfortunately, the 00 circuit is called a NAND gate in data books, because they assign logical 1 to H voltage and because the 00 could be used as an AND or OR or NAND or NOR or four other operators. This is why our names for gates are the logic family member's number.

Mixed-Logic Analysis

This procedure has three basic steps. The number of times any step is used depends on the circuit.

STEP 1   Replace gates by their physical duals to minimize mismatches of active levels at the two ends of any wire.

STEP 2   Start at the circuit's input(s) and work towards the output(s). Ignore the bubbles and execute the $*$ operator for AND gates. Ignore the bubbles and execute the $+$ operator for OR gates.

STEP 3   Take the bubbles into account: Execute the ' operator for active level mismatches at OR gate inputs and outputs. Execute the ' operator for active level mismatches at AND gate inputs and outputs. Mark the gate output pins with the output equation.

Mixed-Logic Synthesis

This procedure goes as follows.

STEP 1   Given a truth table, a K map, an equation, or some combination thereof, obtain the equation to be synthesized. Put the equation into two-level format (sum of products or product of sums).

STEP 2   Rewrite the equation deleting all ' operators. Do not consolidate terms. Synthesize a two-level AND, OR circuit directly from the re-written equation.

STEP 3   Show the active levels of gate terminals. Convert each AND and OR to the specified, or desired, gate symbol by adding bubbles. Use the AND equivalent gate symbol for AND's and the OR equivalent gate symbol for OR's (see Figure 4.27). Show active levels of all input variables. Add a bubble at each input terminal corresponding to active low variables.

STEP 4   Obtain the equation of the newly converted circuit by analysis.

STEP 5   Compare the original equation to the new equation to ascertain which variables, if any, require active low sources and destinations. The active low sources and destinations are implemented by adding inverters or bubbles to input and output circuit nodes. This process introduces the correct terminal conditions for each input and output variable.

REFERENCES   Breeding, K. J. 1989. *Digital Design Fundamentals*. Englewood Cliffs, N.J.: Prentice-Hall.

Fletcher, W. L. 1979. *An Engineering Approach to Digital Design*. Englewood Cliffs, N.J.: Prentice-Hall.

TTL Logic Data Book. 1990. Dallas: Texas Instruments.

Winkel, D., and F. Prosser. 1987. *The Art of Digital Design*. Englewood Cliffs, N.J.: Prentice-Hall.

Name _____   SID# _____   Section _____
           (last)   *(initials)*

Approved by _____   Date _____

Grade on report _____

Mixed-Logic Circuit Design 1

1. Four-output truth table.

    1.1   Check the logic equations for the four o_j outputs.

        _____ o_0
        _____ o_1
        _____ o_2
        _____ o_3

    1.2   Design and build the circuit.

        _____ circuit documents

    1.3   Use the truth table signal generator to test the circuit.

        _____ input/output timing diagram
        _____ o_0
        _____ o_1
        _____ o_2
        _____ o_3

2. One-output truth table

    2.1   Check the logic equation for the z output.

        _____ z

    2.2   Make short-form truth table

        _____ truth table

    2.3   Design and build the circuit.

        _____ circuit documents

2.4 Use the truth table signal generator to test the circuit.

_____ input/output timing diagram
_____ z when d_{3210} = LHLH
_____ z when d_{3210} = HHLL
_____ z when d_{3210} = HLLH

1. Consider the followng truth table.

INPUTS			OUTPUTS				VOLTAGE ASSIGNMENTS
g	b	a	o_3	o_2	o_1	o_0	
L	L	L	H	H	H	L	g T := L
L	L	H	H	H	L	H	b T := H
L	H	L	H	L	H	H	a T := H
L	H	H	L	H	H	H	o_j T := L j = 0, 1, 2, 3
H	—	—	H	H	H	H	

1.1 Write the logic equations for the four o_j outputs.

1.2 Design and build a combinational logic circuit implementing the equations. Use 74LS00 and 74LS10.

1.3 Connect signal generator outputs q_2, q_1, q_0 to the g, b, a inputs, respectively. Plot the four o_j over a cycle of eight clock periods to prove your circuit is correct.

2. Consider the following truth table.

			INPUTS				OUTPUT	VOLTAGE ASSIGNMENTS
g	b	a	d_3	d_2	d_1	d_0	z	
L	L	L	X	X	X	L	L	g T := L
L	L	L	X	X	X	H	H	b T := H
L	L	H	X	X	L	X	L	a T := H
L	L	H	X	X	H	X	H	d_j T := H
L	H	L	X	L	X	X	L	z T := H
L	H	L	X	H	X	X	H	
L	H	H	L	X	X	X	L	
L	H	H	H	X	X	X	H	
H	X	X	X	X	X	X	L	

2.1 Write the logic equation for the z output.

2.2 Make a "short-form" five-row truth table where d_0, d_1, d_2, d_3 are table-entered variables replacing H and L in the z column.

2.3 Design and build a combinational logic circuit implementing the z equation. Use 74LS04 and 74LS20.

2.4 Connect signal generator outputs q_2, q_1, q_0 to the g, b, a inputs, respectively. Connect dip switches to the d_j inputs. Plot the z output for the switch combinations

$$d_{3210} = \text{LHLH}$$
$$= \text{HHLL}$$
$$= \text{HLLH}$$

Are the results what you expect? Why?

Name _____   SID# _____   Section _____
      (last)   *(initials)*

Approved by _____   Date _____

Grade on report _____

Mixed-Logic Circuit Design 2

5.1   Put the assigned equation into two-level format.

_____ equation

5.2   Delete all ' operators. Do not consolidate terms. Synthesize an OR of ANDs circuit.

_____ OR of ANDs circuit

5.3   Convert each AND and OR to the specified gate symbols by adding bubbles. Mark active low input variables with a bubble.

_____ mixed-logic circuit

5.4   Derive the equation for the new circuit.

_____ new equation

5.5   Design and build the circuit.

_____ circuit documents

5.6   Use the truth table signal generator to test the circuit.

_____ input/output timing diagram

Assignment: logic _____ , equation _____

A.   NAND NAND logic.
B.   NOR NOR logic.

Assume variables n_0, n_2 are active high, and that the remaining variables are active low.

1. $f = n_3'[n_1 + (n_2 \text{ xor } n_0)'] + n_3 n_2 n_1'$

2. $f = n_3' n_2' + n_2[n_1' n_0' + n_0(n_3 \text{ xor } n_1)]$

3. $f = n_3'(n_1' + n_0) + n_2(n_1' + n_3)$

4. $f = (n_3 \text{ xor } n_0) + n_3'[n_2'(n_1 + n_0') + n_1 n_0']$

5. $f = n_0'(n_3 \text{ xor } n_2 n_1')$

6. $f = n_2(n_1' + n_2' n_0') + n_2' n_1' n_0'$

7. $f = n_2 n_1' + n_3' n_1(n_0' + n_2')$

5.1   Put the assigned equation into two-level format.

5.2   Delete all ' operators. Do not consolidate terms. Synthesize an OR of ANDs circuit.

5.3   Convert each AND and OR to the specified gate symbols by adding bubbles. Mark active low input variables with a bubble.

5.4   Derive the equation for the new circuit.

5.5   Design and build the circuit.

5.6   Use the truth table signal generator to test the circuit.

PROBLEMS

4.1 Construct a logic truth table for the circuit in Figure P4.1.

4.2 Construct a logic truth table for the circuit in Figure P4.2.

4.3 Construct a logic truth table for the circuit in Figure P4.3.

4.4 Construct a logic truth table for the circuit in Figure P4.4.

Refer to Figure 4.17 for Problems 4.5 to 4.8.

4.5 Show that $g = x'y$ when the 00 assignments are $x\,(--)$, $y\,(++)$, $g\,(--)$.

4.6 Show that $g = x + y'$ when the 00 assignments are $x\,(--)$, $y\,(++)$, $g\,(++)$.

4.7 Show that $g = xy'$ when the 00 assignments are $x\,(++)$, $y\,(--)$, $g\,(--)$.

FIGURE P4.1

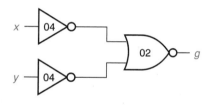

FIGURE P4.2

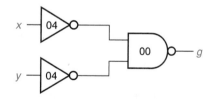

FIGURE P4.3

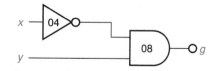

FIGURE P4.4

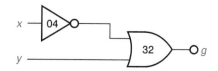

4.8 Show that $g = x' + y$ when the 00 assignments are x $(++)$, y $(--)$, g $(++)$.

Refer to Figure 4.21 for Problems 4.9 to 4.12.

4.9 Show that $g = x' + y$ when the 02 assignments are x $(--)$, y $(++)$, g $(--)$.

4.10 Show that $g = xy'$ when the 02 assignments are x $(--)$, y $(++)$, g $(++)$.

4.11 Show that $g = x + y'$ when the 02 assignments are x $(++)$, y $(--)$, g $(--)$.

4.12 Show that $g = x'y$ when the 02 assignments are x $(++)$, y $(--)$, g $(++)$.

4.13 Use algebra to prove $g_0 = g_2$ (see Figure 4.23).

4.14 Use algebra to prove $g_1 = g_2$ (see Figure 4.23).

4.15 Use algebra to prove $g_4 = g_5$ (see Figure 4.24).

4.16 Use algebra to prove $g_6 = g_7$ (see Figure 4.24).

4.17 Prove Theorems X1 and X1D. Use axioms and theorems in Table 3.1.

4.18 Prove Theorem X6. Use axioms and theorems in Table 3.1.

4.19 Prove Theorem X8. Use axioms and theorems in Table 3.1.

4.20 Prove Theorem X11. Use axioms and theorems in Table 3.1.

4.21 Mixed-Logic and DeMorgan's Theorems

Using all possible product combinations of two variables and their complements, eight functions are generated as follows. DeMorgan's theorems generate the eight complementary functions. Draw the circuits for each function pair.

(a) $z = ba = (b' + a')'$
(b) $z = ba' = (b' + a)'$
(c) $z = b'a = (b + a')'$
(d) $z = b'a' = (b + a)'$

(e) $z = (ba)' = b' + a'$
(f) $z = (ba')' = b' + a$
(g) $z = (b'a)' = b + a'$
(h) $z = (b'a')' = b + a$

4.22 Use mixed-logic circuit analysis to find g (for the circuits in the listed figures) as follows:

Use algebra to find an expression for g' (*not* g).

Write g' to the K-map for g (not g').

Read g from the K-map.

(a) Figure P4.5
(b) Figure P4.6
(c) Figure P4.7
(d) Figure P4.8
(e) Figure P4.9
(f) Figure P4.10
(g) Figure P4.11

FIGURE P4.5

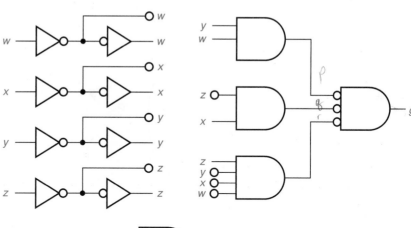

FIGURE P4.6

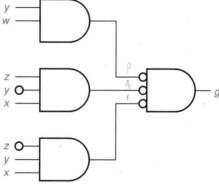

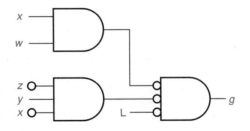

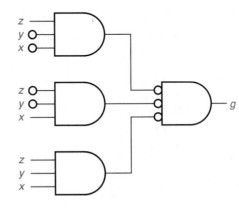

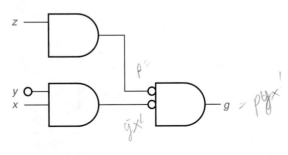

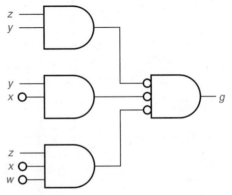

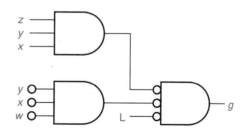

4.23 Assume all variables are active high. Draw a mixed-logic circuit for the following equations. Do not use 08 or 32 type gates. Do not modify the equations.

(a) $f = n_3'[n_1 + (n_2 \text{ xor } n_0)'] + n_3 n_2 n_1'$
(b) $f = n_3'n_2' + n_2[n_1'n_0' + n_0(n_3 \text{ xor } n_1)]$
(c) $f = n_3'(n_1' + n_0) + n_2(n_1' + n_3)$
(d) $f = (n_3 \text{ xor } n_0) + n_3'[n_2'(n_1 + n_0') + n_1 n_0']$
(e) $f = n_0'(n_3 \text{ xor } n_2 n_1')$
(f) $f = n_2(n_1' + n_3'n_0') + n_3'n_1'n_0'$
(g) $f = n_2 n_1' + n_3'n_1(n_0' + n_2')$

4.24 Assume all variables are active high. Draw a mixed-logic circuit for the following equations. Do not use 08 or 32 type gates. Use NAND NAND logic.

(a) $f = n_3'[n_1 + (n_2 \text{ xor } n_0)'] + n_3 n_2 n_1'$
(b) $f = n_3'n_2' + n_2[n_1'n_0' + n_0(n_3 \text{ xor } n_1)]$
(c) $f = n_3'(n_1' + n_0) + n_2(n_1' + n_3)$
(d) $f = (n_3 \text{ xor } n_0) + n_3'[n_2'(n_1 + n_0') + n_1 n_0']$

4.25 Assume all variables are active high. Draw a mixed-logic circuit for the following equations. Do not use 08 or 32 type gates. Use NOR NOR logic.

(e) $f = n_0'(n_3 \text{ xor } n_2 n_1')$
(f) $f = n_2(n_1' + n_3'n_0') + n_3'n_1'n_0'$
(g) $f = n_2 n_1' + n_3'n_1(n_0' + n_2')$

5 COMBINATIONAL BUILDING BLOCKS

Now that we know about logic theory and how discrete gates implement logical operators, we turn to complex logic operators. Complex logic operators are used in most if not all digital systems. Their need is so pervasive you do not have to design them, because they are available as standard integrated circuits. This is why at this point we set aside discrete gate logic design and investigate modern logic design.

Modern logic design is based on complex logical operators, not discrete gate logical operators. Complex logical operators are available as standard medium-scale and large-scale integrated circuits known as MSI and LSI circuits, respectively. Nevertheless, as you will see, discrete gate logic still has a place in modern design processes.

After introducing timing diagrams, we discuss the differences of steady-state and transient responses of combinational-logic circuits. We learn that gate outputs respond to changes at the inputs after propagation delays, that the delays can create unexpected outputs, and that timing diagrams are a graphical means for combinational-logic circuit timing analysis and synthesis.

The first standard logic combination encountered is the multiplexer, also known as the *mux*. We learn the equation that defines the mux and the related truth table. Then, by examining the equation from various points of view, we discover several specific applications for the mux: function generators, paraller-to-serial converters, and data routing. We show how the Karnaugh map expedites circuit analysis and synthesis. Then the process is repeated for decoders, priority encoders, comparators, parity generators, adders, and finally the arithmetic-logic unit. We show how each complex function is designed, as well as how each one is used. The study of complex logical operators also enhances our knowledge of concepts used repeatedly in digital designs.

INTRODUCTION

Combinational building blocks are assemblies of gates that implement complex logical operators. Complex logical operators replace assemblies of gates that appear repeatedly in discrete gate designs. Complex operators are needed often enough to justify their inclusion in a standard logic family of parts. These blocks allow a designer to move up the ladder of design complexity. An important consideration for such a move up is the time required to perfect a design. Designers quickly learn that building blocks reduce project design time. (*Programmable logic*, a design alternative to combinational building blocks, is explored in Chapter 10.)

Nevertheless, discrete gates make possible nonstandard applications of MSI and LSI parts. This extra circuit logic is called *glue logic*. Examples show that glue is the logic that allows the MSI and LSI chips to work together in a variety of applications.

Active high—no bubble; active low—bubble.

Active high and active low assignments are implied by the absence and presence of bubbles in the figures. The active high input or output variable is True when the voltage is high (H). The active low input or output variable is True when the voltage is low (L). Any variable, active high or active low, is said to be asserted when it is True.

When variables are true they are asserted.

Asserted is used to emphasize action without the distraction of saying high or low. We do *not* want to say asserted high or asserted low (even though it is so easy to do so). We say p is asserted, or we say p is active low. For example, in the equation $g = xy'$, g is asserted when xy' is True. The equation $g = xy'$ does not reveal the active levels until it is written, for example, as $/g = x \times /y'$ or as $g_L = x_H \times y'_L$. In this way we learn that g and y are active low variables and x is active high. In schematics, the bubble can substitute for the / or the subscript $_L$.

Note that the figures in this chapter illustrate blocks representing logic family members whose complete logic circuits are found in data books. Also, many enable inputs in the figures are asserted in order to reduce the size of truth tables. Only typical members of each type of block are mentioned. However, all members are found in data books. Complete IC schematics and package block diagrams are also found in data books.

We start with a discussion of circuit timing. Timing is important because switching equations do not represent or predict the dynamic performance of circuits.

5.1 Timing

Combinational-logic circuit performance is measured by how quickly results are achieved. This is a major reason why time is an important issue in logic circuit design. Another major reason is verifying correct transient behavior. Timing appears in the form of timing diagrams, which reveal propagation delays and unexpected outputs called *glitches*. Boolean equations represent steady-state behavior. Boolean equations reveal very little, if anything, about transient response.

Before proceeding to timing diagrams, let's learn about propagation delay.

5.1.1 Propagation Delay

Propagation delay cannot be avoided.

Nothing in the physical world changes instantaneously. This reality is symbolized by light's finite velocity in free space: 3×10^8 meters per second. In electronics a more useful form for the velocity of light is employed: 30 centimeters/nanosecond. (A *nanosecond (ns)* is a unit of time equal to 10^{-9} seconds.) The velocity of light is important for two reasons that may not be obvious. In an integrated circuit, digital signals are represented by electromagnetic waves, which propagate at the velocity of light in the semiconductor medium, or by a moving electron "gas," whose velocity is a very small fraction of the velocity of light. Also, integrated circuits are constructed on pieces of silicon that have finite dimensions. Therefore a signal changing from H to L, or vice versa, at one chip input terminal will cause a change at an output terminal only after a finite amount of time elapses. The *propagation delay* of a signal path is the time required for a change in an input signal to produce a change in an output signal.

5.1.2 Timing Diagrams

Timing diagrams are essential for understanding actual circuit behavior.

A *timing diagram* is a graphical illustration of a circuit's logical behavior as a function of time. A timing diagram's horizontal axis represents time. The vertical axis presents H and L voltage levels. A *digital waveform* is a plot of the H and L levels a variable takes as a function of time (Figure 5.1a). Transitions from L to H require a finite time called *rise time*. An H-to-L transition duration is called *fall time* (Figure 5.1b). As a practical matter, finite (nonzero) rise and fall times are shown in timing diagrams only when performance limits are being examined.

A circuit's timing diagram includes digital waveforms representing circuit behavior at various circuit nodes. The waveforms at these circuit nodes are usually related. When one waveform impacts another waveform, cause and effect are at work. *Causality* is shown (Figure 5.1c) by arrows from one waveform's transitions (*cause*) to other waveforms' transitions (*effects*). Cause produces effect after a delay. Industrial practice measures delays from the cause's 50% level to the effect's 50% level.

The timing diagram is supplemented by a *timing table* specifying time values, a symbol for each delay, and the conditions under which

FIGURE 5.1
Timing diagrams

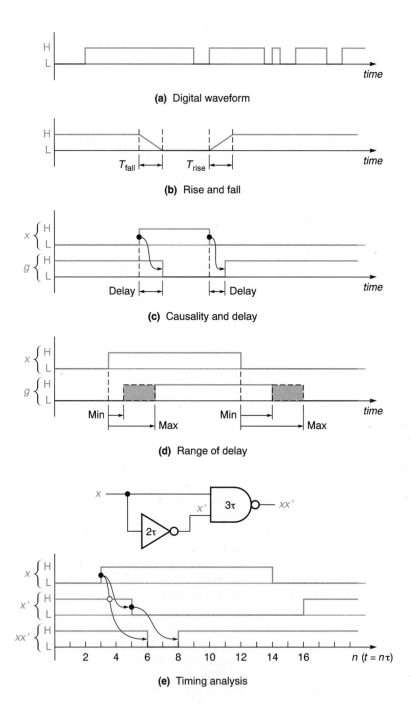

(a) Digital waveform

(b) Rise and fall

(c) Causality and delay

(d) Range of delay

(e) Timing analysis

they apply. Timing tables are included in specifications found in data books. Tables show minimum, typical, and maximum values for each timing parameter. The range of possible delays arises from variations in manufacturing processes, circuit supply voltage, and temperature (Figure 5.1d). A prudent designer works with the appropriate max or min values.

The expression $g = xx'$ represents the steady-state behavior of the circuit in Figure 5.1e. Here is how to ascertain the transient behavior: As shown in Figure 5.1e, we assume that input x transitions from L to H, remains H for some time, and then transitions back to L. We also assume that the delay through the 04 inverter is 2τ, and that the delay through the 00 gate is 3τ. This is why the time markers on the t axis are units of τ.

> *Transient behavior may cause glitches.*

Notice that the x' waveform is an inverted x waveform delayed by 2τ. The xx' waveform starts at level H (logical 0). The first xx' transition to L (logical 1) is caused by, and delayed 3τ from, the first x transition. This xx' transition to L occurs because both of the 00 gate's inputs are at H for 2τ. (In effect, x' is not-x' for 2τ because the 04 inverter delay is 2τ.) The x' transition to L starts the 3τ delay time of its effect on xx'. When the delay time expires, the output xx' returns to H (logical 0). The inverter delay has caused an unexpected output fall to L for 2τ. This is known as a *glitch*. Finally, when x returns to L, no glitch occurs. The glitch is one example of a hazard actually occurring (see upcoming Section 5.1.3).

5.1.3 Hazards

A *hazard* exists when there is the possibility of one or more unexpected outputs. Hazards occur because propagation delays exist; a change in any variable affects outputs after a finite time delay. These delays occur within circuits and on wires (about 1 ns delay per 25 cm of wire). Logic equations represent the *steady-state behavior* of logic circuits. Changes in inputs induce *transient behavior* in logic circuits.

> *Steady-state behavior represents equations.*

The transient behavior may be different from the steady-state behavior. The major difference is occurrence of unexpected outputs (glitches). Glitches can occur at combinational network outputs one or more propagation times after one or more inputs change from high to low, or vice versa. Whether they actually occur depends on actual combinations of delays for the gates in each one of, say, 100 production copies of a circuit.

> *Static hazards arise from xx' or $x + x'$ terms.*

Static hazards   Static hazards arise from xx' or $x + x'$ terms. For example, in Figure 5.2a the transient behavior of an AND of ORs circuit with output g_0 can generate a momentary 1-glitch because the

FIGURE 5.2
Static hazards

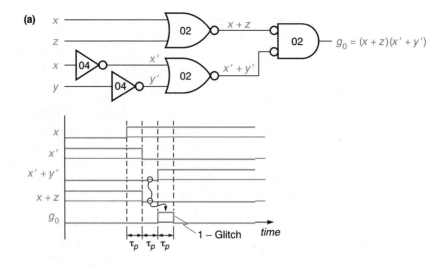

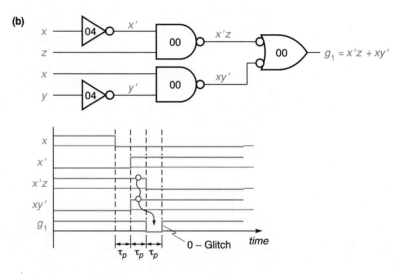

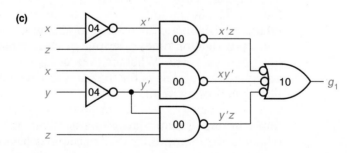

steady-state behavior of g_0 is a 0 output when $z = 0$ and $y = 1$:

$$g_0 = (x + z)(x' + y')$$

and when $z = 0$ and $y = 1$,

$$g_0 = xx' = 0$$

On the other hand, the transient behavior of the OR of ANDs circuit with output g_1 in Figure 5.2b can generate a momentary 0-glitch because the steady-state behavior of g_1 is a 1 output when $z = 1$ and $y = 0$:

$$g_1 = x'z + xy'$$

and when $z = 1$ and $y = 0$,

$$g_1 = x + x' = 1$$

Prudence dictates that we assume and-or-not circuits will create hazards. Many tricks are alleged to eliminate hazards. Perhaps they do; however, we use a K map to verify that they do. Here is how to do it.

The $x'z$ waveform in Figure 5.2b represents the transient behavior of the m_1/m_3 cluster in the K map of Example 3.23, which is reproduced below. The xy' waveform in Figure 5.2b represents the transient behavior of the m_4/m_5 cluster. When x switches from 1 to 0, the m_4/m_5 cluster terms go to zero before the m_1/m_3 cluster terms go to one, thereby creating the 0-glitch, because both clusters are at 0 (H) momentarily. The key observation is that the glitch results because the two clusters do not overlap. Intuition tells us that an additional overlapping cluster will eliminate the hazard: this is true in general. Furthermore, overlapping clusters do not change the function because they do not add new 1 entries in the K map. The overlapping clusters are consensus terms. Overlapping clusters provide the solution. The price paid in this case is an additional AND gate and replacement of the two-input OR gate by a three-input OR gate (Figure 5.2c).

Consensus terms eliminate static hazards.

EXAMPLE 3.23 (Repeated)

The function g with minterms m_1, m_3, m_4, and m_5 has a map whose 1's indicate when $g = 1$ and whose 0's indicate when $g = 0$. The lines around the adjacent pairs of 1's account for all the 1's. The clusters show that $g = x'z + xy'$. The redundant consensus term $y'z$ clearly stands out as a cluster overlapping the other clusters.

$g(x, y, z)$
number $= xyz$

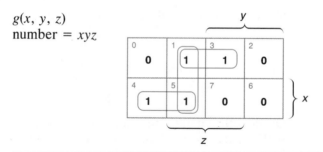

Dynamic hazards A *dynamic hazard* produces a glitch when a change in output from 0 to 1 is actually 0 1 0 1 or 1 0 1 0. That is, the output changes more than once for a single change in the input. Such hazards exist when three or more paths for the same variable have different delay times. For example, in Figure 5.3 three paths for c have different delay times in a real network. (McClusky [see references] explains how to discover dynamic hazards.)

$$f = (ac + bc') \text{ xor } (c)$$

FIGURE 5.3
Dynamic hazards

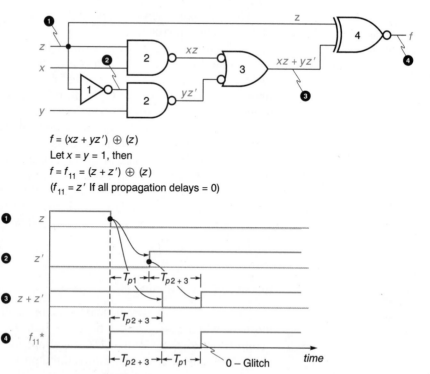

$f = (xz + yz') \oplus (z)$
Let $x = y = 1$, then
$f = f_{11} = (z + z') \oplus (z)$
$(f_{11} = z'$ If all propagation delays $= 0)$

*Assume $T_{p4} = 0$.
{Observe: $f_{11} \to z'$ If $T_{p2+3} \to 0$ and $T_{p1} \to 0$}

5.2 Multiplexing

Output is a selection of one of many inputs.

Multiplexing is the process of selecting one of a group of N sources and then routing the selected source to one destination. A multiplexing network is a combinational network; presentation of selection bits connects the selected source to the destination after propagation times elapse.

5.2.1 Definition and Implementation

Use data inputs as table-entered variables.

The 157 logic family member, such as 74LS157, is a two-to-one multiplexer. The 157 selects one of two sources—a_0 and a_1—with one select input z (Figure 5.4). Actually, the 157 consists of four two-to-one multiplexers controlled by a common select input z, as shown in a data book. The long-form truth table is derived by treating minterm coefficients a_j as well as z as variables in the given defining equation. Next, we interpret a_0 and a_1 as coefficients of the minterms of the z variable in the given defining equation to derive a short-form truth table with a_0 and a_1 as table-entered variables. We call z a *multiplexing variable*.

The 153 logic family member, such as 74LS153, is a four-to-one multiplexer. The 153 selects one of four sources—a_0, a_1, a_2, a_3—with two select inputs—y and z (Figure 5.5). The 153 package contains two

FIGURE 5.4
Two-to-one multiplexer

$$g = a_0 z' + a_1 z \quad \text{Defining equation}$$

$$= a_0 m_0 + a_1 m_1 \quad \text{Minterm format}$$

z	a_0	a_1	g	Row	z	g
0	0	—	0	0	0	a_0
0	1	—	1	1	1	a_1
1	—	0	0			
1	—	1	1			

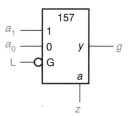

FIGURE 5.5
**Four-to-one
multiplexer**

$$g = a_0 y'z' \quad + a_1 y'z + a_2 yz' + a_3 yz \quad \text{Defining equation}$$

$$= a_0 m_0 \quad + a_1 m_1 + a_2 m_2 + a_3 m_3 \quad \text{Minterm format}$$

$$g = e[a_0 y'z' + a_1 y'z + a_2 yz' + a_3 yz] \quad \text{Enable variable added}$$

Row	G	y	z	g
0	0	0	0	0
1	0	0	1	0
2	0	1	0	0
3	0	1	1	0
4	1	0	0	a_0
5	1	0	1	a_1
6	1	1	0	a_2
7	1	1	1	a_3

Row	G	y	z	g
0–3	0	—	—	0
4	1	0	0	a_0
5	1	0	1	a_1
6	1	1	0	a_2
7	1	1	1	a_3

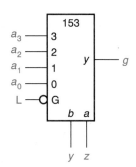

four-to-one multiplexers controlled by common select inputs y and z and separate enable inputs. The truth table uses sources a_0, a_1, a_2, and a_3 as table-entered variables. The multiplexing variables are y and z. The function of the enable input G is illustrated in Figure 5.5. When $G = 0$, inputs x and y become ''don't cares'' (they are disabled) and the output $g = 0$. Two forms of the 153 truth table are shown in Figure 5.5. The enable input G is ignored in subsequent figures ($G = 1$).

The 151 logic family member is an eight-to-one multiplexer selecting one of eight sources—a_0 to a_7—using three inputs—x, y, and z (Figure 5.6). There is one 151 circuit in a package.

A larger multiplexer is created by cascading multiplexers. A twelve-to-one configuration is shown in Figure 5.7. If the L at the number-3 input of the output multiplexer is replaced by a four-to-one multiplexer, the configuration becomes a sixteen-to-one multiplexer. The twelve-to-one multiplexer also illustrates the fact that inputs need not be used. Unused inputs are not left unconnected; they are connected to an appropriate input, such as L, as shown in Figure 5.7.

FIGURE 5.6

Eight-to-one multiplexer

$$g = a_0 x'y'z' + a_1 x'y'z + a_2 x'yz' + a_3 x'yz \qquad \text{Defining equation}$$

$$+ a_4 xy'z' + a_5 xy'z + a_6 xyz' + a_7 xyz$$

$$g = a_0 m_0 + a_1 m_1 + a_2 m_2 + a_3 m_3 + a_4 m_4 + a_5 m_5 + a_6 m_6 + a_7 m_7$$

Row	x	y	z	g
0	0	0	0	a_0
1	0	0	1	a_1
2	0	1	0	a_2
3	0	1	1	a_3
4	1	0	0	a_4
5	1	0	1	a_5
6	1	1	0	a_6
7	1	1	1	a_7

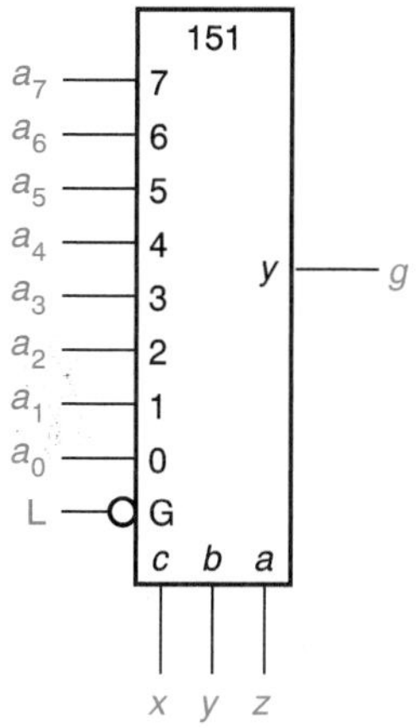

5.2.2   Devices with Three-State Outputs

Connecting two gate outputs together usually produces a malfunction. If one gate has an H output and the other an L output, the result is indefinite when the gate outputs are connected together. Furthermore, physical damage to the gates can also occur. However, many designs are facilitated if gate outputs can be connected together. This need was a major reason why devices that have three-state output gates became available. Three-state outputs are also known as tri-state[1] outputs.

The output of a three-state device has three values: high, low, and disconnected (Figure 5.8a). The output is connected, and equals the input, when the enable input is active. The output is disconnected when the enable input is inactive. As a practical matter, a disconnected three-state device output behaves as if there were an open switch in series with the output (Figure 5.8b). The disconnected output is referred to as the *Hi-Z state* instead of the open state, because the switch is imperfect (a small leakage current can flow through the z circuit element shown in Figure 5.8c).

Can build multiplexers with 3-state devices.

Three-state elements are used to build a multiplexer when physically separated sources are to be connected to a wire one at a time. In

---

[1] Tri-state is a trademark of National Semiconductor Corp.

FIGURE 5.7
Twelve-to-one multiplexer

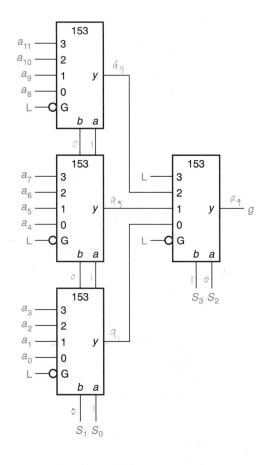

Figure 5.8d the sources and outputs are groups of *n* bits. Wires marked with *n* above a slash (\) represent a group of *n* wires. The 241 and 541 are typical family members.

Cascading multiplexers increases the propagation time in the data paths. This increase can be avoided by the configuration in **Figure 5.9**. The 253 multiplexers have three-state outputs. Each 139 decoder output (see upcoming Section 5.3) enables one multiplexer to connect its output to the wire representing output *g*.

5.2.3   Function Generators

Mux inputs are coefficients of output equation's terms.

A function of *n* variables has 2^n minterms. The output column of a truth table for a function of *n* variables is a list of 2^n 1's and 0's. A multiplexer controlled by *n* multiplexing variables selects one of 2^n sources. If we equate the list of 2^n 1's and 0's to the 2^n input sources, we realize a multiplexer can be used as a function generator. This

FIGURE 5.8
Tri-state multiplexer

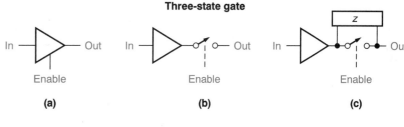

Three-state gate

(a)   (b)   (c)

Multiplexer built with three-state gates

(d)

process is illustrated in Figure 5.10. The key is to equate minterm coefficients to source variables a_3, a_2, a_1, a_0.

A four-to-one multiplexer can be used in a straightforward manner to generate the $2^n = 2^4 = 16$ functions of two variables. If $a_3 a_2 a_1 a_0 =$ LHHL, the corresponding truth table when H := 1 is the XOR truth table (Figure 5.11). Next we show how the process can be made more efficient; a mux with n multiplexing variables can generate functions of $n + 1$ variables.

More efficient function synthesis The key question is: How many terms do we need to represent a function of 2^n variables? Table 3.4 (page 101) shows by evaluating all cases that a two-variable function has at most two terms. From this can we infer that a function of n variables requires at most $2^n/2 = 2^{n-1}$ terms? The answer is yes. We offer the following heuristic proof of this assertion by using K maps.

First we fill K maps with checkerboard patterns of 1's (Figure 5.12). Since, as we can see, there are 2^{n-1} 1's in the three cases portrayed, there are 2^{n-1} terms in the function. We readily generalize that this is true for an n-dimensional K map. Replacement of any 0 by a 1 adds a term to the function. Any such 1 is adjacent to some 1 in

FIGURE 5.9

**Faster
sixteen-to-one
multiplexer**

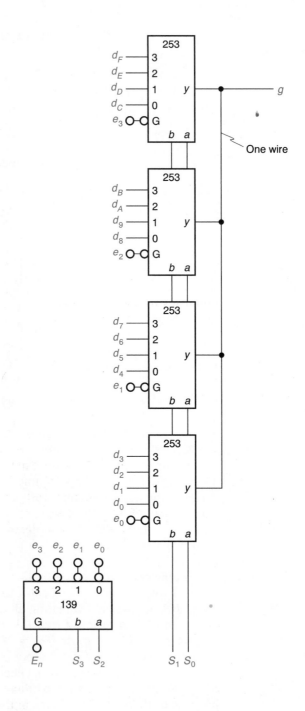

FIGURE 5.10

**Generating
functions**

y	z	g
0	0	a_0
0	1	a_1
1	0	a_2
1	1	a_3

$$g = a_3yz + a_2yz' + a_1y'z + a_0y'z'$$

$$g = a_3m_3 + a_2m_2 + a_1m_1 + a_0m_0$$

If $a_j = 1$, we include minterm j in the equation for g; otherwise (i.e., if $a_j = 0$), we omit minterm j from the equation for g.

the checkerboard pattern (try it). The new adjacent pair of two terms collapses to one term in an algebraic expression. Therefore the number of terms never increases beyond 2^{n-1}, even though the number of 1's can increase to 2^n.

When the coefficients of the 2^{n-1} terms are mux inputs, only $n - 1$ select lines are required. The number of terms does not increase if an

FIGURE 5.11

**Generating the
XOR function**

$$g = y \text{ xor } z = yz' + y'z$$

y	z	g	
0	0	0	(a_0)
0	1	1	(a_1)
1	0	1	(a_2)
1	1	0	(a_3)

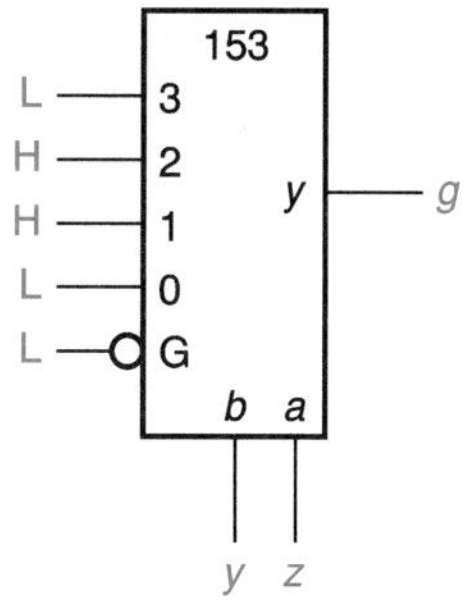

$$g = 0yz + 1yz' + 1y'z + 0y'z'$$

$$g = a_3yz + a_2yz' + a_1y'z + a_0y'z'$$

Equate coefficients:

$$a_3 = 0 \qquad a_2 = 1 \qquad a_1 = 1 \qquad a_0 = 0$$

FIGURE 5.12
Checkerboard patterns in K maps

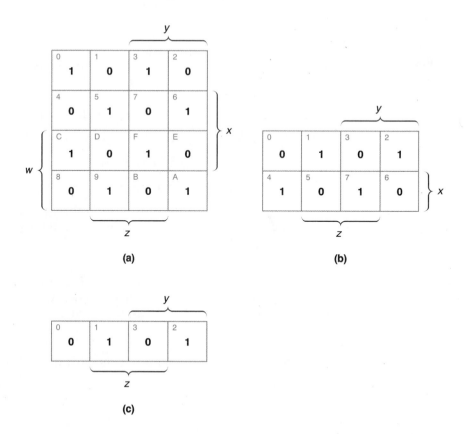

(a)

(b)

(c)

nth variable is treated as a coefficient. This implies that the nth variable is a literal at one or more mux inputs. This process is more efficient because a function of n variables can be realized by a mux with $n - 1$ select lines. Immediate practical consequences are:

A two-to-one mux generates $g(y, z)$ (see Example 5.1).

A four-to-one mux generates $g(x, y, z)$ (see Example 5.2).

An eight-to-one mux generates $g(w, x, y, z)$ (see Example 5.4).

And so forth.

Examples 5.1, 5.2, and 5.3 illustrate how the sources are calculated by using algebra or truth tables or K maps, respectively. Examples 5.4 and 5.5 illustrate conversion of a K map into a reduced-dimension map so that a mux with $n - 1$ select lines can be used. The reduced-dimension map usually has map-entered variables.

EXAMPLE 5.1 XOR Function

The XOR function is synthesized by equating the XOR function to the mux equation. The fundamental theorem of algebra permits solving for a_0, a_1 by equating coefficients of the variables.

$$g = yz' + y'z$$
$$= a_0z' + a_1z$$

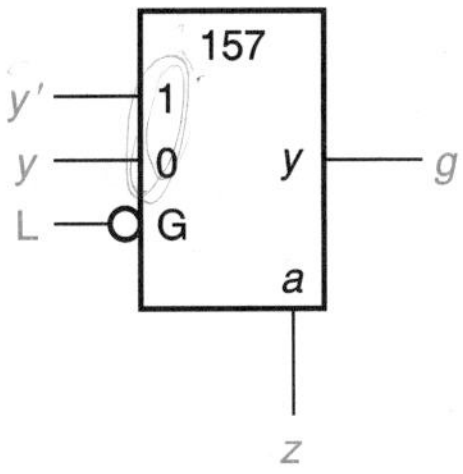

Equate coefficients:

$$a_0 = y \qquad\qquad a_1 = y'$$

Truth table:

$$g = m_1 + m_2$$

Row	z	g	a_j
0	0	y	a_0
1	1	y'	a_1

K map with map-entered variable:

Let z be the mux select input.

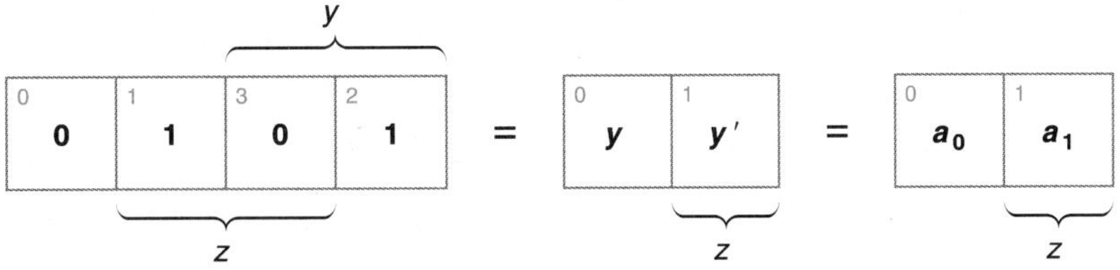

Let y be the mux select input.

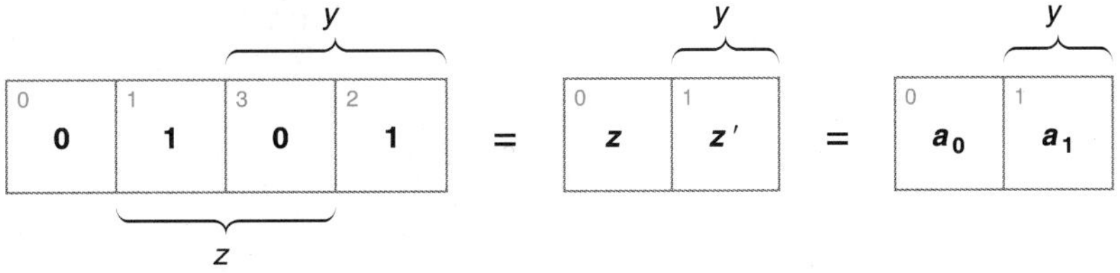

EXAMPLE 5.2   G(x, y, z) Function Generator

Given $g(x, y, z) = x$ xor y xor z, here is one way to calculate the
coefficients a_0, a_1, a_2, a_3.

$g = x$ xor y xor z

$g = x$ xor $(y$ xor $z)$

$g = x$ xor $(yz' + y'z)$

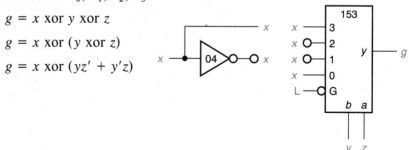

$$g = x'(yz' + y'z) \quad + x(yz' + y'z)'$$
$$g = x'(yz' + y'z) \quad + x(y'z' + yz)$$
$$g = xyz \quad + x'yz' + x'y'z + xy'z'$$
$$g = a_3 yz \quad + a_2 yz' + a_1 y'z + a_0 y'z'$$

Equate coefficients:

$$a_3 = x, \qquad a_2 = x', \qquad a_1 = x', \qquad a_0 = x$$

Truth table:

$$g = m_1 + m_2 + m_4 + m_7$$

Row	y	z	g	a_j
0	0	0	x	a_0
1	0	1	x'	a_1
2	1	0	x'	a_2
3	1	1	x	a_3

K map with map-entered variable:

Let yz be the mux select inputs.

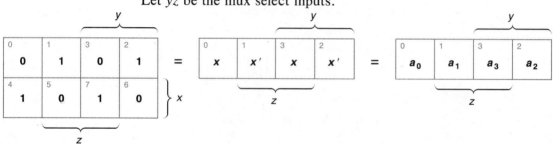

Let xy be the mux select inputs.

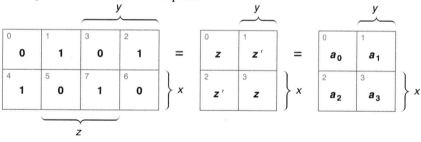

EXAMPLE 5.3   From Minterms to Generated Function

$g(x, y, z) = \Sigma\, m(1, 2, 3, 5, 6, 7)$

$g = xyz + xyz' + xy'z + x'yz + x'yz' + x'y'z$

      7   6   5   3   2   1   ← Minterm number

$$g = xy(z + z') + xy'(z) + x'y(z + z') + x'y'(z)$$

$$g = xy(1) + xy'(z) + x'y(1) + x'y'(z)$$

$$g = xy(a_3) + xy'(a_2) + x'y(a_1) + x'y'(a_0)$$

Equate coefficients:

$$a_3 = 1, \qquad a_2 = z, \qquad a_1 = 1, \qquad a_0 = z$$

Truth table:

$$g = m_1 + m_2 + m_3 + m_5 + m_6 + m_7$$

Row	x	y	g	a_j
0	0	0	z	a_0
1	0	1	1	a_1
2	1	0	z	a_2
3	1	1	1	a_3

K map with map-entered variable:

Let *xy* be the mux select inputs.

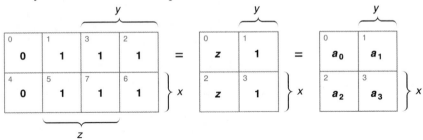

Let *yz* be the mux select inputs.

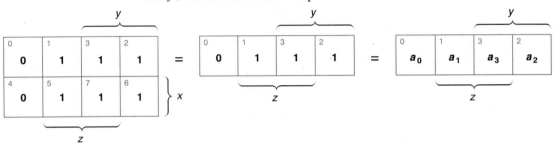

EXAMPLE 5.4   **From K Map to Generated Function g(w, x, y, z)**

Given:

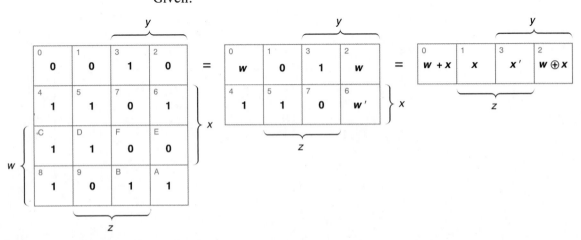

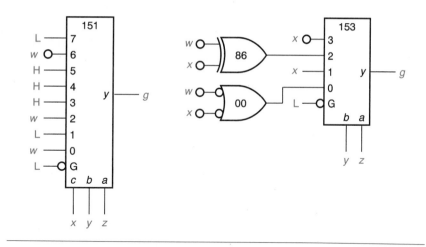

EXAMPLE 5.5 **From K Map to Generated Function**
g(a, w, x, y, z)

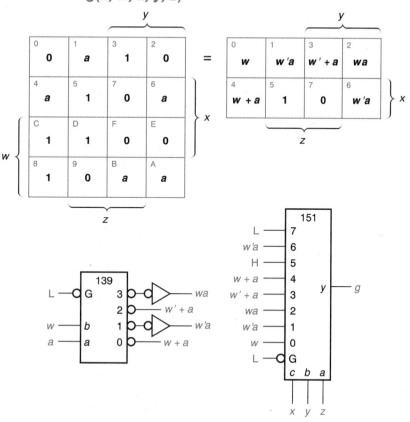

EXERCISE 5.1

Write $g = y + z$ to an xyz K map (see Figure 3.3, page 127). Cluster four pairs of squares corresponding to the four xy minterms. Deduce the coefficient of each xy minterm.

Answer:

xy minterm	Coefficient
m_0	z
m_1	1
m_2	z
m_3	1 ∎

EXERCISE 5.2

Write $g = x \oplus y \oplus z$ to an xyz K map (see Figure 3.3, page 127). Cluster four pairs of squares corresponding to the four yz minterms. Deduce the coefficient of each yz minterm.

Answer:

yz minterm	Coefficient
m_0	x
m_1	x'
m_2	x'
m_3	x ∎

5.2.4   Conversion from Parallel to Serial

Inputs are available in parallel.

Assume that bits $d_0, d_1, \ldots d_6, d_7$ are available simultaneously, that is to say, in parallel. A *parallel-to-serial conversion* occurs when the bits are made to occupy sequential time slots (Figure 5.13a). Suppose three variables q_2, q_1, and q_0 generate the binary number sequence 000, 001, 010, . . . , 110, 111. Furthermore, suppose each number is present for the same amount of time, i.e., for one time slot (Figure 5.13a). Also, suppose q_2, q_1, and q_0 are connected to multiplexer select lines as shown in Figure 5.13b. When the term $q_2q_1q_0$ has the value 110, multiplexer input six is selected and bit d_6 appears at the output. As the value of $q_2q_1q_0$ sequences from 000 to 111 bits, d_0 to d_7 appear at the output in the same sequence. Each bit is present for a time interval equal to the time the corresponding number is present. This is why each bit occupies one time slot.

FIGURE 5.13

Parallel-to-serial conversion

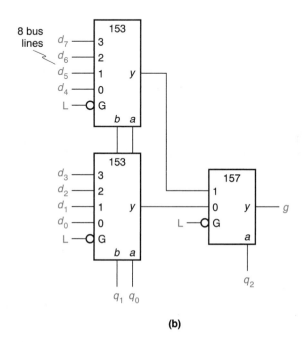

(a)

(b)

5.2.5   Data Routing

If today's air temperature is automatically measured once per hour, we will have 24 numbers at the end of the day. Each number is one piece of data we could use, for example, in a plot of today's temperature versus time. And, if the measuring instrument converts the value of each piece of data into a binary number, the result is 24 groups of, say, four bits called *data words*.

Now assume there are two instruments measuring temperature but only one data plotter. That is, there are two sources and one destination. One way each instrument can deliver data is via four wires so that all bits of each data word are available simultaneously. The group of four wires is called a *bus,* because the simultaneous signals on the bus are related to each other—they are a data word.

A *bus* is a group of two or more wires carrying related signals.

FIGURE 5.14

Data routing

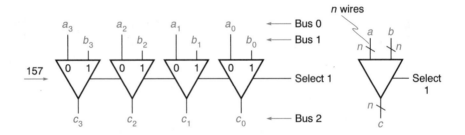

There are many occasions when source bus_0 or source bus_1 is to be connected to destination bus_2. A group of two-to-one multiplexers provides this capability (Figure 5.14).

5.3 Decoding

Decoding is a process of activating one of 2^n outputs y_j for each combination of values of the n inputs. A decoding network is a combinational network; presentation of input bits activates the selected output line after propagation times elapse. It is useful to remember that decoders are minterm generators. On the other hand, decoding can be considered the inverse of multiplexing (that's why it is also known as *demultiplexing*): routing one source g to one of 2^n selected outputs y_j. Presentation of n selection bits connects the source to the selected output destination after propagation times elapse.

Decoders generate minterms.

5.3.1 Definition and Implementation

One binary digit, or *bit,* has two states. Each decoded state activates one output. A one-bit decoder has two outputs, y_0 and y_1, selected by input a (Figure 5.15). This is not a standard logic family member. The equations in Figure 5.15 show that when the decoder is enabled ($g = 1$), then $a = 0$ activates output y_0 and $a = 1$ activates output y_1. Or, if the decoder is interpreted as a demultiplexer, the equations show how g is routed to output y_1 or output y_0 by selection bit a. Observe the use of g as a table-entered variable.

The next level of complexity uses two bits to define four states. Again each decoded state activates one output. A two-bit decoder has four outputs—y_3, y_2, y_1, and y_0—selected by two bits, ba (Figure 5.16). This decoder is realized by the 139 (74LS139) logic family member, which has two decoder circuits per package. The enable input g is a factor in each output equation y_j. When g is zero (False), all outputs are zero. When $g = 1$, any output can be enabled by the select lines.

FIGURE 5.15
One-bit decoder/ demultiplexer

The defining equations are:

$$y_{1L} = ga \qquad y_{0L} = ga'$$
$$\phantom{y_{1L}} = gm_1 \qquad \phantom{y_{0L}} = gm_0$$

	DECODER					DEMULTIPLEXER			
Row	g	a	y_1	y_0		Row	a	y_1	y_0
0–1	0	—	0	0		0	0	0	g
2	1	0	0	1		1	1	g	0
3	1	1	1	0					

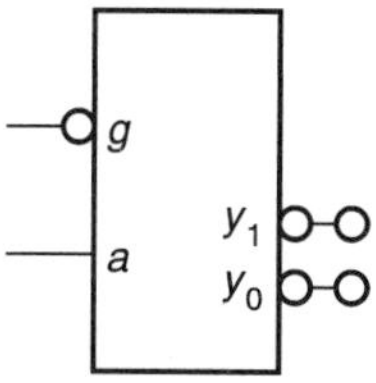

(Not a logic family member)

The level of complexity continues to increase by a power of two. The three-bit decoder defines eight states. Once again each decoded state activates one output. A three-bit decoder has eight outputs—y_7, y_6, y_5, y_4, y_3, y_2, y_1, and y_0—selected by three bits, cba (Figure 5.17). This decoder is implemented by the 138 logic family member, which has one circuit per package. The 138 has three enable inputs, g_1, g_2,

FIGURE 5.16
Two-bit decoder/ demultiplexer

The defining equations and truth table are:

$$y_{3L} = gba \qquad y_{2L} = gba' \qquad y_{1L} = gb'a \qquad y_{0L} = gb'a'$$

	DECODER							DEMULTIPLEXER					
Row	g	b	a	y_3	y_2	y_1	y_0	Row	b	a	y_3	y_2	y_1	y_0
0–3	0	—	—	0	0	0	0	0	0	0	0	0	0	g
4	1	0	0	0	0	0	1	1	0	1	0	0	g	0
5	1	0	1	0	0	1	0	2	1	0	0	g	0	0
6	1	1	0	0	1	0	0	3	1	1	g	0	0	0
7	1	1	1	1	0	0	0							

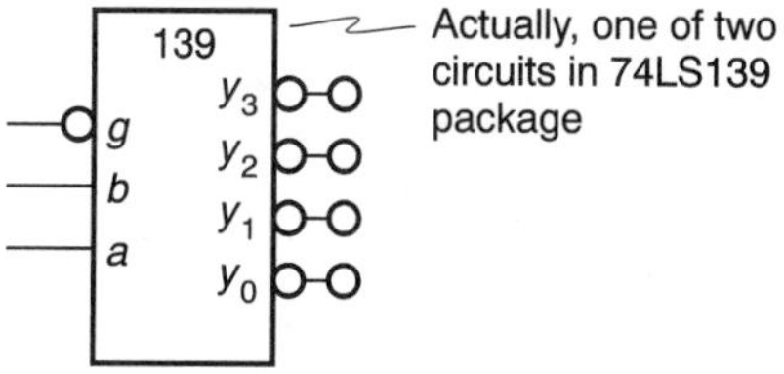

Actually, one of two circuits in 74LS139 package

FIGURE 5.17
**Three-bit
decoder/
demultiplexer**

The defining equations are (where $g = g_1g_2g_3$):

$$y_{7L} = gcba \qquad y_{6L} = gcba'$$

$$\cdot\; y_{5L} = gcb'a \qquad y_{4L} = gcb'a'$$

$$y_{3L} = gc'ba \qquad y_{2L} = gc'ba'$$

$$y_{1L} = gc'b'a \qquad y_{0L} = gc'b'a'$$

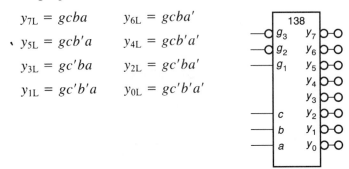

and g_3. The 138 outputs are enabled only if all three enable inputs are True. This fact is emphasized by the equation $g = g_1g_2g_3$. Three enable inputs make the 138 a versatile building block.

Expanding a decoder is a matter of using variables as inputs to one or more of the 138 enable inputs $g_1g_2g_3$. The circuit (Figure 5.18) recognizes that when the $a_3a_2a_1a_0$ variables' combinations of values cycle from 0000 to 1111, the variables $a_2a_1a_0$ cycle from 000 to 111

FIGURE 5.18
**Four-bit decoder:
Four inputs to
sixteen outputs**

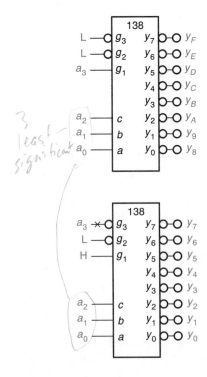

when variable $a_3 = 0$ and again when $a_3 = 1$. The equations that demonstrate these properties are as follows.

Given the four variables $a_3 a_2 a_1 a_0$, typical output equations y_4 and y_A take the following forms for the circuit in Figure 5.18:

The enable equation for the decoder with outputs 0 to 7 is $g_1 = 1$, $g_2 = 1$, $g_3 = a_3'$, so

$$y_4 = gcb'a' = a_3'a_2a_1'a_0'$$

The enable equation for the decoder with outputs 8 to F is $g_1 = a_3$, $g_2 = 1$, and $g_3 = 1$, so

$$y_A = gc'ba' = a_3 a_2'a_1 a_0'$$

5.3.2   Decoders as Function Generators

Decoder plus OR gates implement sum of minterms equations.

Each decoder output y_j represents one minterm m_j when the enable input(s) $g = 1$. In fact, $y_j = m_j$ when $g = 1$.

Any function is some sum of minterms. This is why we may add one or more OR gates to a decoder to build any OR of ANDs function we please. In example 5.6, the 30 is an eight-input version of the 00 (positive true logic) NAND gate.

EXAMPLE 5.6   Function Generator Using a Decoder and OR Gates

Given a function, we convert it to canonical sum of products form to find the minterms. (Note how unused inputs to the 30 gate are connected to H.)

$$g_1 = yz \ = (x + x')yz = xyz + x'yz = m_7 + m_3$$

$$g_2 = xz \ = x(y + y')z = xyz + xy'z = m_7 + m_5$$

$$g_3 = xy \ = xy(z + z') = xyz + xyz' = m_7 + m_6$$

$$g_4 = xyz' + x'y + z = m_7 + m_6 + m_5 + m_3 + m_2 + m_1$$

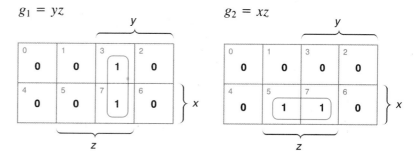

$$g_3 = xy \qquad\qquad\qquad g_4 = xyz' + x'y + z$$

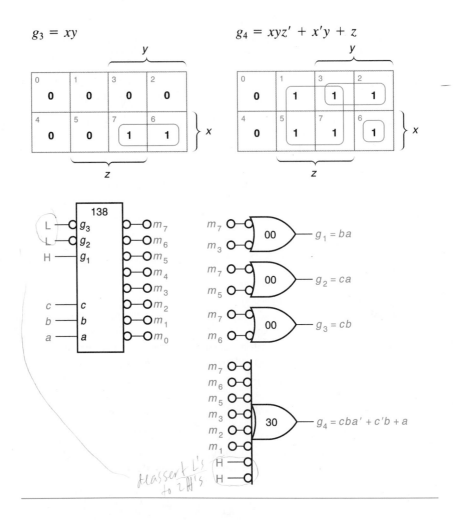

EXAMPLE 5.7   Common Factors and the Decoder Enable Input

The enable inputs allow introduction of factors common to all min-terms. Variables a, b, x, y, f_2, and f_3 are active high. The subscript $_L$ marks the variable f_1 as active low.

$$f_{1L} = gm_1 = (x + y)b'a$$

$$f_2 = gm_3 = (x + y)ba$$

$$f_3 = g(m_3 + m_2)$$

$$= (ba + ba')(x + y)$$

$$= b(a + a')(x + y)$$

$$= b(x + y)$$

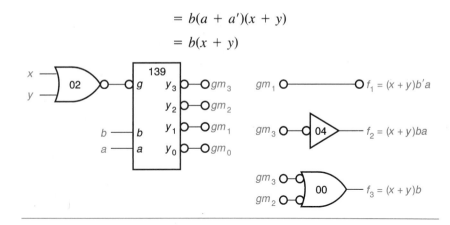

See g_1 circuit in Example 5.6.

EXERCISE 5.3

Write $g(c, b, a) = ba$ to a cba K map (see Figure 3.3, page 127). Use a decoder and gates to design a circuit for this function.

Answer:

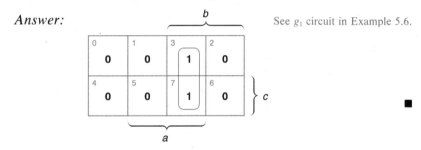

EXAMPLE 5.8   9's complement function

The 9's complement function is a more unusual application of decoders as a function generator. In decimal, the 9's complement of s is $r = 9 - s$. In binary, this is more easily shown by truth tables. All variables are active high in the following.

s	s_3	s_2	s_1	s_0	r_3	r_2	r_1	r_0	r
0	0	0	0	0	1	0	0	1	9
1	0	0	0	1	1	0	0	0	8
2	0	0	1	0	0	1	1	1	7
3	0	0	1	1	0	1	1	0	6
4	0	1	0	0	0	1	0	1	5
5	0	1	0	1	0	1	0	0	4
6	0	1	1	0	0	0	1	1	3

s	s_3	s_2	s_1	s_0	r_3	r_2	r_1	r_0	r
7	0	1	1	1	0	0	1	0	2
8	1	0	0	0	0	0	0	1	1
9	1	0	0	1	0	0	0	0	0
A	1	0	1	0	–	–	–	–	–
B	1	0	1	1	–	–	–	–	–
C	1	1	0	0	–	–	–	–	–
D	1	1	0	1	–	–	–	–	–
E	1	1	1	0	–	–	–	–	–
F	1	1	1	1	–	–	–	–	–

By inspection:

$$r_0 = s_0{}'$$

$$r_1 = s_1$$

$$r_2 = s_3{}'(s_2 \text{ xor } s_1) = s_3{}'(m_2 + m_1) \quad \text{Minterms of } s_2, s_1$$

$$r_3 = s_3{}'s_2{}'s_1{}' = s_3{}'m_0$$

5.3.3   Decoders as Demultiplexers

A decoder g_j input is a common factor in all y_j output equations (see Section 5.3.1 and Example 5.7). When we emphasize this factor, interpretation of the decoder as a demultiplexer is straightforward. Perhaps the multiplexer-demultiplexer subsystem block diagram in Figure 5.19 can make this clear. Input lines $s_1 s_0$ select source d_j and deliver it to destination y_j.

FIGURE 5.19
**Multiplexer-
demultiplexer
subsystem**

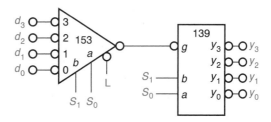

Priority Encoding

Logic family member 148 is an eight-line priority encoder. There are eight active low input lines I_j, numbered from 0 to 7. There are three active low output lines $a_2 a_1 a_0$. The three other lines, EI, EO, and GS, are discussed shortly.

The 148 is an *encoder* because input line I_j produces the binary number j at outputs $a_2 a_1 a_0$. When inputs I_0 to I_7 are active low one at a time, the output $a_2 a_1 a_0$ is an active low binary number equal to the asserted input's decimal number. Asserting line I_6 places 110_2 in the form LLH at outputs $a_2 a_1 a_0$.

The 148 is a *priority encoder* because the number at the output, $a_2 a_1 a_0$, is the binary number corresponding to the highest-numbered input when two or more input lines are asserted simultaneously. Asserting inputs I_3 and I_5 places 101_2 in the form LHL at outputs $a_2 a_1 a_0$.

Three 148's are cascaded together in Figure 5.20 to illustrate how 24 input lines are priority encoded. The enable input EI activates the 148. When all inputs are inactive, the enable output EO is active low. When any input is asserted, the EO is not asserted. The idea is to have the EO deactivate (not assert) lower-priority 148s. Conversely, an active EO passes priority to the next 148. This is another interpretation of the EO's role in a hierarchy of 148 chips.

The request–report equation is GS = EI × EO'. The GS output is asserted if the 148 is enabled *and* at least one input is asserted to make EO not true. The GS reports that one or more inputs to the enabled 148 are asserted. One application of the GS report is to ask the system to identify the active 148 as well as to check the identified 148's output binary number $a_2 a_1 a_0$.

The GS output is used in conjunction with the other GS outputs to identify the active 148. The identification circuit is based on the fact that only one GS report is active at a time in a system, because the EO disables lower-priority 148 chips. Therefore, if the GS_j are connected to inputs of yet another 148, the output number $e_2 e_1 e_0$ identifies the

FIGURE 5.20
Priority encoder

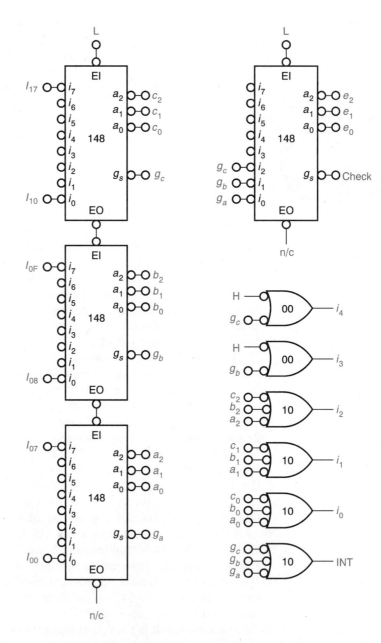

active 148. The system check output tells the system to read the number $e_2 e_1 e_0$ (Figure 5.20). The 10 in figure 5.20 is a three input version of the 00. The five i_j circuits generate a five-bit number that reports the highest-priority request. The INT circuit reports that a request has been made.

Priority encoders are used to assign priority to interrupt inputs in microprocessor systems where multiple simultaneous requests are possible. They are also used to manage complex input/output systems, and in code conversion schemes.

5.5 ━━━━━━━━ **Comparing**

Let us compare the decimal numbers 7592 and 8392. We literally see that 7592 is the smaller number; hardware, however, cannot see. In hardware what tests do we need to make? The simplest test compares the most significant digits (MSDs) first. That test tells us that $8 > 7$. This is sufficient, because $8000 > 7999 > 7592$. Digit 8 times its weight is greater than the other number. (The process assumes both numbers have the same number of digits, which is true in most systems. If it is not true, the comparator's MSB inputs are wired as zeros to fill out the missing digits.) Now let us compare 7592 and 7392. Clearly, $7592 > 7392$ because $7 = 7$ *and* $5 > 3$. Or, given that both MSDs are 7, we move right to the next digit pair.

Comparison by digit pairs reduces truth table size.

The algorithm for comparing two positive numbers x and y starts with the most significant digit.

STEP 1   Test a digit pair: the digits are equal or not equal.

STEP 2   IF the digits are equal,
THEN IF   this is the last digit pair, assert the $x=y$ output, and exit to
             end the process,
          ELSE move right to the next digit pair and go to step 1
    ELSE (the digits are not equal)
             assert the $x>y$ or the $x<y$ output lines, according to which is
             greater, and exit to end the process.

This algorithm requires two circuits: one to detect equality and the other to detect which is greater than or less than (Figure 5.21). Assemblies of these circuits are used to build four-bit comparators for $x = y$ and $x > y$. A less-than circuit is the same as the greater-than circuit with x and y exchanged. The 85 is the logic family four-bit magnitude comparator (Figure 5.22). The 85 reports $A < B$, $A = B$, and $A > B$. The 85's $<$, $=$, and $>$ inputs provide for cascading 85's in order to compare n-bit words. An eight-bit comparator is shown in Figure 5.23.

2's complement numbers require signed comparison.

Comparing signed numbers involves the same process as magnitude comparison except for the most significant digits, which are the sign digits. In magnitude comparisons, $1 > 0$; but in sign comparisons, $1 < 0$ (negative is less than positive). Therefore only the circuit comparing the first binary digit pair is changed by exchanging the x and y

FIGURE 5.21
**Comparing two
unsigned digits**

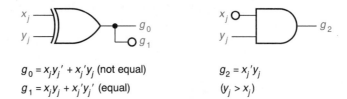

$g_0 = x_j y_j' + x_j' y_j$ (not equal)
$g_1 = x_j y_j + x_j' y_j'$ (equal)

$g_2 = x_j' y_j$
$(y_j > x_j)$

FIGURE 5.22
**74LS85 four-bit
comparator**

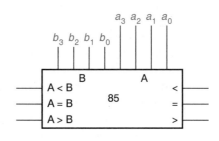

FIGURE 5.23
**Eight-bit
comparator
(requires two 85's)**

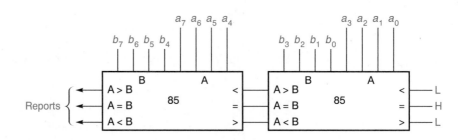

bits. If the numbers have been sign extended, the result is not changed.

The 518 through 522 and 682 through 689 are eight-bit magnitude comparators with various features.

EXERCISE 5.4   Derive the equation for $x = y$, where $x = x_1 x_0$ and $y = y_1 y_0$.

Answer: $x=y = (x_1 \oplus y_1)'(x_0 \oplus y_0)'$ ∎

EXERCISE 5.5   Derive the equation for unsigned $x > y$, where $x = x_1 x_0$ and $y = y_1 y_0$.

Answer: $x>y_U = x_1 y_1' + (x_1 \oplus y_1)'(x_0 y_0')$ ∎

EXERCISE 5.6   Derive the equation for signed $x > y$, where $x = x_1 x_0$ and $y = y_1 y_0$.

Answer: $x>y_S = x_1' y_1 + (x_1 \oplus y_1)'(x_0 y_0')$ ∎

5.6 | Generating and Checking Parity

Parity bit allows detection of one-bit errors.

Single-parity-check codes of n digits contain only one check digit, that is, the modulo-2 sum of the code's $n - 1$ information digits or its complement. The check digit is the nth digit. The digits are added according to the binary rules: $0 + 0 = 0$, $0 + 1 = 1$, $1 + 0 = 1$, and $1 + 1 = 1$. The binary sum of $n - 1$ binary digits is 0 or 1 when the number of ones in the group is even or odd, respectively. Therefore, when the check digit is the modulo-2 sum, the number of ones in an n-digit codeword is even. And when the check digit is the complement of the modulo-2 sum, the number of ones in an n-digit codeword is odd. Single-parity-check codes enable detection of single-bit errors.

If the number of 1 bits in a group is odd, the parity of the group is odd. If the number of 1 bits in a group is even, the parity of the group is even. Odd parity is called *parity 1*. Even parity is called *parity 0*. Furthermore, the parity of any two groups is the sum of their parities modulo 2.

EXAMPLE 5.9 Calculating Single-Parity-Check Digits

A single-parity-check digit is calculated from the information digits and appended to those digits to form the codeword. Assume codewords have nine digits ($n = 9$).

Calculate the single-parity-check digit for the information digits 01000111_2. Append the single-parity-check digit to the information digits, making it the most significant digit in the codeword. Form codewords with odd and even parity.

The modulo-2 sum of 01000111_2 is calculated by applying the binary rules.

$$s = 0 + 1 + 0 + 0 + 0 + 1 + 1 + 1 = 0$$

$$s' = 1$$

even-parity codeword $s01000111_2 = 001000111_2$

odd-parity codeword $s'01000111_2 = 101000111_2$

Parity is calculated in hardware with XOR gates. The function of the XOR truth table can be interpreted as addition modulo 2.

y	z	g	
0	0	0	$0 + 0 = 0$
0	1	1	$0 + 1 = 1$
1	0	1	$1 + 0 = 1$
1	1	0	$1 + 1 = 0$

And $g = x$ xor y xor z "sums" the digits in a three-bit group. Proceeding heuristically, we note that

$$g = x \text{ xor } (y \text{ xor } z) = x \text{ xor } (yz \text{ "sum"}) = xyz \text{ "sum"}$$

Generalizing, we can say the output g of an n-input XOR is 1 if parity is odd and 0 if parity is even. The 74LS280 in effect is a nine-input XOR with parity 1 output plus one inverter providing a parity 0 output: all lines are active high (Figure 5.24). A parity bit can be added to an eight-bit bus with one 280 (Figure 5.24). The parity of the nine-bit bus can be checked, or calculated, by connecting the nine bus lines to the nine inputs of another 280.

5.7 Adding and Carrying

Binary arithmetic using 2's complements consists essentially of addition operations, as was shown in Chapter 1. Therefore the only arithmetic operator we need to implement in hardware is the addition operator. Addition generates carries, so the addition operator must include the carry function. In fact, the hardware partitions into two parts. One part implements the addition function, and the other part implements the carry function.

The binary number system with radix 2 has two digits, 0 and 1, which for purposes of addition we make correspond to the logical

FIGURE 5.24

Adding a parity bit to a bus

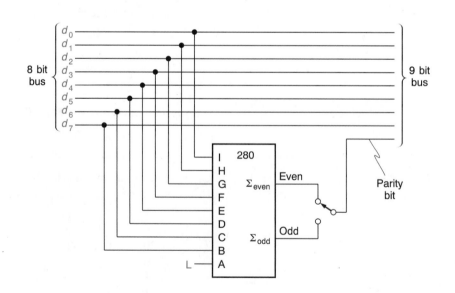

symbols 0 and 1 (False and True). For given inputs we write a logic truth table with the two outputs sum and carry corresponding to the binary number addition table. From the truth table we design the hardware addition operator.

Reduce truth table size by processing bit pairs.

The question of what are the given inputs remains. If we add two six-bit numbers, there are 12 inputs and the table has 2^{12} rows! This apparently impossible situation is resolved by processing only two inputs at a time to implement one-bit adders. Then we interconnect as many one-bit adders as are needed to build an adder for the actual numbers we wish to deal with.

5.7.1 Adding

A sum may generate a carry.

An adder with no carry input is called a *half-adder*. We build the truth table for a half-adder by adding two one-bit numbers. In Figure 5.25 we add x to y to get sum s and carry c.

Processing bit pairs requires carry input and output.

A *full-adder* adds an input carry from the prior digit pair to the current digit pair as well as providing a sum-and carry output. The use of two half-adders is implied by the two addition operations. In Figure 5.26 we add input carry c_i to x to y to get sum s and output carry c. Inspection of columns x, y, and s in the first four rows reveals that s equals x xor y when $c_i = 0$. In the last four rows, s equals x xnor y when $c_i = 1$. The equations for s and c are:

$$s = c_i'(x \text{ xor } y) + c_i(x \text{ xor } y)'$$
$$= c_i \text{ xor } (x \text{ xor } y)$$
$$= c_i \text{ xor } x \text{ xor } y$$

$$c = c_i'xy + c_i(x'y + xy' + xy)$$
$$= xy(c_i' + c_i) + c_i(x'y + xy')$$
$$= xy + c_i(x \text{ xor } y)$$

FIGURE 5.25

Half-adder truth table and circuit

x	y	s	c
0	0	0	0
0	1	1	0
1	0	1	0
1	1	0	1

$s = x$ xor y

$c = xy$

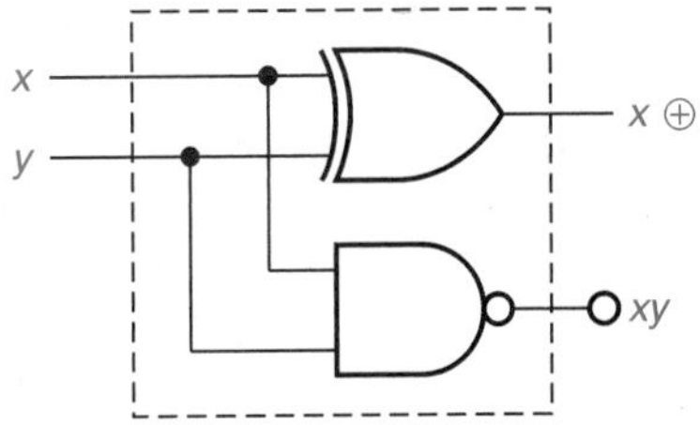

FIGURE 5.26
**Full-adder truth
table and circuit**

c_i	x	y	s	c
0	0	0	0	0
0	0	1	1	0
0	1	0	1	0
0	1	1	0	1
1	0	0	1	0
1	0	1	0	1
1	1	0	0	1
1	1	1	1	1

$s = c_i$ xor x xor y

$c = xy + c_i(x$ xor $y)$

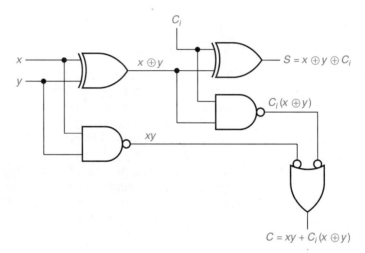

or

$$c = c_i'xy + c_i(x'y + xy' + xy)$$
$$= xy(c_i' + c_i) + c_i(x'y + xy' + xy) \quad \text{Using } p + p = p$$
$$= xy + c_i(x + y) \qquad\qquad \text{Alternate form for this equation}$$

The physical interpretation of the xy term in the carry output equation
is that a carry is generated when the digit pair bits x and y equal 1 (1
plus $1 = 0$ with carry 1). The physical interpretation of the $c_i(x + y)$
term in the carry output equation is that a carry is propagated when
there is a carry into the pair and either bit x or bit y equals 1.

Whether we use the verb *generate* or *propagate* depends on the
source of the new carry. A new carry is *generated* when both digits
are 1, whereas an input carry *propagates* through to the carry output
when either digit is 1. These terms have special symbols—g and p—
and the property $g = gp$.

Bit pair may generate
or propagate a carry.

Proof:

$$g = xy$$

$$p = x + y$$

$$pg = (x + y)xy = xxy + yxy = xy + yx = xy = g$$

$$pg = g \qquad \text{Q.E.D.}$$

Important: $g = gp$ and
$p' = p'g'$.

The special property is not true when the alternate p form (x xor y) is used.

The complement of this equation is $p' = p'g'$.

Proof:

$$g' = (pg)'$$

$$g' = p' + g'$$

$$p'g' = p'(p' + g') = p' + p'g' = p'(1 + g') = p'$$

$$p'g' = p' \qquad \text{Q.E.D.}$$

EXERCISE 5.7   Derive $f = xy + c_i(x \text{ xor } y)$ from $f = xy + c_i(x + y)$.   ■

EXERCISE 5.8   Use algebra to convert $s = c_i \oplus x \oplus y$ and $c = xy + c_i(x + y)$ to sums of minterms (OR of ANDs).

Answer: $s = m_1 + m_2 + m_4 + m_7$

$c = m_3 + m_5 + m_6 + m_7$ ■

5.7.2 Carrying

The carries in arithmetic operations either pass through a cascade of bit adders or they do not. When the carries are cascaded, they implement what is called a *ripple carry* scheme, in which the propagation

FIGURE 5.27
**Four-bit adder with
ripple carry**

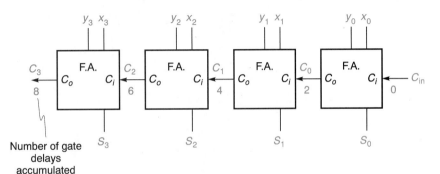

How much delay is the issue.

delay accumulates as the carry passes through each full-adder (Figures 5.27 and 5.28). When carries are not cascaded, they implement what is called a *look-ahead carry* scheme, in which delays do not accumulate. This faster scheme is preferred in practice. Our notation for addition is as follows:

Let

c_{in} = carry into the least significant bits of the two words

c_{j-1} = carry into bit-pair j

$sum_j = b_j$ plus a_j plus c_{j-1} sum_j out of full-adder j

c_j = carry out of bit-pair j

g_j = generate out of bit-pair j

p_j = propagate out of bit-pair j

FIGURE 5.28
Ripple carry circuit emphasizing gate delays

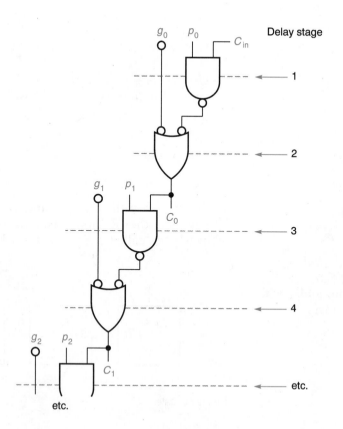

Bit 0:

$$\text{sum}_0 = x_0 \text{ xor } y_0 \text{ xor } c_{\text{in}}$$

$$c_0 = g_0 + p_0 c_{\text{in}}$$

$$g_0 = x_0 y_0$$

$$p_0 = x_0 + y_0$$

Bit j:

$$\text{sum}_j = x_j \text{ xor } y_j \text{ xor } c_{j-1}$$

$$c_j = g_j + p_j c_{j-1}$$

$$g_j = x_j y_j$$

$$p_j = x_j + y_j$$

Ripple carry accumulates delays.

Ripple carry The output carry equations c_j are in ripple carry format because the carry output of stage $j - 1$ is the carry input to stage j. The number of gate delays required for c_j to be available is listed to the left of the c_j equations. These are the equations when four full-adders are connected in ripple carry format (Figure 5.27).

Ripple carry format:

Cumulative number
of gates
introducing delay

0	
	$c_0 = g_0 + p_0 c_{\text{in}}$
2	
	$c_1 = g_1 + p_1 c_0$
4	
	$c_2 = g_2 + p_2 c_1$
6	
	$c_3 = g_3 + p_3 c_2$
8	

Figure 5.28 portrays how the number of gate delays accumulates in a ripple carry scheme. Looking back at Figure 5.26 we see explicitly that input carry c_i passes through two gates before exiting as output carry c, with or without modification. Gate pairs abstracted from the full-adders are shown as a cascade of abstracted pairs in Figure 5.28.

Look-ahead carry does
not accumulate delays.

Look-ahead carry   In a look-ahead carry scheme the cumulative delay is independent of the number of bit-pairs to be added. The look-ahead carry concept is implemented by retaining in the equations only the input carry c_{in} while eliminating the bit-pair to bit-pair intermediate carries c_{j-1}. Furthermore, the cumulative delay is minimized if the equations are physically realized as two level AND/OR logic circuits.

   The ripple carry equations are in the desired form; however, they include the intermediate carries c_0, c_1, c_2, and c_3. The ripple carry equations are converted to look-ahead carry format in two steps. First each intermediate carry c_{j-1} is replaced with its equivalent equation in each ripple carry output equation c_j. Then all factors are multiplied out. The final equations are in the form of sum of product terms. The price paid for look-ahead carry capability is a more complex set of carry output circuits. Such a set is illustrated in Figure 5.29.

*Cumulative
delay*

0

$$c_0 = g_0 + p_0 c_{in}$$

2

$$c_1 = g_1 + p_1 c_0$$
$$= g_1 + p_1(g_0 + p_0 c_{in})$$
$$= g_1 + p_1 g_0 + p_1 p_0 c_{in}$$

2

$$c_2 = g_2 + p_2 c_1$$
$$= g_2 + p_2(g_1 + p_1 g_0 + p_1 p_0 c_{in})$$
$$= g_2 + p_2 g_1 + p_2 p_1 g_0 + p_2 p_1 p_0 c_{in}$$

2

$$c_3 = g_3 + p_3 c_2$$
$$= g_3 + p_3(g_2 + p_2 g_1 + p_2 p_1 g_0 + p_2 p_1 p_0 c_{in})$$
$$= g_3 + p_3 g_2 + p_3 p_2 g_1 + p_3 p_2 p_1 g_0 + p_3 p_2 p_1 p_0 c_{in}$$

2

5.7.3   283 Four-Bit Adder

The 283 four-bit adder (Figure 5.29) adds four bit-pairs using a look-ahead carry scheme. Analysis of the 283 is instructive for several

FIGURE 5.29

283 four-bit adder in mixed-logic format

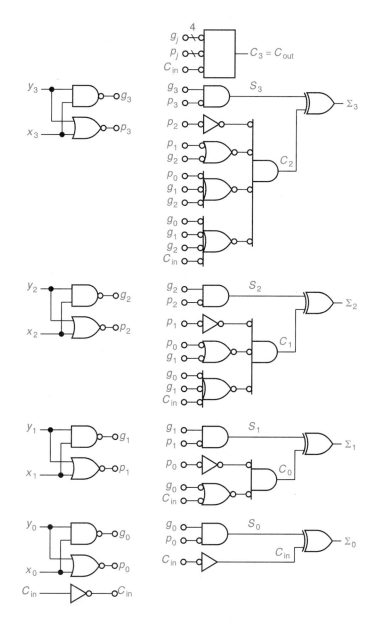

reasons: the 283 represents a substantial application of mixed logic, the 283 uses the universal-logic circuit, and the 283 uses look-ahead carry circuits. (Figure 5.29 is the 283 data book schematic converted into mixed-logic format for ease of analysis, as shown shortly.)

The designers minimized gate count by manipulating the sum and carry equations. In some data books, unfortunately, the 283 bits are numbered from 1 to 4, not 0 to 3. Nevertheless, we number them 0 to 3, as illustrated in Figure 5.29 and in the analysis of the second 283 bit-pair circuit (subscript 1). (The relevant fragment from the data book schematic is found in Figure 5.30.) The analysis is divided into three parts: calculate s_1, calculate c_0, and calculate Σ_1.

$$s_1 = p_1 g_1'$$
$$= (x_1 + y_1)(x_1 y_1)'$$
$$= (x_1 + y_1)(x_1' + y_1')$$
$$= x_1 y_1' + x_1' y_1$$
$$= x_1 \text{ xor } y_1$$

The s_1 analysis reveals another application of the p_j and g_j functions: The sum of two bits is $p_j g_j'$.

The logical mismatch from the carry circuit output c_0 into the XOR gate shown in Figure 5.30 implies that we first calculate c_0' and then use De Morgan's Theorem to find c_0.

$$c_0' = p_0' + g_0' c_{in}'$$
$$= (x_0 + y_0)' + (x_0 y_0)'(c_{in})'$$

$$c_0 = (x_0 + y_0)''[(x_0 y_0)'' + (c_{in})''] \quad \text{Using De Morgan's Theorem}$$
$$= (x_0 + y_0)(x_0 y_0 + c_{in})$$
$$= p_0(g_0 + c_{in})$$
$$= p_0 g_0 + p_0 c_{in})$$
$$= g_0 + p_0 c_{in} \quad\quad\quad \text{Using } g_0 = p_0 g_0$$
$$= x_0 y_0 + (x_0 + y_0) c_{in}$$

The output XOR gate adds the carry to the bit-pair sum.

$$\Sigma_1 = s_1 \text{ xor } c_0$$
$$= x_1 \text{ xor } y_1 \text{ xor } c_0$$
$$= x_1 \text{ xor } y_1 \text{ xor } [x_0 y_0 + (x_0 + y_0) c_{in}]$$

FIGURE 5.30

74LS283 circuit for second bit pair, x_1, y_1.

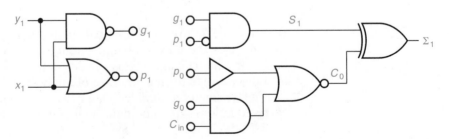

FIGURE 5.31
**74LS283 mixed-logic
circuit for carry c_0**

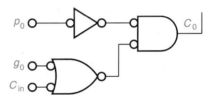

Here is how to avoid the complexities of the foregoing analysis and proceed smartly. We convert the data book carry circuit part of the Σ_1 circuit into mixed-logic format (Figure 5.31, see also Figure 5.30). Now we can go directly to the result in the c_0 analysis.

$$s_1 = p_1 g_1' = x_1 \text{ xor } y_1 \quad \text{As before}$$

$$c_0 = p_0(g_0 + c_{in})$$
$$= p_0 g_0 + p_0 c_{in}$$
$$= g_0 + p_0 c_{in}$$
$$= x_0 y_0 + (x_0 + y_0)c_{in}$$

$$\Sigma_1 = x_1 \text{ xor } y_1 \text{ xor } [x_0 y_0 + (x_0 + y_0)c_{in}] \quad \text{As before}$$

EXERCISE 5.9

From Figure 5.29, show that $c_1 = p_1(p_0 + g_1)(g_0 + g_1 + c_{in})$. Then use algebra to convert this form to a sum of product terms (OR of ANDs).

Answer: $c_1 = g_1 + p_1 g_0 + p_1 p_0 c_{in}$ ∎

EXAMPLE 5.10 **Four-Bit Addition with the Sum and
Carry Equations**

Pencil and paper:

$$x_3 x_2 x_1 x_0 = x = 1011_2 \qquad y_3 y_2 y_1 y_0 = y = 1010_2 \qquad c_{in} = 1$$

$$
\begin{array}{rl}
1011 & \text{Carries} \\
1 & c_{in} \\
1010 & y \\
\underline{1011} & x \\
10110 & x \text{ plus } y \text{ plus } c_{in}
\end{array}
$$

Inside the 283 (see Figure 5.29):

Reference: $g_j = x_j y_j \qquad p_j = x_j + y_j \qquad j = 0, 1, 2, 3$

in c_{in}

0    $\Sigma_0 = s_0 \text{ xor } c_{in}$
          $= x_0 \text{ xor } y_0 \text{ xor } c_{in}$
          $= 1 \text{ xor } 0 \text{ xor } 1 = 0$

1    $c_0 = g_0 + p_0 c_{in}$
          $= (1 \times 0) + (1 + 0) \times 1 = 1$

     $\Sigma_1 = s_1 \text{ xor } c_0$
          $= x_1 \text{ xor } y_1 \text{ xor } c_0$
          $= 1 \text{ xor } 1 \text{ xor } 1 = 1$

2    $c_1 = g_1 + p_1 g_0 + p_1 p_0 c_{in}$
          $= (1 \times 1) + (1 + 1)(1 \times 0) + (1 + 1)(1 + 0) \times 1 = 1$

     $\Sigma_2 = s_2 \text{ xor } c_1$
          $= x_2 \text{ xor } y_2 \text{ xor } c_1$
          $= 0 \text{ xor } 0 \text{ xor } 1 = 1$

3    $c_2 = g_2 + p_2 g_1 + p_2 p_1 g_0 + p_2 p_1 p_0 c_{in}$
          $= (0 \times 0) + (0 + 0)(1 \times 1) + (0 + 0)(1 + 1)(1 \times 0)$
          $\qquad\qquad\qquad + (0 + 0)(1 + 1)(1 + 0) \times 1 = 0$

     $\Sigma_3 = s_3 \text{ xor } c_2$
          $= x_3 \text{ xor } y_3 \text{ xor } c_2$
          $= 1 \text{ xor } 1 \text{ xor } 0 = 0$

out $c_3 = g_3 + p_3 g_2 + p_3 p_2 g_1 + p_3 p_2 p_1 g_0 + p_3 p_2 p_1 p_0 c_{in}$
          $= (1 \times 1) + (1 + 1)(0 \times 0) + (1 + 1)(0 + 0)(1 \times 1)$
          $\qquad\qquad + (1 + 1)(0 + 0)(1 + 1)(1 \times 0)$
          $\qquad\qquad\qquad + (1 + 1)(0 + 0)(1 + 1)(1 + 0) \times 1 = 1$

5.7.4   Conversion from BCD to Binary

The conversion process illustrates eight-bit addition using 283 chips, subtraction in the form of 2's complement addition (add x' plus 1 in lieu of subtraction), and multiplication of a number to be added by positioning of the bits. You may want to reread section 1.5.2 (BCD Numbers) before proceeding.

Add BCD digit pairs as groups of 4 bits.

A two-digit BCD number ranges from 0 to 99 decimal in value. A seven-digit binary number ranges from 0 to 127 decimal. Allowing for a carry out, we had better use eight bits. As we will literally see, the wiring plays a significant role in setting up values. Consider a two-digit BCD number n with digits $x = x_3 x_2 x_1 x_0$ and $y = y_3 y_2 y_1 y_0$ whose polynomial is $n = 10x + 1y$. Anticipating a conversion to binary we put n in a "powers of two" format.

$$n = (16 - 6)x + y$$
$$= 16x + y - 4x - 2x$$
$$= 16x + y + 4(-x) + 2(-x)$$
$$= (16x + y) + 4(x' + 1) + 2(x' + 1)$$

Position digits with wiring arrangement.

We form $16x$ by physically positioning x with respect to y. Term $(x' + 1)$ has a multiplier of either 2 or 4 established by positioning. Often

FIGURE 5.32

BCD-to-binary conversion circuit

$bcd = xy$ $\quad x = x_3x_2x_1x_0$ $\quad y = y_3y_2y_1y_0$

binary $= b = b_7b_6b_5b_4b_3b_2b_1b_0$

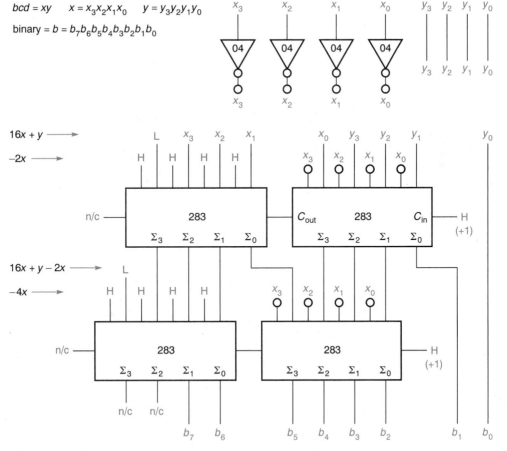

n/c = no connection

$bcd = xy$ $x = x_3x_2x_1x_0$ $y = y_3y_2y_1y_0$

$\text{binary} = b = b_7b_6b_5b_4b_3b_2b_1b_0$

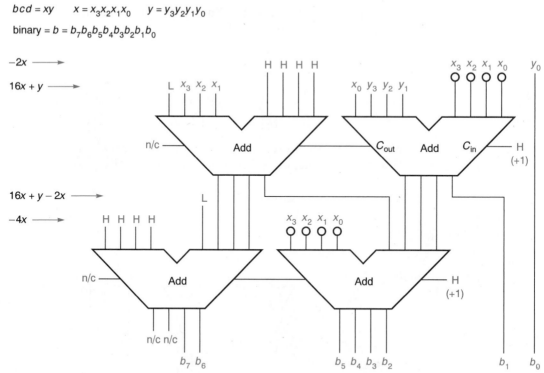

FIGURE 5.33
BCD-to-binary conversion logic schematic

overlooked is sign extension of negative numbers. On paper, positioning shows why the wiring is as it is (Figure 5.32). The conversion circuit also has a symbolic form (Figure 5.33). The 283 implements the equations in a straightforward manner; however, there are faster conversion circuits. (The ♦ symbol in the following means zeros are not used, because no wires are involved.)

	$= x_3$	x_2	x_1	x_0	y_3	y_2	y_1	y_0	
$16x + y$	x_3	x_2	x_1	x_0	y_3	y_2	y_1	y_0	
$2(x' + 1) =$	1	1	1	$x_{3'}$	$x_{2'}$	$x_{1'}$	$x_{0'}$	♦	With sign extension
								1	
$4(x' + 1) =$	1	1	$x_{3'}$	$x_{2'}$	$x_{1'}$	$x_{0'}$	♦	♦	With sign extension
						1	♦	♦	
$\text{binary sum} =$	b_7	b_6	b_5	b_4	b_3	b_2	b_1	b_0	

Example: $xy = 99 = 10011001$

$$
\begin{array}{llllllllll}
16x + y = & & 1 & 0 & 0 & 1 & 1 & 0 & 0 & 1 \\
2(x' + 1) = & & 1 & 1 & 1 & 0 & 1 & 1 & 0 & \blacklozenge \\
& & & & & & & & 1 & \blacklozenge \\
\hline
\text{partial sum} = & & 1 & 1 & 0 & 0 & 0 & 0 & 1 & 1 & 1 \quad \text{First add} \\
4(x' + 1) = & & 1 & 1 & 0 & 1 & 1 & 0 & \blacklozenge & \blacklozenge \\
& & & & & & 1 & \blacklozenge & \blacklozenge \\
\hline
\text{binary sum} = 1 & 0 & 0 & 1 & 1 & 0 & 0 & 0 & 1 & 1 \quad \text{Second add} \\
\text{8-bit sum} = & & 0 & 1 & 1 & 0 & 0 & 0 & 1 & 1 & = 99 \text{ decimal} \quad \text{Q.E.D.}
\end{array}
$$

5.8 Arithmetic-Logic Unit (ALU)

Five bits program the ALU function.

An n-bit arithmetic-logic unit (ALU) performs one operation at a time on two n-bit words. A block diagram of a four-bit ALU is illustrated in Figure 5.34. Five input lines specify the function to be performed by an ALU. Four lines—s_3, s_2, s_1, and s_0—select one of 16 operations. The fifth line, M, defines the operation as logic (carry circuits disabled) or arithmetic (carry circuits enabled). The ALU has carry input and carry output lines to accommodate word sizes greater than four bits. Internal carry lines implement a look-ahead carry scheme. External input and output carry lines can implement ripple carry or look-ahead carry schemes (with the assistance of additional chips). An ALU is an assembly of universal-logic circuits, look-ahead carry circuits, and adders. These are discussed next.

5.8.1 Universal-Logic Circuit (ULC)

ULC generates 16 functions of 2 variables.

The Universal-logic circuit (ULC) in Figures 5.35 and 5.36 generates the 16 functions z_j of two variables (Tables 5.1 and 5.2). Four selection inputs, $s_3 s_2 s_1 s_0$, select one function at a time (Table 5.1). In addition to the z_j function output, the ULC has the generate and propagate function outputs g_j and p_j. Subscript j ranges from 0 to 15.

FIGURE 5.34
ALU block diagram

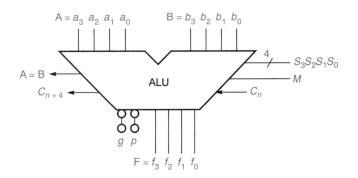

FIGURE 5.35

Universal-logic circuit block diagram

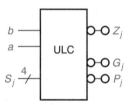

The actual universal-logic circuit is well known (Figure 5.36). The subscript j on the single line outputs z_j, g_j, and p_j refers to the different interpretations selected by the four s_j lines.

For any value of $S_j = s_3 s_2 s_1 s_0$, the outputs are:

$$g_j = s_3 \times ba + s_2 \times b'a$$

$$p_j = s_1 \times b' + s_0 \times b + a$$

$$z_j = g_j \text{ xor } p_j$$

The g_j, p_j also have the following useful property: $g_j = g_j p_j$.

TABLE 5.1   Universal-Logic Circuit Equations

$z_j = (s_3 \times ba + s_2 \times b'a) \text{ xor } (s_1 \times b' + s_0 \times b + a) = g_j \text{ xor } p_j$		
s_{3210}	z_j	
0000	$z_0 = 0 \text{ xor } a$	$= a$
0001	$z_1 = 0 \text{ xor } (b + a)$	$= b + a$
0010	$z_2 = 0 \text{ xor } (b' + a)$	$= b' + a$
0011	$z_3 = 0 \text{ xor } (b' + b + a)$	$= 1$
0100	$z_4 = b'a \text{ xor } a$	$= ba$
0101	$z_5 = b'a \text{ xor } (b + a)$	$= b$
0110	$z_6 = b'a \text{ xor } (b' + a)$	$= b' \text{ xor } a$
0111	$z_7 = b'a \text{ xor } (b' + b + a)$	$= b + a'$
1000	$z_8 = ba \text{ xor } a$	$= b'a$
1001	$z_9 = ba \text{ xor } (b + a)$	$= b \text{ xor } a$
1010	$z_A = ba \text{ xor } (b' + a)$	$= b'$
1011	$z_B = ba \text{ xor } (b' + b + a)$	$= b' + a'$
1100	$z_C = (ba + b'a) \text{ xor } a$	$= 0$
1101	$z_D = (ba + b'a) \text{ xor } (b + a)$	$= ba'$
1110	$z_E = (ba + b'a) \text{ xor } (b' + a)$	$= b'a'$
1111	$z_F = (ba + b'a) \text{ xor } (b' + b + a)$	$= a'$

FIGURE 5.36
Universal-logic circuit

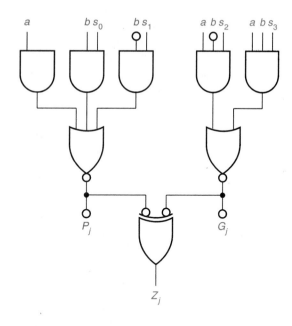

TABLE 5.2 **Universal-Logic Circuit Functions**

z	s_j	z'
0	C	1
1	3	0
a'	F	a
a	0	a'
b'	A	b
b	5	b'
b'a'	E	b + a
b'a	8	b + a'
ba'	D	b' + a
ba	4	b' + a'
b' + a'	B	ba
b' + a	2	ba'
b + a'	7	b'a
b + a	1	b'a'
b' xor a	6	b xor a
b xor a	9	b' xor a

The universal-logic circuit g_j and p_j differ from the g and p used in the full-adder circuit in detail but not in principle. Example 5.11 shows that $S_9 = 1001$ programs the ULC to be an adder. Tables 5.1 and 5.2 list all 16 cases in two ways.

EXAMPLE 5.11

$S_9 = 1001 \Rightarrow S_3 = 1, S_2 = 0, S_1 = 0, S_0 = 1$

$$g_9 = s_3 \times ba + s_2 \times b'a$$
$$= 1 \times ba + 0 \times b'a$$
$$= ba$$

$$p_9 = s_1 \times b' + s_0 \times b + a$$
$$= 0 \times b' + 1 \times b + a$$
$$= b + a$$

$$z_9 = g_9 \text{ xor } p_9$$
$$= ba \text{ xor } (b + a)$$
$$= ba(b + a)' + (ba)'(b + a)$$
$$= bab'a' + (b' + a')(b + a)$$
$$= 0 + ba' + b'a$$
$$= b \text{ xor } a$$

5.8.2   The 74LS181 Four-Bit ALU

The 181 arithmetic-logic unit uses universal-logic circuits, look-ahead carry circuits, and adders to implement arithmetic and logic functions on the two four-bit input words B and A (Figure 5.37).

The 181 is a combinational-logic circuit.

The ALU internal look-ahead carry equations are the same as the look-ahead carry equations developed for the 283. The look-ahead carry circuits are enabled when the M line is low, and disabled when the M line is high. Logic functions are performed when the M line is high. The M and S_j lines select the specific logic or arithmetic functions (Table 5.3). Ripple carry input and output lines c_n and c_{n+4} are

FIGURE 5.37
Four-bit 181 ALU package

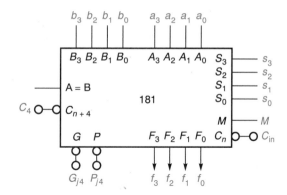

available. In addition, look-ahead carry logic outputs g_{j4} and p_{j4} allow a look-ahead carry scheme with multiple 181 chips if the 182 is used.

Figure 5.38 shows the 181 as an abstract black box for four-bit pairs. Four 181 packages wired as shown in Figure 5.39 constitute a sixteen-bit ALU using ripple carry format.

TABLE 5.3 **181 ALU Functions**

SELECT:				M = H	M = L: ARITHMETIC OPERATIONS	
S_3	S_2	S_1	S_0	Logic operations	$C_n = $ H	$C_n = $ L
L	L	L	L	$f = a'$	$f = a$	$f = a$ plus 1
L	L	L	H	$f = (b + a)'$	$f = b + a$	$f = (b + a)$ plus 1
L	L	H	L	$f = ba'$	$f = b' + a$	$f = (b' + a)$ plus 1
L	L	H	H	$f = 0000$	$f = 1111 = -1$	$f = 1111$ plus 1 = 0
L	H	L	L	$f = (ba)'$	$f = a$ plus $b'a$	$f = a$ plus $b'a$ plus 1
L	H	L	H	$f = b'$	$f = (a + b)$ plus $b'a$	$f = (a + b)$ plus $b'a$ plus 1
L	H	H	L	$f = b$ xor a	$f = a$ minus b minus 1	$f = a$ minus b
L	H	H	H	$f = b'a$	$f = b'a$ minus 1	$f = b'a$
H	L	L	L	$f = b + a'$	$f = a$ plus ba	$f = a$ plus ba plus 1
H	L	L	H	$f = (b$ xor $a)'$	$f = b$ plus a	$f = b$ plus a plus 1
H	L	H	L	$f = b$	$f = (b' + a)$ plus ba	$f = (b' + a)$ plus ba plus 1
H	L	H	H	$f = ba$	$f = ba$ minus 1	$f = ba$
H	H	L	L	$f = 1111$	$f = a$	$f = a$ plus a plus 1
H	H	L	H	$f = b' + a$	$f = (b + a)$ plus a	$f = (b + a)$ plus a plus 1
H	H	H	L	$f = b + a$	$f = (b' + a)$ plus a	$f = (b' + a)$ plus a plus 1
H	H	H	H	$f = a$	$f = a$ minus 1	$f = a$

FIGURE 5.38
Four-bit ALU abstraction

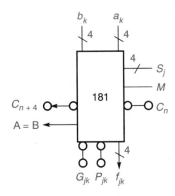

5.8.3   The 74LS182 Sixteen-Bit Look-Ahead Carry

The 182 look-ahead carry generator (Figure 5.40) provides look-ahead carry operation across 181 ALU's set up as four-bit binary adders. The 181 look-ahead carry outputs g_k and p_k are connected directly to the 182 g_k, p_k inputs. The 182 equations are

$$c_{n+x} = g_0 + p_0 c_{in}$$

$$c_{n+y} = g_1 + p_1 c_{n+x}$$
$$= g_1 + p_1 g_0 + p_1 p_0 c_{in}$$

$$c_{n+z} = g_2 + p_2 c_{n+y}$$
$$= g_2 + p_2 g_1 + p_2 p_1 g_0 + p_2 p_1 p_0 c_{in}$$

$$g_{out} = g_3 + p_3 c_{n+z}$$

$$p_{out} = p_3 p_2 p_1 p_0 c_{in}$$

FIGURE 5.39
Sixteen-bit ALU with ripple carry every fourth bit

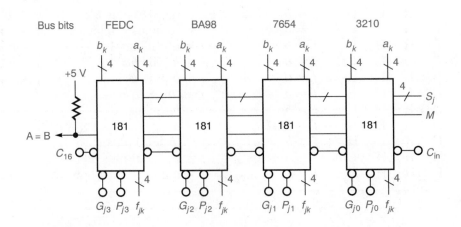

FIGURE 5.40
**182 Look-ahead
carry generator**

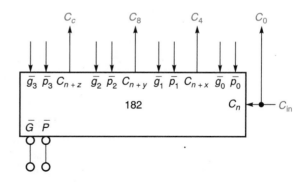

FIGURE 5.40
**182 Look-ahead
carry generator**

A valid, perhaps useful, point of view is that the 182 treats the 181 chips as if they are single-bit adders. The 182 does not know that the 181s' g and p outputs are from every fourth bit-pair.

Figure 5.41 shows how multiple 182 parts are wired to implement look-ahead carry across nine and more 181 ALUs.

FIGURE 5.41
**182 thirty-two-bit
(and more)
look-ahead carry
generator**

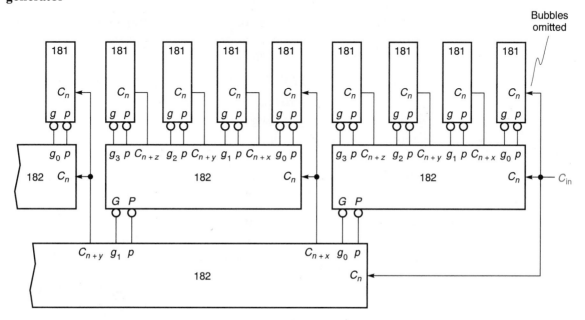

SUMMARY

Early designers worked with discrete parts: resistors, diodes, vacuum tubes, and so forth. The number of gates had to be small or the product would fill a room. When the transistor replaced the vacuum tube, more complex designs (still using parts) became commercially feasible. The first integrated circuits (ICs) replaced assemblies of parts implementing gates. IC availability accelerated the trend to more complex designs. But designers still found themselves using gates to form the same complex building blocks (mux, decode, arithmetic-logic unit, and so forth) over and over again. When the complex function IC appeared, design complexity surged. At that point discrete gates were relegated to the secondary role of glue logic supplementing the complex building blocks. This is why we set aside discrete gate logic design here and investigate modern logic design that uses complex logical operators.

Timing

A timing diagram is a graphical illustration of a circuit's logical behavior as a function of time. A timing diagram's horizontal axis represents the time dimension, the vertical axis presents the voltage dimension. It is used for H and L voltage levels. A digital waveform is a plot of the H and L levels a variable takes as a function of time.

Multiplexing

Multiplexing is the process of selecting one of a group of 2^n sources and then routing that source to one destination. A multiplexing network is a combinational network; presentation of selection bits connects the selected source to the destination after propagation times elapse. A typical multiplexer equation shows that the multiplexer inputs a_j are the minterm coefficients.

$$g = a_0 y'z' + a_1 y'z + a_2 yz' + a_3 yz \quad \text{Defining equation}$$

$$= a_0 m_0 + a_1 m_1 + a_2 m_2 + a_3 m_3 \quad \text{Minterm format}$$

Important multiplexer applications are derived by making different interpretations of the defining equations.

Decoding

Decoding is a process of activating one of 2^n outputs y_j for each combination of values of the n inputs. A decoding network is a combinational network; presentation of input bits activates the selected output line after propagation times elapse. On the other hand, decoding can be considered to be demultiplexing: routing one source g to one of 2^n selected outputs y_j. Presentation of n selection bits connects the source to the selected output destination after propagation times elapse.

The defining equations for a two-bit decoder show that decoders are minterm generators. A multitude of decoder applications are based on various uses of the minterms and the enabling coefficient g.

$$y_3 = gba \qquad y_2 = gba' \qquad y_1 = gb'a \qquad y_0 = gb'a'$$

Priority Encoding, Comparing, and Generating and Checking Parity

These important functions have many uses in digital systems. Priority encoders allow assignment of importance to devices requiring service from a computer. Compare functions implemented with hardware execute faster than software compare functions. Generating and checking parity is the tip of the iceberg of error detection and correction functions that make many digital systems possible in difficult environments.

Adding and Carrying

The only arithmetic operator we need to implement in hardware is the addition operator. Addition generates carries, so the addition operator must include the carry function. Addition is carried out one bit-pair at a time, which makes the design process tractable. This is why addition hardware is essentially an assembly of full-adder circuits. A full-adder consists of a sum circuit (s) and a carry circuit (c) derived from the equations for s and c.

$$s = c_i \text{ xor } x \text{ xor } y \qquad c = xy + c_i(x + y)$$

The look-ahead carry concept eliminates accumulation of delays from a chain of full-adders executing the addition of two n-bit words.

Arithmetic-Logic Unit (ALU)

An arithmetic-logic unit (ALU) is a programmable combinational-logic block. Input lines program the ALU selecting one of n logical or arithmetic operations on two multibit words. The universal-logic circuit provides one way to implement an ALU's programming capability. The universal-logic circuit's equation is

$$z_j = (s_3 \times ba + s_2 \times b'a) \text{ xor } (s_1 \times b' + s_0 \times b + a) = g_j \text{ xor } p_j$$

REFERENCES

Blakeslee, T. R. 1979. *Digital Design with Standard MSI and LSI,* 2nd ed. New York: Wiley.

Breeding, K. J. 1989. *Digital Design Fundamentals*. Englewood Cliffs, N.J.: Prentice-Hall.

Mano, M. M. 1991. *Digital Design,* 2nd ed. Englewood Cliffs, N.J.: Prentice-Hall.

McClusky, E. J. 1986. *Logic Design Principles*. Englewood Cliffs, N.J.: Prentice-Hall.

Seidensticker, R. B. 1986. *The Well-Tempered Digital Design*. Reading, Mass.: Addison-Wesley.

Wakerly, J. F. 1990. *Digital Design Principles and Practices*. Englewood Cliffs, N.J.: Prentice-Hall.

Name _____ SID# _____ Section _____
      (last)   *(initials)*

Approved by _____ Date _____

Grade on report _____

Greater-Than Design

1. Two-bit greater than

   1.1 _____ four-bit Truth table
        _____ four-bit K map
        _____ z output waveform
   1.2 Design and build the circuit.
        _____ circuit documents
   1.3 Use the truth table signal generator to test the circuit.
        _____ input/output timing diagram

2. Four-bit greater than or equal to

   2.1 _____ Equal$_j$ truth table and equation for bit-pair j
        _____ Greater$_j$ truth table and equation for bit-pair j
   2.2 _____ $z>$ equation
        _____ $z=$ equation
   2.3 Design and build the signed $a > b$ circuit.
        _____ circuit documents
   2.4 Design and build the signed $a = b$ circuit.
        _____ circuit documents
   2.5 Use the truth table signal generator to test the circuit.
        _____ input/output timing diagram, $a_{3210} = $ LHLH
        _____ input/output timing diagram, $a_{3210} = $ HHLL
        _____ input/output timing diagram, $a_{3210} = $ HLLH
        _____ Are the results what you expect? Why?

1. Two-bit greater-than function

Define a two-bit greater-than function for unsigned numbers. The output z is True when the two-bit number $a = a_1 a_0$ is greater than $b = b_1 b_0 (a > b)$.

    1.1   Create a truth table, a K map, and an output equation for z as a function of inputs b_1, b_0, a_1, and a_0.

    1.2   Design and build a combinational-logic circuit implementing the greater-than function. Use 74LS04 and 74LS10.

    1.3   Connect signal generator outputs q_3, q_2, q_1, and q_0 to the b_1, b_0, a_1, a_0 inputs. Plot the z output to prove that the circuit is correct.

2. Four-bit greater-than function

Is it practical to use K maps to design these functions with eight input variables? *Hint:* create bit-pair greater-than (>) and equal-to (=) functions.

    2.1   Define four bit-pair greater-than and equal-to functions for signed numbers b, a, where $b = b_3 b_2 b_1 b_0$ and $a = a_3 a_2 a_1 a_0$.

    2.2   Write the equations for outputs $z>$ and $z=$ in terms of bit-pair $>$ and $=$ functions.

    2.3   Design and build a combinational-logic circuit implementing the four-bit signed equal-than function.

    2.4   Design and build a combinational-logic circuit implementing the four-bit signed equal-to function.

    2.5   Connect signal generator outputs q_3, q_2, q_1, and q_0 to the b_3, b_2, b_1, b_0 inputs. Connect H/L dip switch circuits to the four a_j inputs. Plot the $z>$ and $z=$ outputs for the following switch combinations:

$$a_{3210} = \text{LHLH}$$
$$= \text{HHLL}$$
$$= \text{HLLH}$$

Are the results what you expect? Why?

Name _____ SID# _____ Section _____
     (last)   *(initials)*

Approved by _____ Date _____

Grade on report _____

Combinational Binary Multiplier Design

1. Multiply on paper:

   _____ $111_2 \times 111_2$

2. Calculate $c = a \times b$.

   _____ symbolic calculation

3. Design and build the circuit.

   _____ circuit documents

4. Use the truth table signal generator to test the circuit.

   _____ Input/output timing diagram.
   _____ Explain why plots should cover at least eight clock periods.
   _____ Prove by the method of perfect induction (Section 3.11.5) the output products to expect for the specified inputs.

5. Use the truth table signal generator to test the circuit.

   _____ input/output timing diagram for $a_{210} = $ _____.
   _____ What products should you obtain, and why?

1. In decimal, $7 \times 7 = 49$. On paper, use binary multiplication to multiply 111×111.

2. For any numbers a, b, c, where

$$a = a_2 a_1 a_0$$
$$b = b_2 b_1 b_0$$
$$c = c_5 c_4 c_3 c_2 c_1 c_0$$

use pencil and paper to calculate $c = a \times b$. Do this very neatly.

3. Design and build a combinational-logic circuit implementing $c = a \times b$. Use 74LS08 and 74LS283.

4. Connect the following signal generator outputs:

$$q_2 \text{ to } a_2 \text{ and } b_2$$
$$q_1 \text{ to } a_1 \text{ and } b_1$$
$$q_0 \text{ to } a_0 \text{ and } b_0$$

Plot the six c_j outputs over an eight-clock-period cycle, or longer, to prove the circuit is correct. Why eight minimum? What products should you obtain, and why?

5. Connect the following signal generator outputs:

$$q_2 \text{ to } b_2$$
$$q_1 \text{ to } b_1$$
$$q_0 \text{ to } b_0$$

Connect H/L dip switch circuits to a_2, a_1, a_0.
The a_{210} input is assigned to be _____. (000 to 111 binary)
Plot the six c_j outputs over an eight-clock-period cycle.
What products should you obtain, and why?

PROBLEMS

5.1 *Static hazard:* Show by replicating Figure 5.2a that the static hazard in $g_0 = (x + z)(x' + y')$ is eliminated when the consensus term $(y' + z)$ is added as a factor to g_0. Show by using a K map that g_0 is not changed by adding $(y' + z)$ to the expression.

5.2 *Static hazard:* Show by replicating Figure 5.2b that the static hazard in $g_1 = x'z + xy'$ is eliminated when the consensus term zy' is added to g_1. Show by using a K map that g_1 is not changed by adding zy' to the expression.

5.3 *Dynamic hazard:* Show by replicating Figure 5.3 that the dynamic hazard is eliminated by adding the consensus term to $(xz + yz')$.

5.4 The 74LS153 schematic is found in any TTL data book. Assume select inputs A (pin 14) and B (pin 2) are at the L level. Assume that data input 1C0 (pin 6) is at the L level, and that 1C1 (pin 5) is at the H level. Let input A switch from L to H. Make a timing diagram showing the ideal digital waveforms, including propagation delays, at the outputs of the four gates in the signal path from A (pin 14) to 1Y (pin 7). Refer to Figure 5.2.

5.5 Use one 4 × 1 mux (153) and supplementary gates to synthesize the following functions.

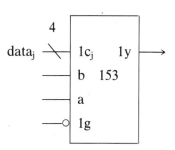

(a) $f = (x \text{ xor } y \text{ xor } z)w$
(b) $f = [xy + z(x \text{ xor } y)]w$
(c) $f = [xy + z(x + y)]w$
(d) $f = x' + y'z$
(e) $f = x' + z + y'$

5.6 Use one 8 × 1 mux (151) and supplementary gates to synthesize the following functions.

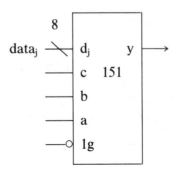

(a) $f = x$ xor y xor zw
(b) $f = xy + z(x$ xor $wy)$
(c) $f = xy + z(x + yw)$
(d) $f = wx' + y'z$
(e) $f = wx' + wz + xy' + w'yz'$

5.7 Use multiplexers and supplementary gates to design Gray-code-to-binary converters defined by the following truth table.

(a) Use four 151 chips plus gates.
(b) Use four 153 chips plus gates.
(c) Use two 151 chips plus gates.
(d) Use two 153 chips plus gates.

INPUTS—GRAY				OUTPUTS—BINARY			
w	x	y	z	g_3	g_2	g_1	g_0
0	0	0	0	0	0	0	0
0	0	0	1	0	0	0	1
0	0	1	1	0	0	1	0
0	0	1	0	0	0	1	1
0	1	1	0	0	1	0	0
0	1	1	1	0	1	0	1
0	1	0	1	0	1	1	0
0	1	0	0	0	1	1	1
1	1	0	0	1	0	0	0
1	1	0	1	1	0	0	1
1	1	1	1	1	0	1	0
1	1	1	0	1	0	1	1

INPUTS—GRAY				OUTPUTS—BINARY			
w	x	y	z	g_3	g_2	g_1	g_0
1	0	1	0	1	1	0	0
1	0	1	1	1	1	0	1
1	0	0	1	1	1	1	0
1	0	0	0	1	1	1	1

5.8 The 74LS139 schematic is found in any TTL data book. Assume select inputs 1A (pin 2) and 1B (pin 3) are at the L level. Assume that enable input 1G (pin 1) is at the L level. Let input 1A switch from L to H. Make a timing diagram showing the ideal digital waveforms, including propagation delays, at the outputs of the three gates in the signal path from 1A (pin 2) to 1Y1 (pin 5). Refer to Figure 5.2.

5.9 Use a 3×8 decoder (138) and supplementary gates to synthesize the following functions.

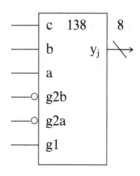

(a) $f = (x \text{ xor } y \text{ xor } z)w$
(b) $f = [xy + z(x \text{ xor } y)]w$
(c) $f = [xy + z(x + y)]w$
(d) $f = x' + y'z$
(e) $f = x' + z + y'$

5.10 Use two 3×8 decoders (138) and supplementary gates to synthesize the following functions.

(a) $f = x \text{ xor } y \text{ xor } zw$
(b) $f = xy + z(x \text{ xor } wy)$
(c) $f = xy + z(x + yw)$
(d) $f = wx' + y'z$
(e) $f = xw' + wz + wy' + w'yz'$

5.11 Use decoders and supplementary gates to design Gray-code-to-binary converters defined by the truth table in Problem 5.7.

(a) Use four 138 chips plus gates.
(b) Use four 139 chips plus gates.

5.12 Use a 3×8 decoder (138) and supplementary gates to decode an eight-bit address a_j ($j = 0$ to 7) by producing asserted low output lines o_j for hex addresses C7, CF, D7, DF, E7, EF, F7, and FF. Write the equation for the E7 output.

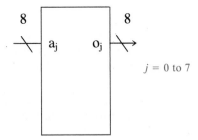

5.13 Use two 3×8 decoders (138) and *no* supplementary gates to synthesize f_0 and f_1.

$$f_0 = a_5' a_4' a_3' a_2' a_1 a_0'$$
$$f_1 = a_5' a_4' a_3 a_2' a_1 a_0$$

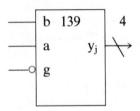

5.14 Use one or two 2×4 decoders (139) and supplementary gates to synthesize the following functions.

(a) $f = x \text{ xor } y \text{ xor } z$
(b) $f = xy + z(x \text{ xor } y)$
(c) $f = xy + z(x + y)$
(d) $f = x' + y'z$
(e) $f = (y + z)(x' + y')$

5.15 The 74LS148 schematic is found in any TTL data book. Assume all inputs are at the H level except input EI, which is at the L level. Let input_6 (pin 3) switch from H to L. Make a timing diagram showing the ideal digital waveforms, including propagation delays, at the outputs of the gates in the signal paths from input_6 (pin 3) to A1 (pin 7). Refer to Figure 5.2.

5.16 Use a priority encoder (148) and supplementary gates to synthesize the following truth table.

In	Address out
r_0	C7
r_1	CF
r_2	D7
r_3	DF
r_4	E7
r_5	EF
r_6	F7
r_7	FF

8 → r_j a_j → 8

$j = 0$ to 7

5.17 Synthesize a less-than function for four-bit signed numbers.

5.18 Can glue logic plus an 85 be used for signed number comparisons? Explain why not, or design a circuit to do so.

5.19 The 74LS283 schematic is found in any TTL data book. Assume all inputs are at the L level. Let A3 (pin 14) switch from L to H. Make a timing diagram showing the ideal digital waveforms, including propagation delays, at the outputs of all the gates in the signal path from A3 (pin 14) to $\Sigma 3$ (pin 13). Refer to Figure 5.2.

5.20 Convert binary 0000 to 1111 (d_j) into BCD 00 to 15 (b_j). Use a 157 mux, a 283 adder, and supplementary gates for this highly specialized code converter.

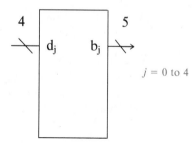

5.21 Synthesize a one-digit BCD adder $s = a$ BCD_add b. Use 283 adders and supplementary gates.

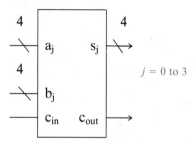

5.22 Use the adder equations for sum_j, and carry out c_j for $j = 0, 1, 2, 3$ to sum $c_{in} = 0$, $a = 1011$, and $b = 0101$. Build a table of values with columns for b_j, a_j, c_j, and sum_j, in that order.

5.23 Use algebra to demonstrate that

$$p_j g_j' \text{ xor } c_{i,j} = b_j \text{ xor } a_j \text{ xor } c_{i,j} = sum_j$$

For $c_{in} = 1$, $a = 1011$, and $b = 0101$, build a table of values with columns for b_j, a_j, p_j, g_j, $p_j g_j'$, c_j, and sum_j, in that order.

5.24 Use the ULC z equation to demonstrate that

$$z_6 = b' \text{ xor } a = b \text{ xor } a'$$

5.25 For $c_{in} = 1$, $a = 1011$, and $b = 0101$, calculate a minus b using 181 s_j code 6. Build a table of values with columns for b_j, a_j, z_{6j}, c_j, and f_j, in that order.

5.26 Synthesize a programmable add/subtract circuit using 86 and 283. The circuit adds when the sub input is False. Sub is a minus b.

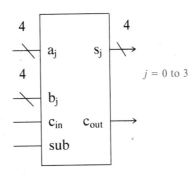

$j = 0$ to 3

5.27 Find $s_j = g_j(w, x, y, z)$ for $j = 0, 1, 2, 3$ (Figure P5.1).

5.28 Find $s_1 = g(w, x, y, z)$ (Figure P5.2).

5.29 Find $s_0 = g(p, q, r, s, t)$ (Figure P5.3).

5.30 Find $f = f(x, y)$ (Figure P5.4).

FIGURE P5.1

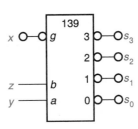

FIGURE P5.2

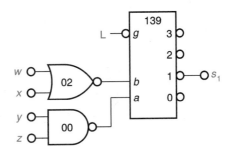

FIGURE P5.3

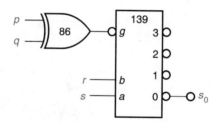

FIGURE P5.4

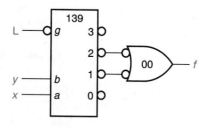

6 SEQUENTIAL CIRCUIT ELEMENTS

In this chapter we adapt our knowledge of combinational-logic circuits and physical gates to create a different type of logic circuit: latches and flip-flops. We create these because we need them to implement sequential-logic designs.

The solution to many problems requires a sequence of events to take place. In this way, time enters logic design. If the solution is available for use on command then it is in storage somewhere. Once stored, the solution can be used over and over again. A circuit that stores information in effect remembers the information.

Combinational-circuit outputs are simply functions of the present inputs. When the inputs change, so do the outputs. We need something new. If the inputs change, remembering requires an output to sustain itself. For this reason, feedback of the output to an input comes into play. We learn that feedback is remembering as we start with one AND gate and one OR gate. We discover this two-gate circuit (latch) has two stable states; that is, it is bistable. An input level becomes active and then inactive as the output goes to H voltage and stays there. Later, another input level becomes active and then inactive as the output goes to L voltage and stays there. And so we learn that latches respond to level changes. The first bistable circuit we study is the RS latch. We learn the defining equation, the related truth table, and the properties of the latch. Then we deal with the bistable latch.

However, latches have their intrinsic problems that make them unusable in many applications. The work that solved these problems gradually transformed the latch circuit into something known as the *edge-triggered flip-flop,* which responds to fast transitions, known as edges, from L to H or vice versa. That work, however, is not used today, so we simply skip over most of it. We study in detail the two types of commercially available flip-flops: the D flip-flop and the JK flip-flop. The T flip-flop is also studied. Then we apply out knowledge to build asynchronous ripple counters using D and JK flip-flops.

INTRODUCTION

In combinational-logic circuits, current outputs are functions of current inputs. There is no remembering or memory of the past. In sequential-logic circuits, there is memory of the past, so that present outputs are functions of the present and past inputs. Another way to put this is that the next state of a sequential circuit depends on both the present state and the present inputs.

"Memory" in a circuit means that the present state values remain constant until a change-to-the-next-state command is given. We need to introduce time as a variable in order to understand what the words *present* and *next* imply. The concept of *state* in a synchronous-state machine means essentially that the circuit will hold present values until the next clock edge occurs. Each clock edge is a change-to-the-next-state command in synchronous-state machines. The concept of *state* in an asynchronous-state machine means essentially that the circuit will hold present values until one or more inputs change. We focus only on synchronous machines at this time.

A modern sequential-circuit design process is set forth in the next chapter. There the business of how one moves from the present state to the next state is resolved. In order to design and build those sequential circuits we need to add to our tool kit the necessary sequential-circuit elements.

Feedback Is Remembering

Remembering requires feedback from output to input.

A combinational circuit acquires memory capability when a circuit output is connected (fed back) to a circuit input. The minimum circuit has one gate, which can be AND (Figure 6.1) or OR (Figure 6.2). The AND circuit does not remember that the input was high, as shown in the timing diagram, because when input h goes low the feedback loop is broken, and it cannot be restored. The AND circuit *does* remember that its input was low. The OR circuit provides a solution for high inputs. The timing diagram shows that q remains high after d goes low, which demonstrates that the loop is not broken. The circuit remembers that d was high in the past. The problem with this simple OR memory circuit is that we cannot reset it to $q = L$. So AND or OR alone are not satisfactory; however, together they are.

FIGURE 6.1
AND gate with feedback

FIGURE 6.2
OR gate with feedback

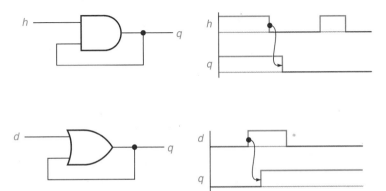

Memory capability:
one OR and one AND
in a feedback loop.

OR provides memory (remembering) in a loop we cannot break, whereas we can break a feedback loop using AND. Let us combine them into one circuit by inserting the AND in the otherwise-unbreakable OR feedback loop (Figure 6.3). This circuit has two inputs, which we call data (d) and hold (h). When h is low, the loop is broken, so that output q follows data: $q = d$. We say the circuit is *transparent;* except for propagation delay, the transparent circuit can be omitted. *Propagation delay* is the time required for input changes to appear at an output. When h is high, the AND-OR circuit is in the hold mode (Figure 6.3) because q remains high after d goes low.

When h goes high, the value of d at that instant is *captured;* that is, the circuit remembers that value. The case when d is low as h goes high is illustrated in Figure 6.3. And later, when d goes high, q is driven high. The circuit can be *set* anytime h is high. When d goes low, the high value of d is remembered because the feedback loop is active. Still later, when h goes low, the feedback loop is broken again and q follows d. The circuit is *reset* because d happens to be low. This elementary circuit has many applications, one of which is the RS latch.

FIGURE 6.3
**AND/OR gates
with feedback**

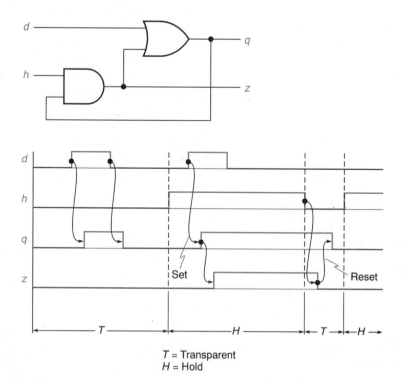

T = Transparent
H = Hold

6.2

RS Latch

The RS (shorthand for "reset/set") latch has two simple forms: cross-coupled 00 gates (Figure 6.4) and cross-coupled 02 gates (Figure 6.5). Observe how mixed logic allows us to redraw the circuits as AND-OR pairs, revealing the function of each gate when we associate the s input with d (for "data") and the r input with h (for "hold") (Figure 6.3). Also note that q and z are complementary outputs. The latch has two stable states: one when $q = H$, and the other when $q = L$. To "prove" this we need only trace levels around the feedback loop. The stable states have the names *set* and *reset*. All of these facts follow from the defining equation.

The R and S input levels reset or set the latch.

In Figure 6.4 or 6.5 observe that

$$q = s + z \quad \text{and} \quad z = r'q$$

Therefore

$$q = s + r'q$$

In this equation q is an input (right side) and an output (left side). What does the equation with q on both sides mean?

Suppose one or more of the input variables r, s change. What is q? One answer uses the "right-now" (present) values of r, s, and q to evaluate $s + r'q$. Then the calculated value is assigned to the output q *after* changes to inputs propagate through the circuit. This new value of q is called q^+ (q-plus). We conclude that the left-side q does not equal the right-side q and that the left-side q is replaced by q^+.

So q^+ is the next state of q—next in the sense that q^+ is the output value when an input changes and after signal propagation times expire. (Signal propagation is created by changes in the inputs r or s.) The next-state equation for q^+ is evaluated for all r, s combinations of values in Figure 6.6. The latch functions listed in the table need to be verified by an H, L voltage analysis.

FIGURE 6.4

RS latch using 00 gates

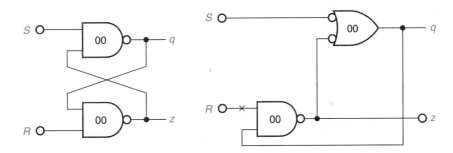

FIGURE 6.5
**RS latch using 02
gates**

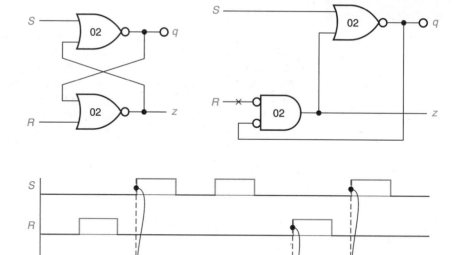

Assumption: Propagation time equals |↔|

FIGURE 6.6
RS latch definition

Defining equation: $q^+ = s + r'q$

If:

$rs = 00$ the next state is q

$rs = 01$ the next state is 1

$rs = 10$ the next state is 0

$rs = 11$ the next state is 1

Excitation table (with q as a table-entered variable):

Row	r	s	q^+	Latch Function
0	0	0	q	hold
1	0	1	1	set
2	1	0	0	reset
3	1	1	1	(do not use)

FIGURE 6.7

**Next-state K map
for the RS latch**

Next-state K map for $q^+ = s + r'q$:

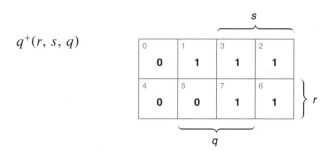

$q^+(r, s, q)$

Consider the 02 form of the RS latch (Figure 6.5). When both r and s are at the inactive L level, then the latch cannot change state. The latch *holds* the present value (and $z = q'$). When s is at the active H level while r is kept at L, the q output is driven active low and z is driven high, so that again $z = q'$. The latch is set. When r is at the active H level while s is kept at L, the z output is driven low and q is driven inactive high, so that once again $z = q'$. The latch is reset. When *both* r and s are at the active H level, then *both* outputs q and z are driven low and z does not equal q'. If *either* r or s falls to L, the complementary relationship of z and q is restored. If *both* r and s fall to low *simultaneously,* the next latch state is not predictable. The next state could be set or reset, or the latch could burst into oscillation. This is why the latch function entry is "do not use." In practice, the final value is determined by the last input to remain active. The r and s input signals provide *excitation* for the latch. The *excitation table* is generated by the defining equation, because it defines the output for various input excitations. The next-state K map for the RS latch is found in Figure 6.7.

State of latch outputs depends upon past levels of the inputs.

Now we modify the circuit to get what we call the data-and-hold form of the RS latch (Figure 6.8). This form is the circuit of standard logic family members 75 and 375.

FIGURE 6.8

**RS latch, DH
format**

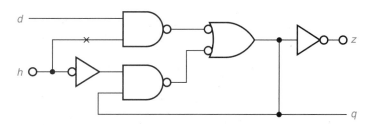

EXERCISE 6.1   Derive the defining equation for the circuit in Figure 6.8. Make a truth table with d and q as table-entered variables.

Answer: $q^+ = hq + h'd$

h	q^+
0	d
1	q

∎

6.3   Bistable Latch

Bistable latch has four modes: hold, set, reset, and toggle.

The RS latch does not have a *toggle function* (that is, the next state becomes the 1 state if the present state is 0, and vice versa). The equation for the toggle function is $q^+ = q'$. The following is one example of a toggle function.

> The caps-lock key on a keyboard is a toggle key. We press it to lock the keyboard in uppercase; we press it again to release to the upper/lowercase mode. Thus, the computer keyboard caps-lock key toggles the keyboard between the two typing modes.

The RS latch equation is $q^+ = s + r'q$. If $rs = 11$, then $q^+ = 1 + 0 \times q = 1$. If we want $q^+ = q'$ when $rs = 11$, then $q^+ = sq' + r'q$. Having made this change we need to verify by evaluating the modified equation that q^+ for the other three rs combinations (00, 01, and 10) has not changed. As shown in Figure 6.9, this is indeed the case. The bistable latch has four functions: hold, set, reset, and toggle. The next-state K map for the bistable latch is given in Figure 6.10.

We did not design the RS latch circuit; we just presented it. Nevertheless, given the defining equation, how *would* we design a bistable latch? The design of any latch circuit requires an asynchronous design

FIGURE 6.9

Bistable latch

Defining equation: $q^+ = sq' + r'q$

Excitation table (with q as a table-entered variable):

Row	r	s	q^+	Latch Function
0	0	0	q	hold
1	0	1	1	set
2	1	0	0	reset
3	1	1	q'	toggle

FIGURE 6.10

FIGURE 6.10
Next-state K map for the bistable latch

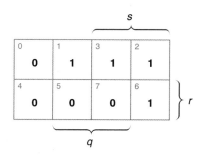

A latch is an asynchronous circuit.

process, which is not covered in detail in this text until Chapter 11. For now we will proceed intuitively.

First we synthesize the bistable q^+ equation as a combinational circuit, assuming there are no feedback connections (Figure 6.11). Then we add feedback lines to provide sources for the required active high q and active low q inputs, creating the bistable latch (Figure 6.12).

6.4 Sequential Circuit Timing

Synchronous circuits require a system clock to define precisely when events can occur. Events can occur once per clock period at a specified point in the clock period known as the *triggering edge*. A synchronous circuit performs reliably if inputs are stable for specified times (setup and hold) before and after every triggering edge. When these conditions are violated, unpredictable events can occur.

6.4.1 Clock

The *clock period* is the time between successive transitions in the same direction in a clock waveform (Figure 6.13a). The *clock frequency* is the number of successive transitions per second in the same direction in the clock waveform. Thus the clock frequency is the

FIGURE 6.11
Bistable circuit without feedback

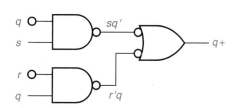

FIGURE 6.12

**Bistable latch, RS
format**

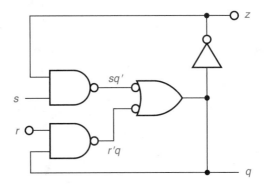

reciprocal of the clock period. A useful abstraction of the clock waveform is a series of up arrows and down arrows representing corresponding clock transitions (Figure 6.13b). In a synchronous system,
only one set of arrows (representing edges) is used to trigger events.
The selected clock edges (up or down but not both) are referred to as
the *active clock edges*.

Clock transitions can occur periodically or aperiodically. Clock
transition rates range from single pulses initiated manually by pressing
a push button to a maximum periodic rate set by hardware limitations
(Figure 6.13c).

6.4.2   Setup Time and Hold Time

Synchronous circuits require input signals to meet *setup time and hold
time* specifications. Failure to meet these specifications usually leads
to uncertain next-state and output results. Reliable results follow
when input variables are stable during setup and hold times.

Any flip-flop is an asynchronous circuit, because it is assembled
from AND and OR gates.* An important assumption in the design of
any flip-flop asynchronous circuit is that only one input changes at a
time. In other words, the circuit must be quiescent before any input
changes. From the point of view of an input, looking backward and
forward in time, all other inputs must be quiescent. For example, if
the d input to the D flip-flop discussed in the next section is changing,
the clock input must be held constant, and vice versa. But for how
long before and after any d input change must the clock input be held

---

* This may surprise you. In effect, setup time and hold time specifications allow us to
think of, and use, edge-triggered flip-flops as if they themselves are synchronous circuits.

FIGURE 6.13
**Clock timing
diagrams**

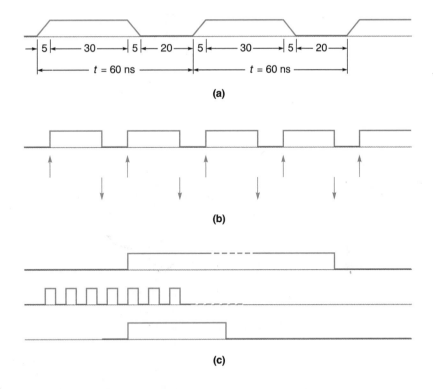

constant? *After* comes into play because an input change must propagate through the flip-flop circuit before another input changes. As it is doing so there can be no interferences from changes in other inputs starting to propagate through the circuit. In data books, *before* and *after* have the names t_{setup} (t_{su}) and t_{hold} (t_h).

The series of periodic up arrows in Figure 6.14 are an abstract representation of the clock waveform L-to-H transitions. The up arrows represent active clock edges. A time window defined by the parameters t_{su} and t_h straddles any active clock edge. A window is illustrated in Figure 6.14.

6.4.3 Synchronous Circuit Timing Diagrams

In a *synchronous sequential circuit* all flip-flops are clocked by the same clock signal. In principle, all flip-flop outputs change some time *after* every active clock edge. Any q waveform transitions occur after the active clock edge. The delay is the time required by the active clock edge to propagate through a flip-flop circuit and do its work. This is the active-clock-edge-to-output-q propagation time.

$f = 1/\tau$

Setup and hold times imply only one input changes at a time.

One clock drives all flip-flops in a synchronous circuit.

Actual setup time is clock period τ minus t_{PHL}.

When a high flip-flop output is switched to low by the active clock edge, the time delay from the active clock edge to the flip-flop output transition is the time to propagate and implement an H-to-L transition, or t_{PHL}. When a low flip-flop output is switched to high by the active clock edge, the time delay from the clock edge to the flip-flop output transition is the time to propagate and implement an L-to-H transition, or t_{PLH}. A realistic (delayed) output waveform is illustrated in Figure 6.15a. Waveforms like these that are synchronized to a clock are known as *synchronous waveforms*. In a sequential circuit, one flip-flop's output is in effect another flip-flop's input (Figure 6.15a). This is why, *within* a sequential circuit, all input and output waveforms are synchronous waveforms. (External inputs may be asynchronous. This complex issue is discussed in Section 6.5.3 and in Chapter 7.)

If we assume that the synchronous waveform in Figure 6.15a is an input to a flip-flop, then we can say that the *actual* hold time for the active clock edge ending clock cycle 1 is the propagation time delay t_{PLH}. And the *actual* hold time for the active clock edge ending clock cycle 2 is the propagation time delay t_{PHL}. The *actual* setup time in clock cycle 2 is the time interval from the L-to-H input waveform edge to the active clock edge ending cycle 2. Actual setup and hold times must exceed the flip-flop setup time and hold time requirements. The unavoidable propagation times t_{PHL} and t_{PLH} allow the system to work by providing holding time. A zero t_{PHL} or t_{PLH} provides zero hold time, which is unsatisfactory for most circuits.

When the clock frequency is increased, the period decreases until the period equals t_{setup} plus t_{hold}. This minimum possible clock period is illustrated in Figure 6.15b.

6.4.4   Metastability

Metastability is important to study because it can yield performance that is unreliable. A *metastable* state is a peculiar state of pseudoequilibrium in which the energy content of the state is either more than or less than that required for a stable state.

FIGURE 6.14
Setup and hold

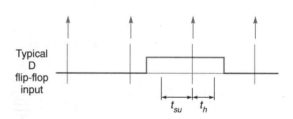

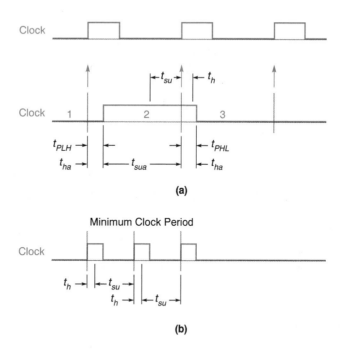

FIGURE 6.15

Synchronous circuit timing diagram

The RS latch assembled with 00 gates is set when the s input is asserted by falling to the L level. Suppose the s input returns to the H level after time T has elapsed. The s waveform is now a *pulse* of duration T. The *energy* E in the pulse is essentially the $dV = H - L$ voltage differential multiplied by the time T. As the pulse duration T decreases, the energy E decreases. At some point (T_1) the energy in the s pulse is insufficient to set the latch.

If the duration T is increased from below T_1, then the energy in the pulse increases; and at some point T_2 (greater than T_1) and beyond, the s pulse will set the latch. Pulses with duration in the range T_1 to T_2 may put the latch in a metastable state. A setup time specification guarantees prior-to-the-clock-edge input pulse durations greater than T_2. This is why asynchronous inputs must be synchronized.

The possibility of a latch's entering a metastable state is revealed by superposition of two inverter transfer functions. In effect, two cascaded inverters represent any latch's feedback loop. The inverter transfer function for inverters 1 and 2 is illustrated in Figure 6.16a. Two cascaded inverters establish relationships between their inputs and outputs: $V_{in1} = V_{out2}$, $V_{in2} = V_{out1}$. This is why their transfer functions may be superimposed, as shown in Figure 6.16b. The two

Asynchronous inputs can induce metastability in a flip-flop.

transfer functions intersect at three points: points 1 and 3 represent stable states set and reset, point 2 represents the metastable state. The intrinsic instability of point 2 is illustrated in Figure 6.16b by a slight perturbation of V_{in2}, which "jumps" the latch to stable point 3 (reset).

Once in a metastable state, the output can "oscillate" for an indefinite time, or suddenly switch after an indefinite time. The output becomes unpredictable. These events can generate what the uninformed perceive as mysterious random failures in a digital system. This is why, in practice, metastability is a source of disasters.

Synchronize all asynchronous inputs.

Two cascaded flip-flops can synchronize an asynchronous input if the clock period is sufficiently long in duration (Figure 6.16c). The parameter *metastable resolution time* (t_{mr}) is the "maximum" time a flip-flop will remain in a metastable state. If a metastable state is

FIGURE 6.16
Stable and metastable states

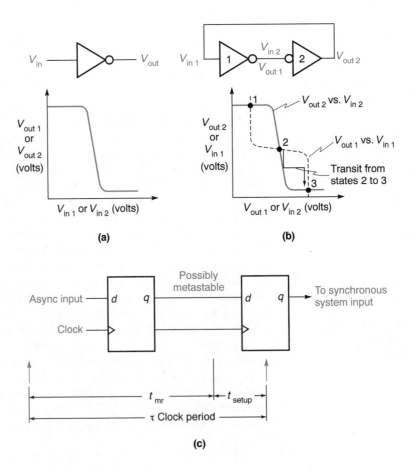

(a)

(b)

(c)

resolved in less than t_{mr} seconds, then a sufficiently long clock period is $t_{meta} + t_{setup}$. For more information on this complex subject, consult Chaney 1983.

6.5

Edge-Triggered Flip-Flops: General Comments

Let us assume that the bistable latch h_L input (see Figure 6.8) is a clock. When the clock is low, the latch holds the present value of q. The latch is in the *hold mode* when the clock is low. When the clock is high, the latch is in the *transparent mode: q* will follow changes in d, because the feedback loop is broken. The circuit is transparent in the sense that the output follows the input when the clock is high. For q not to change while the clock is high, input d must be held constant. Furthermore, any glitches (see Section 5.1.3, page 227) at the d input while the clock is high will be captured by q according to the glitch time duration and direction. The time constraint on d and the vulnerability to glitches are undesirable. This is why latches are replaced in many applications by edge-triggered flip-flops, which capture the value of d when the clock transitions from H to L (or vice versa). The vulnerability to glitches decreases enormously, and d need only have the correct value in a suitable time window that includes the transition.

Any flip-flop has two states, known as set and reset, and the symbol for the output is usually labeled q. The output q reports the state of the flip-flop, and so q is called a *state variable*. Clocked flip-flops are driven by a clock, which is a periodic digital waveform of frequency f and period $t = 1/f$. A 25-megahertz (MHz) waveform has a period of 40 nanoseconds (ns). A digital waveform has only levels H and L. Waveform transitions from H to L or L to H are called *waveform edges* or simply *edges*.

Edge-triggered flip-flops change state only at active clock edges.

Edge-triggered flip-flops are set and reset by either the negative-going (H to L) or the positive-going (L to H) clock waveform edges, but not both. Hence a clocked flip-flop is triggered only once per clock cycle. A 25-MHz clock triggers a flip-flop once per 40 ns. *Every* triggering edge (positive- or negative-going but not both) stores logical 0 or 1 in the flip-flop according to rules derived from the flip-flop defining equation. So a flip-flop state is changed only once per clock period, and the state is constant during each clock period. Edge-triggered flip-flops do not have transparent modes. This is basically why they are different from latches. The present inputs and the present state in the current clock period determine the next state in the next clock period. Finally, we note that flip-flop names are taken from their input terminals, such as D, JK, and T.

Edge-Triggered D Flip-Flop

The edge-triggered D flip-flop has one input called d (for "data") and another input called *clock*. Some D flip-flops have one output line q. Others have two output lines for q that are voltage complements; if one is H, the other is L, and vice versa (Figure 6.17). The key fact to know about a flip-flop is the defining equation: everything is derived from that equation. Flip-flop inputs provide *excitation* for the flip-flop at each clock edge. The defining equation $q^+ = d$ is the source of the flip-flop excitation table in Figure 6.17.

In practice, flip-flops have additional (asynchronous) inputs that preset and clear the flip-flop. Preset (PRE) and clear (CLR) are aliases for set and reset. The commercial function (truth) table for the 74 includes these inputs and the clock, as shown in Figure 6.18. The bar over PRE, CLR, and one of the q symbols does not make these the complement operator. In data books the bar means the variable is given the active low ($--$) assignment. (Data books make extensive use of mixed logic.) We do not use the bar in our circuit diagrams. The bubble at the corresponding input replaces the bar. This statement is verified in the first line of the table, which shows that the 74 is set to ($q =$) H when PRE is active low. The second line shows that an active low CLR resets the flip-flop to ($q =$) L. The third line reveals the indefinite RS latch behavior (see Figure 6.6) when both PRE and CLR are active low. The dashes mean "don't care." The clock (CLK) and the D input are disabled when either PRE or CLR is active. In the next two lines, with PRE and CLR inactive, the up arrow in the CLK column means the $q^+ = d$ equation is executed when the clock transitions from L to H (Q is q^+ in this table).

The equation $q^+ = d$ implies that q is a copy of the d waveform delayed by one clock period (Figure 6.19a). (A one-clock-period slip

D flip-flop defining equation is q$^+$ = d.

FIGURE 6.17
D flip-flop (74LS74)

Defining equation: $q^+ = d$

Excitation table:

d	q^+	Function
0	0	reset
1	1	set

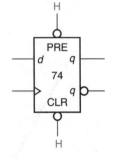

FIGURE 6.18
D flip-flop commercial function table

	INPUTS			OUTPUTS		
$\overline{PRE}$	$\overline{CLR}$	CLK	D	Q	$\overline{Q}$	
L	H	–	–	H	L	
H	L	–	–	L	H	
L	L	–	–	H[1]	H[1]	
H	H	↑	H	H	L	set
H	H	↑	L	L	H	reset
H	H	L	–	Q_0	$\overline{Q_0}$	hold

[1] This configuration is not stable.

D flip-flop output follows input with a one clock period delay.

assumes the d input is synchronous with the clock.) Observe that there is no q^+ waveform, because in the next state q^+ becomes the present-state q. The value of d in clock period n determines the value of q in the next clock period, $n + 1$. Therefore, during clock period n, q^+ is shorthand for the value of q in clock period $n + 1$. The next-state q^+ is determined by the value of the d input at the time the L-to-H clock edge occurs. (The clock edge executes the q^+ equation.) The coincidence of these values and the clock edge are marked by circles in Figure 6.19a. They are the cause, and the arrows show the effect. (Some flip-flops are designed to have the negative-going H-to-L edge trigger the flip-flop.)

If t_{PHL} and t_{PLH} are zero, then the d and q waveform transitions are coincident with the clock edge (Figure 6.19b). The equation $q^+ = d$ in effect asks the question "What is the value of d at the clock edge?" Suppose d transitions from L to H coincident with the clock edge. What value does the circuit use when executing the q^+ equation? Is d low, or is it high? When propagation delays equal zero there is no definite answer. Showing the d and q transitions coincident with the clock edge has no meaning. This is why nonzero propagation delays are necessary for synchronous operation.

EXAMPLE 6.1   **Divide-by-2 Circuit**

One application of any flip-flop is dividing by 2, or toggling. To divide by 2 means to divide the clock frequency by 2, thereby doubling the clock period. Here is one way to deduce what connections must be

FIGURE 6.19
**D flip-flop *d* and *q*
waveforms**

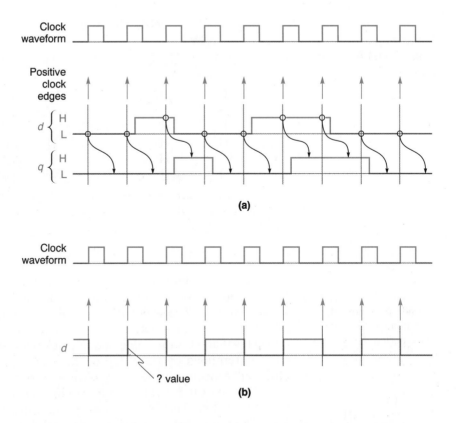

made so that a D flip-flop divides by 2. We start by drawing the desired digital waveform for q with a period twice the clock period. Given q and the defining equation $q^+ = d$, we deduce that the d waveform must be the same q waveform slipped back one clock period. (A one-clock-period slip assumes the d input is synchronous with the clock.) This conclusion is verified by inspection of each clock period (as a present state) of the d and q waveforms shown below. In each clock period the waveforms show that $d = q'$. Substituting for d, the defining equation

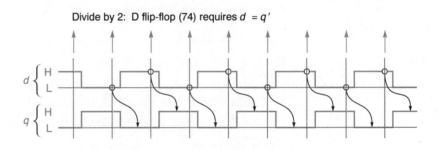

Divide by 2: D flip-flop (74) requires $d = q'$

becomes $q^+ = q'$. That is, q-next is the complement of q-present, which is what we want.

EXAMPLE 6.2 Divide by 2: Calculating the Minimum Clock Period

The minimum clock period for reliable division by 2 is derived from the 74LS74 switching characteristics and from the clock, setup, and hold specifications. A drawing of the clock and q waveforms starts the process.

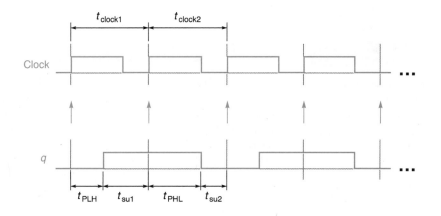

From the data book:

$t_{PLH} = 25\,\text{ns max},\quad t_{PHL} = 40\,\text{ns max},\quad t_{su} = 20\,\text{ns min},\quad t_{h} = 5\,\text{ns min}$

From the waveforms:

$$t_{clock1} = t_{PLH} + t_{su1} \qquad \text{or} \qquad t_{clock2} = t_{PHL} + t_{su2}$$
$$= 25 + 20 \qquad\qquad\qquad = 40 + 20$$
$$= 45 \qquad\qquad\qquad\qquad = 60$$

Therefore t_{clock} must be greater than or equal to 60 ns in a divide-by-2 circuit.

Check:

$$t_{clock\text{-}min} = 60\text{ ns} \Rightarrow f_{max} = 16.667\text{ MHz}$$

FIGURE 6.20
JK flip-flop

Defining equation: $q^+ = jq' + k'q$

Excitation table:

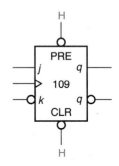

				← Assignments for 73, 112
	+ +	+ +		← Assignments for 109
	+ +	− −		
Row	j	k	q^+	Function
0	0	0	q	hold
1	0	1	0	reset
2	1	0	1	set
3	1	1	q'	toggle

6.7 Edge-Triggered JK Flip-Flop

JK flip-flop defining
equation is
$q^+ = jq' + k'q$.

The JK flip-flop has two data inputs called j and k and a third input called *clock* (Figure 6.20). Some JK flip-flops have one output line q. Others have two output lines for q that are voltage complements; if one is H the other is L, and vice versa. The 109 is our preferred version of the JK flip-flop because input circuits are simplified in many applications when the active levels of j and k are different.

EXAMPLE 6.3 Divide by 2 with the JK Flip-Flop

The divide-by-2 equation is $q^+ = q'$. This is derived from the JK defining equation, $q^+ = jq' + k'q$, by setting $j = k = 1$ so that

$$q^+ = 1 \times q' + 1' \times q = q'$$

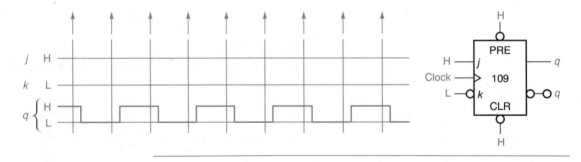

The JK truth table (Figure 6.21) is rearranged to emphasize other JK properties, which are applied in Chapter 8. When $q = 0$, the k input is disabled:

$$q^+ = j0' + k'0 = j1 + k'0 = j$$

When $q = 1$, the j input is disabled:

$$q^+ = j1' + k'1 = j0 + k'1 = k'$$

EXERCISE 6.2

Let $x = q_0'$ in the circuit in Figure P6.2 (page 326). Make a present-state, next-state (PS, NS) truth table where $q_1 q_0$ is the present state and $q_1^+ q_0^+$ is the next state. Derive the in-circuit defining equations.

Answer:

q_1	q_0	q_1^+	q_0^+		
0	0	0	1	$q_1^+ = q_0$	$q_0^+ = q_0'$
0	1	1	0		
1	0	0	1		
1	1	1	0		∎

EXERCISE 6.3

Refer to the circuit in Figure P6.8 (p. 327). Make a present-state, next-state (PS, NS) truth table where $q_1 q_0$ is the present state and $q_1^+ q_0^+$ is the next state. Derive the in-circuit defining equations.

Answer:

q_1	q_0	q_1^+	q_0^+		
0	0	0	1	$q_1^+ = q_1'$ xor q_0	$q_0^+ = q_0'$
0	1	1	0		
1	0	1	1		
1	1	0	0		∎

FIGURE 6.21
JK flip-flop revisited

Row	j	k	q	q^+	Function
0	0	0	0	0	$q^+ = j$
2	0	1	0	0	
4	1	0	0	1	
6	1	1	0	1	
1	0	0	1	1	$q^+ = k'$
3	0	1	1	0	
5	1	0	1	1	
7	1	1	1	0	

FIGURE 6.22
T flip-flop

Defining equation: $q^+ = tq' + t'q = t \text{ xor } q$

Excitation table:

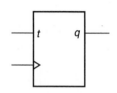

Row	t	q	q^+	Function
0	0	0	0	hold value of q when t is False
1	0	1	1	$q^+ = q$
2	1	0	1	toggle q when t is True
3	1	1	0	$q^+ = q'$

With q as a table-entered variable:

Row	t	q	q^+	Function
0	0	0	q	hold
1	1	1	q'	toggle

6.8 Edge-Triggered T Flip-Flop

The T flip-flop is the special case of the JK flip-flop where $t = j = k$ (Figure 6.22). There is no T flip-flop in the standard logic family. T stands for "toggle."

EXAMPLE 6.4   Divide-by-Two with the T Flip-Flop

The divide-by-2 equation, $q^+ = q'$, is derived from the T defining equation, $q^+ = tq' + t'q$, by setting $t = 1$.

$$q^+ = 1 \text{ xor } q = q' \quad \text{Toggle mode when } t = 1$$

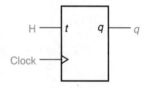

6.9 Converting Flip-Flops

The standard logic family includes only edge-triggered D and JK flip-flops, because the D and JK are readily converted to other types. The conversions are guided by the defining equations.

Converting the JK flip-flop equation to the D flip-flop equation:

$$q^+ = jq' + k'q \qquad \text{and} \qquad q^+ = d = d(q' + q) = dq' + dq$$

imply

$$jq' + k'q = dq' + dq$$

Equating coefficients of q and q' we get $j = d$ and $k' = d$ (Figure 6.23a).

Converting the JK flip-flop equation to the T flip-flop equation:

$$q^+ = jq' + k'q \qquad \text{and} \qquad g^+ = tq' + t'q$$

imply

$$jq' + k'q = tq' + t'q$$

Equating coefficients of q and q' we get $j = t$ and $k' = t'$. Therefore $t = j = k$ (Figure 6.23b).

Converting the D flip-flop equation to the JK flip-flop equation:

$$q^+ = d \qquad \text{and} \qquad q^+ = jq' + k'q$$

imply

$$d = jq' + k'q$$

A conventional AND/OR circuit implements this equation for d. The AND/OR circuit is the circuit of a two-to-one multiplexer with q as the multiplexing variable (Figure 6.24a). Also, note that the d equation is in the form of a defining equation for a two-to-one mux with $a_0 = j$ and $a_1 = k'$.

FIGURE 6.23
Converting a JK flip-flop to T and D flip-flops

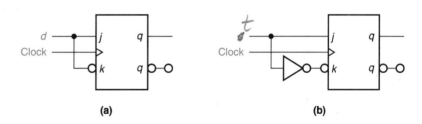

(a)   (b)

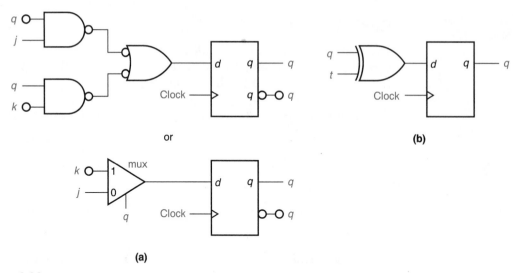

FIGURE 6.24
Converting a D flip-flop to JK and T flip-flops

Converting the D flip-flop equation to the T flip-flop equation:

$$q^+ = d \quad \text{and} \quad q^+ = tq' + t'q$$

imply

$$d = tq' + t'q$$

$$d = t \text{ xor } q$$

Here d is the xor output (Figure 6.24b).

6.10   Asynchronous Ripple Counters

Although synchronous counters (see Chapter 8) are far superior in performance, you should become aware of ripple counters.

A *digital counter* is a group of flip-flops wired to count clock edges, and to store in the flip-flops the number of clock edges counted. The counter number can be wired to be in any radix. We, however, are interested in the case in which the number is a radix-2 binary number. How do we wire a group of flip-flops to count up or down in binary? Chapter 7 presents a general method for answering such questions. At this point, however, let us proceed intuitively.

First we write down a sequence of binary numbers:

q_{3210}	Count by 1's		Count by 0's	
0000	0		15 (decimal)	
0001	1		14	
0010	2		13	
0011	3		12	
0100	4	up	11	down
0101	5	↓	10	↓
0110	6		9	
0111	7		8	
1000	8		7	
1001	9		6	
1010	10		5	
1011	11		4	
1100	12		3	
1101	13		2	
1110	14		1	
1111	15 (decimal)		0	

Count increments when a variable changes from 1 to 0.

The list assigns $q_3 q_2 q_1 q_0$ to be active high variables to count by 1's. The count increments by one as each clock edge is counted. The counter counts clock edges. However, if we use active low variables, in effect we focus on the zeros, and the count decrements by one as each clock edge is counted. Observe that the least significant digit q_0 toggles as clock edges are counted. Also, each time q_0 changes from 1 to 0, digit q_1 changes state (0 to 1 or 1 to 0). Furthermore, q_0's period is twice the clock period, and q_1's period is twice that of q_0. Observe also that each time q_1 changes from 1 to 0, digit q_2 changes state (0 to 1 or 1 to 0). Furthermore, q_2's period is twice q_1's period, just as q_1's period is twice that of q_0. Finally, observe that q_3's period is twice q_2's period.

So we say that

clock divided by 2 produces q_0

q_0 divided by 2 produces q_1

q_1 divided by 2 produces q_2

q_2 divided by 2 produces q_3

and we conclude that one way to wire a group of flip-flops as a binary

FIGURE 6.25
Ripple counters

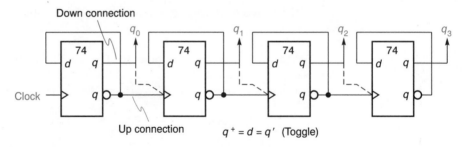

$$q^+ = d = q'\ \text{(Toggle)}$$

(a) *D* flip-flop ripple counter (asynchronous)

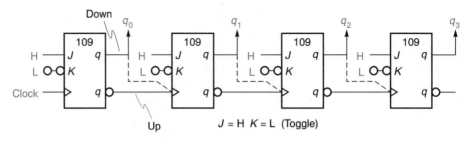

$$J = \text{H}\quad K = \text{L}\ \text{(Toggle)}$$

(b) *JK* flip-flop ripple counter (asynchronous)

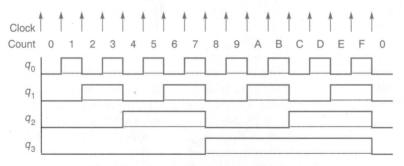

(c) Up waveforms

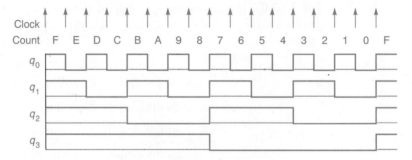

(d) Down waveforms

counter is to use a cascade of divide-by-2 circuits. The cascade of dividers may be wired to count up or down. A cascade may use D flip-flops (Fig. 6.25a) or JK flip-flops (Figure 6.25b). The cascade of dividers is called a ripple counter because higher-order digits cannot change state until the lower-order digits change state and the change of state propagates through the cascade. Now, are we counting by 1's or by 0's?

The q_{j+1} waveforms (Figure 6.25c) for an up counter change state when the q_j waveforms transition from H to L, but the D flip-flops trigger on the L-to-H transitions. The active low q_j flip-flop outputs transition from L to H when the active high q_j outputs transition from H to L. Thus the active low q of any bit should be the clock input to the next bit. The input clock is not changed.

Count decrements when a variable changes from 0 to 1.

Down-counter waveforms (Figure 6.25d) change state when the q_j waveforms go from L to H. Thus the active high q output of any bit should be the clock input to the next bit. Figures 6.25a and 6.25b show the up and down ripple counter connections for D and JK ripple counters, respectively.

SUMMARY

Latch

Defining equations: RS $q^+ = s + r'q$

Bistable $q^+ = sq' + r'q$

Timing

Any flip-flop has two states, known as set and reset; the symbol for the output is usually labeled q. Since the output q reports the state of the flip-flop, q is called a state variable.

Synchronous circuits require input signals to meet setup time and hold time specifications. A time window defined by the parameters t_{setup} and t_{hold} straddles any clock edge. A window is illustrated in Figure 6.14. Actual setup and hold times must exceed the flip-flop setup and hold time requirements.

In a synchronous sequential circuit all flip-flops are clocked by the same clock signal. In principle, all flip-flop outputs change some time after every clock edge. All q waveform transitions occur after the clock edge. The delay is the time required by the clock edge to propagate through a flip-flop circuit and do its work. This is the clock-edge-to-output-q propagation time. The unavoidable propagation times t_{PHL} and t_{PLH} allow the system to work by providing holding time. A zero t_{PHL} or t_{PLH} provides zero hold time, which is unsatisfactory for most circuits.

Edge-Triggered Flip-flops

The defining equation is the next-state equation. The defining equation calculates the next state given both the present state and the inputs.

Defining equations:

$$D \quad q^+ = d$$

$$JK \quad q^+ = jq' + k'q$$

$$T \quad q^+ = tq' + t'q = t \text{ xor } q$$

Excitation tables:

D

Row	d	q^+	Function
0	0	0	reset
1	1	1	set

JK

Row	j	k	q^+	Function
0	0	0	q	hold
1	0	1	0	reset
2	1	0	1	set
3	1	1	q'	toggle

T

Row	t	q^+	Function
0	0	q	hold
1	1	q'	toggle

Asynchronous Ripple Counters

A digital counter is a group of flip-flops wired to count rising or falling clock edges. The counter's purpose is to store the number of clock edges counted. The counter number can be wired to be in any radix.

Ripple counters exist because their circuits are simple. This rationale no longer justifies designing them, however, because synchronous counter design is now very straightforward (see Chapter 7) and because synchronous counter performance is far superior.

REFERENCES

Breeding, K. J. 1989. *Digital Design Fundamentals*. Englewood Cliffs, N.J.: Prentice-Hall.

Fletcher, W. L. 1979. *An Engineering Approach to Digital Design*. Englewood Cliffs, N.J.: Prentice-Hall.

Mano, M. M. 1991. *Digital Design*, 2nd ed. Englewood Cliffs, N.J.: Prentice-Hall.

Roth, C. H. 1992. *Fundamentals of Logic Design*, 4th ed. St. Paul, Minn.: West.

Shiva, S. G. 1988. *Introduction to Logic Design*. Glenview, Ill.: Scott, Foresman.

Wakerly, J. F. 1990. *Digital Design Principles and Practices*. Englewood Cliffs, N.J.: Prentice-Hall.

Metastability:

Brueninger, R., and K. Frank. *Metastable Characteristics of Texas Instruments Advanced Bipolar Logic Families*. Texas Instruments Pub No. SDAA004.

Chaney, T. C. 1979. Comments on "A note on synchronizer or interlock maloperation." *IEEE Transactions on Computers* C-28(10) (Oct.): 802.

Chaney, T. C. 1983. Measured Flip-Flop Responses To Marginal Triggering. *IEEE Transactions on Computers* C-32(12) (Dec.): 1207–1209.

Kleeman, L., and A. Cantoni. 1986. Can Redundancy and Masking Improve the Performance of Synchronizers? *IEEE Transactions on Computers* C-35(7) (July): 643–646.

Kleeman, L., and A. Cantoni. 1987. On the Unavoidability of Metastable Behavior in Digital Systems. *IEEE Transactions on Computers* C-36(1) (Jan.): 109–112.

Marino, L. R. 1977. The Effect of Asynchronous Inputs on Sequential Network Reliability. *IEEE Transactions on Computers* C-26(11) (Nov.): 1082.

Name _____   SID# _____   Section _____
     (last)   *(initials)*

Approved by _____   Date _____

Grade on report _____

Counter Design with D and JK Flip-Flops

1. Design and build two divide-by-2 circuits.

_____ D flip-flop circuit documents
_____ JK flip-flop circuit documents

2. Use the truth table signal generator to test the circuit.

_____ D flip-flop input/output timing diagram
_____ JK flip-flop input/output timing diagram
_____ D flip-flop circuit t_{PHL}, t_{PLH}, t_{rise}, t_{fall} data
_____ JK flip-flop circuit t_{PHL}, t_{PLH}, t_{rise}, t_{fall} data

3. Design and build a one-clock-period-delay circuit.

_____ circuit documents

4. Use the truth table signal generator to test the circuit.

_____ input/output timing diagram

5. Design and build a three-bit q_0, q_1, q_2 JK flip-flop ripple down counter with a count control input p.

_____ circuit documents

6. Use the truth table signal generator to test the circuit.

_____ input/output timing diagram
_____ q_0
_____ q_1 and delay from q_0
_____ q_2 and delay from q_1
_____ q_3 and delay from q_3
_____ circuit change to up counter
_____ circuit change to reset counter

1. Design and build two divide-by-2 circuits: one using a D flip-flop and the other using a JK flip-flop. Use 74LS74 and 74LS109.

2. Connect signal generator output clk to the clock inputs. Plot the triggering waveform, clk, and two cycles of each output waveform. Measure t_{PHL}, t_{PLH}, t_{rise}, and t_{fall} for each circuit.

3. Build a one-clock-period-delay circuit using a D flip-flop. Use 74LS74.

4. Connect signal generator outputs R_{co} to the D input and clk to the clock input. Plot the D flip-flop's input D and output Q waveforms.

5. Design and build a three-bit q_0, q_1, q_2 JK flip-flop ripple down counter with a count control input p. The counter counts when p is True, and holds the present count value when p is False. Do not gate the clock. Use 74LS109 and glue gates as required.

6. Connect signal generator outputs R_{co} to the scope trigger input and clk to the clock input. Plot q_0, q_1, and q_2. Show the ripple delays. What circuit design changes do you make to convert this to an up counter? What circuit additions/changes do you make to reset the counter.

PROBLEMS

6.1 Apply the *d* and *h* (inverted) waveforms of Figure 6.3 to the *s* and *r* bistable latch inputs (Figure 6.12). Plot the *q* and *z* outputs.

6.2 Use the RS latch in Figure 6.4. Plot *q* for the s_L, r_L inputs. Mark each interval between events as *hold, clear, set,* or *toggle*.

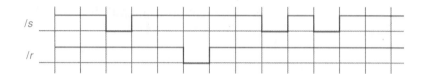

6.3 The circuit in Figure P6.1 is a switch debouncer. Redraw the circuit by replacing the 00 box in P6.1 with the circuit in Figure 6.4. The s_L, r_L waveforms imply that the switch arm is either down, in transit, or up. The switch arm bounces when it hits the destination contact. A consequence is the multiple s_L or r_L input "pulses." Mark each interval between events as *down, in transit,* or *up*. Plot *q* for the given s_L, r_L input waveforms. Explain why *q* does not bounce.

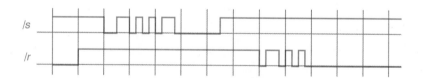

6.4 Using the 74LS109 JK, plot *q*. Mark each clock period as hold, clear, set, or toggle.

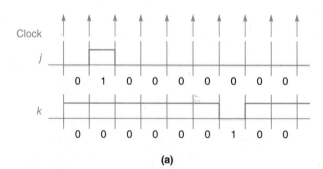

(a)

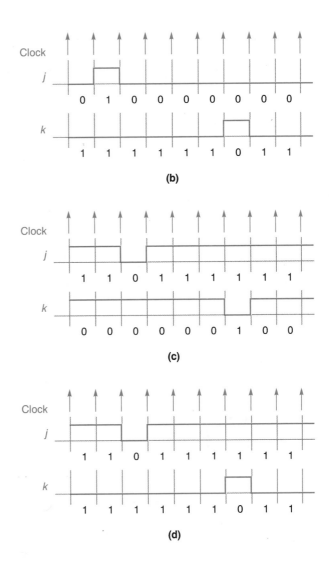

(b)

(c)

(d)

6.5 Plot q_0, q_1 for the x, y inputs to the circuit in Figure P6.2.

6.6 Plot q_0, q_1 for the x, y inputs to the circuit in Figure P6.3.

6.7 Starting with all flip-flop outputs at the L level, plot q_0, q_1, q_2 over nine clock periods for the circuit in Figure P6.4.

6.8 Starting with all flip-flop outputs at the H level, plot q_0, d_1, q_1, q_2 over nine clock periods for the circuit in Figure P6.5.

FIGURE P6.1

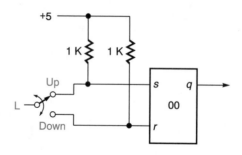

FIGURE P6.2

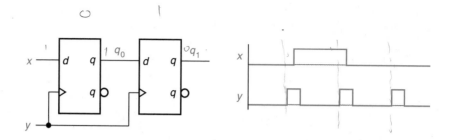

FIGURE P6.3

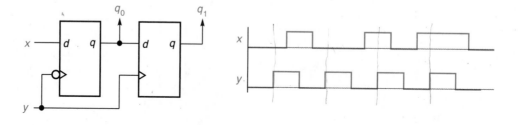

FIGURE P6.4

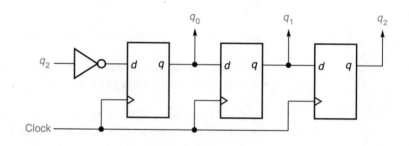

FIGURE P6.5

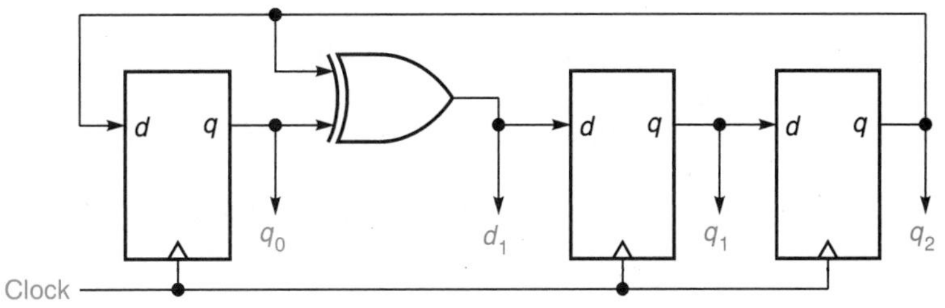

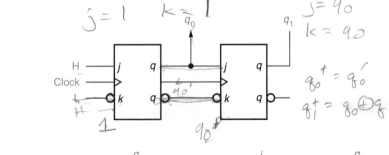

FIGURE P6.6

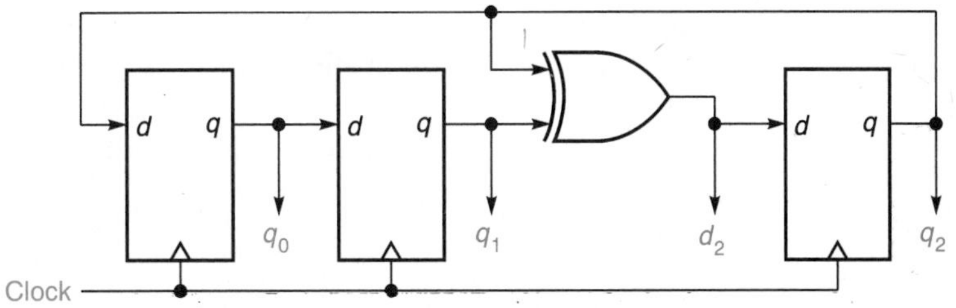

FIGURE P6.7

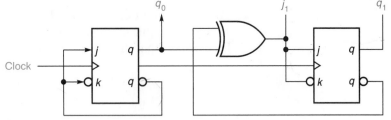

FIGURE P6.8

6.9 Starting with all flip-flop outputs at the H level, plot q_0, q_1, d_2, q_2 over nine clock periods for the circuit in Figure P6.6.

6.10 Starting with all flip-flop outputs at the L level, plot q_0, q_1 over six clock periods for the circuit in Figure P6.7.

6.11 Starting with all flip-flop outputs at the L level, plot q_0, j_1, q_1 over six clock periods for the circuit in Figure P6.8.

6.12 Starting with all flip-flop outputs at the L level, plot q_0, q_1, j_2, k_2, q_2 over nine clock periods for the circuit in Figure P6.9.

6.13 Can a T flip-flop be converted into a D flip-flop?

6.14 The next-state equations for a set of logic functions are as follows:

$$q_0^+ = 1 \qquad q_1^+ = q_0 \qquad q_2^+ = q_1 \qquad q_3^+ = q_2$$

(a) Design input circuits for D flip-flops.
(b) Design input circuits for JK flip-flops.

6.15 The next-state equations for a set of logic functions are as follows:

$$q_0^+ = q_0 \operatorname{xor} 1 \qquad q_1^+ = q_1 \operatorname{xor} q_0 \qquad q_2^+ = q_2 \operatorname{xor} q_1 q_0 \qquad q_3^+ = q_3 \operatorname{xor} q_2 q_1 q_0$$

(a) Design input circuits for D flip-flops.
(b) Design input circuits with one gate delay for JK flip-flops.
(c) Design input circuits with one gate delay for T flip-flops.

6.16 The truth table for a logic function is as follows:

q	q^+
0	1
1	0

(a) Design an input circuit for a D flip-flop.
(b) Design an input circuit for a JK flip-flop.
(c) Design an input circuit for a T flip-flop.

FIGURE P6.9

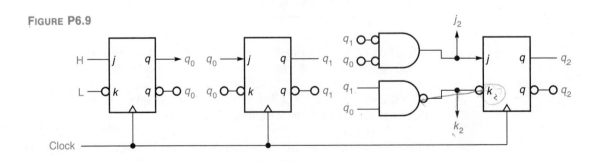

6.17 The truth table for a logic function is as follows:

q_1	q_0	q_1^+	q_0^+
0	0	0	1
0	1	1	0
1	0	1	1
1	1	0	0

(a) Design input circuits for D flip-flops.
(b) Design input circuits with one gate delay for JK flip-flops.
(c) Design input circuits with one gate delay for T flip-flops.

6.18 The truth table for a logic function is as follows:

q_1	q_0	q_1^+	q_0^+
0	0	1	1
0	1	0	0
1	0	0	1
1	1	1	0

(a) Design input circuits for D flip-flops.
(b) Design input circuits with one gate delay for JK flip-flops.
(c) Design input circuits with one gate delay for T flip-flops.

6.19 Derive the equations for flip-flop parameters d_0, d_1, q_0^+, q_1^+ of the circuit in Figure P6.2. List all present states defined as $q_1 q_0$, and all corresponding next states $q_1^+ q_0^+$.

6.20 Derive the equations for flip-flop parameters d_0, d_1, d_2, q_0^+, q_1^+, q_2^+ of the circuit in Figure P6.4. List all present states defined as $q_2 q_1 q_0$, and all corresponding next states $q_2^+ q_1^+ q_0^+$.

6.21 Derive the equations for flip-flop parameters d_0, d_1, d_2, q_0^+, q_1^+, q_2^+ of the circuit in Figure P6.5. List all present states defined as $q_2 q_1 q_0$, and calculate the corresponding next states $q_2^+ q_1^+ q_0^+$.

6.22 Derive the equations for flip-flop parameters d_0, d_1, d_2, q_0^+, q_1^+, q_2^+ of the circuit in Figure P6.6. List all present states defined as $q_2 q_1 q_0$, and calculate the corresponding next states $q_2^+ q_1^+ q_0^+$.

6.23 Derive the equations for flip-flop parameters j_0, k_0, j_1, k_1, q_0^+, q_1^+ of the circuit in Figure P6.7. List all present states defined as $q_1 q_0$, and calculate the corresponding next states $q_1^+ q_0^+$.

6.24 Derive the equations for flip-flop parameters j_0, k_0, j_1, k_1, q_0^+, q_1^+ of the circuit in Figure P6.8. List all present states defined as $q_1 q_0$, and calculate the corresponding next states $q_1^+ q_0^+$.

6.25 Derive the equations for flip-flop parameters j_0, k_0, j_1, k_1, j_2, k_2, q_0^+, q_1^+, q_2^+ of the circuit in Figure P6.9. List all present states defined as $q_2 q_1 q_0$, and calculate the corresponding next states $q_2^+ q_1^+ q_0^+$.

7 ALGORITHMIC STATE MACHINES

OVERVIEW

Now that we know how flip-flops work, we turn to the major object of digital design—the state machine. The state machine is found everywhere, because all controller implementations are state machines and every nontrivial application requires a controller. Our goal is to learn the modern algorithmic state machine (ASM) method after a brief look at the traditional state diagram method.

We start by discussing the style of the ASM method, because it implements top-down design, then discuss the temptations offered by the risky bottom-up design style. Next we present the basic concepts of state machines: state variable, state, state assignment, branch, unconditional output, conditional output, asynchronous machine, and synchronous machine. We discuss traditional state machine diagrams, notation, and methods for analysis and synthesis. We cover the Moore state machine, which has only unconditional outputs, as well as the Mealy state machine, which has only conditional outputs. The generality is the combined Mealy and Moore circuit, which allows for an easy transition to modern ASM notation and methods.

Next we present the vocabulary and notation of algorithmic state machines. Using this notation, we assemble ASM charts. Given such a chart and timing waveforms for input variables, we determine the sequence(s) of states and the corresponding outputs.

Given an algorithm, we learn how to synthesize a circuit implementing that algorithm. A direct synthesis method always yields a circuit that works. Another method analyzes a circuit to produce the ASM chart representing the circuit. This method is also direct. These methods allow us to analyze and design state machines of any complexity. Specific worked examples provide experience with the methods.

The chapter closes with discussions of control loops in ASM charts and the very important linked ASM charts.

INTRODUCTION

What we are about to discuss are *synchronous state machines,* machines that change state only when a clock edge occurs. Machines without clocks are known as *asynchronous state machines;* they change state when an input changes state. The RS latch is an asynchronous state machine, and so are D and JK flip-flops, in the sense that they are designed by asynchronous methods. (In practice, however, the flip-flops are used and thought of as synchronous devices.) Synthesis and analysis methods for asynchronous machines are totally different from methods for synchronous machines. Chapter 11 is devoted to asynchronous state machines.

The algorithmic state machine (ASM) method uses straightforward procedures to create reliable sequential digital designs known as *sequential state machines*. By the ASM method, we can develop designs with style and elegance. However, neither the ASM method nor any other method eliminates the need for creativity. Creativity is required to construct the basic algorithm that solves the current problem. At the same time, the ASM method is an excellent assistant during the creative phase. The issues of creativity and the conversion of words into ASM charts are addressed in this chapter's examples and in later chapters.

The ASM chart represents the controller.

Good design style is intrinsic to the ASM method. One important attribute of the method is *top-down design*. Top-down design starts with a careful study of the problem. Details are ignored at this stage. The relevant questions include the following, amongst others.

1. Is the problem statement clear?
2. Is the problem statement definitive?
3. Can the problem statement be simplified by restatement?
4. Is the problem part of a larger problem?

The data path processes the data.

The ASM state machine controls the data path.

Another attribute of the ASM method is explicit separation of the controller from the data path controlled by the controller. The *controller* is the state machine—the master. The *data path*—the slave—is the hardware that processes information. The controller manipulates the data path so it executes the required operations on data at the appropriate time. The ASM method emphasizes the creation of a clear and detailed definition of the control algorithm prior to implementing a detailed hardware design.

Good design style does not exclude bottom-up considerations. A typical example of a *bottom-up* process is a selection of "the solution," i.e., some integrated circuit, from an IC data book. The bottom-up process is continued by grafting other parts onto "the solution." In this way we can find an (apparent) solution. The usual justifications are the chip(s) saved, the reputedly short design time, and the presumed higher performance. These are fallacious, however. The contribution of chips to the cost of most products is less than 25%. And the cost of field repairs can exceed the cost of a poorly designed product over the product's lifetime. Finally, a properly executed top-down design yields comparable, if not increased, performance. There is no valid justification for the short-term point of view bottom-up design represents.

There are a great many difficulties with bottom-up designs, including the following.

1. They require a great deal of artistic skill unavailable to most bottom-up designers. (Invariably the result is a high-risk mix of asynchronous and synchronous methods.)
2. No systematic theoretical method is used that in effect certifies the design as reliable. (Explain to management how you are going to resolve the mysterious problems that invariably arise during production.)
3. There is an incomplete understanding of the circuit behavior that forecloses clarity in documentation, such as test, repair, and instruction manuals. ("Just how does this work so I can fix it?")

Nevertheless, an intimate knowledge of hardware influences any design decisions. The challenge is to use this knowledge in a way that avoids wasting time on unproductive choices. Discipline keeps bottom-up design in perspective.

The Controller and the Data Path An important attribute of the ASM method is *explicit* separation of the data path from the controller that controls the data path. Controller outputs manipulate the data path so it can perform the required data operations at the appropriate times.

The top-down design process begins with the selection of the data processing algorithm and its translation into a data path circuit. The algorithm might be a known procedure or one of our own creation. The selected algorithm is translated into a hardware configuration that processes the data as desired. One example is an algorithm multiplying two 12-bit binary numbers using a shift-and-add process, which is translated into a data path consisting of an adder, a counter, and shifters. Another example is a memory system whose data path consists of address decoders and an array of memory chips, a case in which data processing is limited to reading data from and writing data to storage.

Once the data path is defined, we list the inputs required to configure it and the inputs required for data. For example, configuration inputs select one of several sources to load into a selected register. Data path configuration inputs become controller outputs. Furthermore, we list the data path outputs required by the task as well as those the controller needs as inputs for confirming execution or ascertaining status of data path operations.

The top-down process ends with the design and implementation of the controller. Controller inputs represent external events or data path outputs reporting data processing results. Controller outputs configure the data path so desired actions take place. If this digital system is a subsystem of another system, there may be additional controller outputs reporting to the higher-level system.

Knowing the inputs, the outputs, and the algorithm, the controller is designed and implemented. Note that the controller cannot be designed until the data path is known.

7.1   Fundamentals

We start by relating everyday activities to the language of state machines. Suppose you are at work reading a technical report and the boss calls and asks you to come into her office. You stop reading and do so. In other words, the boss's input forces you to branch from your present activity (reading the report) to the next activity (going to the boss's office). Until the input was asserted, you were in your present activity. New inputs transition you into the next activity. *Present state* is your activity before inputs change. *Next state* is your activity after you have reacted to changes in inputs. Your behavior is asynchronous. Moreover, you are storing the present state in a part of your mind, which is a *state register* that remembers.

A clock synchronizes activities. Suppose there is a clock in the background that emits a timing mark periodically (e.g., a high-pitched chirp emitted once a minute). Furthermore, suppose you agreed not to change activity until inputs change *and* the next timing mark occurs. Now your behavior has become synchronous. In a sequential digital design, a clock emits pulses periodically. Each clock pulse orders a move to the next state, which can be the same as the present state or a different state. You were dwelling for many clock periods in the read-the-report present state until the telephone rang *and* a timing mark occurred.

Many tasks require events to take place sequentialy over time. When the present event ends, the next event starts immediately thereafter. Or the present event cannot end until some input variable changes state. In the meantime, the task dwells in the present state. And there may be more than one next event possible. Or the present event is repeated *n* times. And so forth.

Let us state all this more precisely.

State variables represent the state of a machine.

State variable: a stored quantity capable of assuming the value 0 or 1.

Example: Flip-flop output q is a state variable.

State: The state of a machine encodes in a memory sufficient past history so that future behavior may be determined. Sufficient information is available to determine from the present state and the present inputs both the present outputs and the next state. The present state of the sequential digital circuit is represented by a set of state variables.

Example: A traffic light controller encodes as a state the fact that the east–west light is red and the north–south light is green.

Example: The state of a sequential digital circuit with three flip-flops is represented by the set of three flip-flop outputs q_j concatenated in some order such as $q_2 q_1 q_0$. (The three flip-flops constitute the state register—see *State assignment.*)

State assignment: The process of state assignment converts each state identifier to a pattern of state variables. The pattern is usually a minterm whose literals are the state variables. In effect the process of state assignment gives each state a number. The present-state number is stored in a group of flip-flops, called the *state register,* which is the physical representation of the state machine's present state. The state assignments can be arbitrary; however, there are preferred choices. Any set of numbers assigned to states affects in an unpredictable manner the complexity of the hardware computing the next-state number and the hardware implementing the output equations. When the number of states is small, the hardware consequences of choices may be investigated in a reasonable amount of time. For larger state machines, computer-aided design (CAD) tools are recommended. There is a large body of knowledge on this complex subject, which is outside the scope of this text. Here we will use a simple binary representation for the state identifier number.

Branch: Associated with each state are none, one, two, or more input variables. The next state is selected by the present (associated) state and the present value of these input variables in some combination. This next-state decision is referred to as a *branch.* The next state is determined without conditions when there are no input variables associated with the present state. Input variables *not* associated with a present state have "don't care" status.

Outputs: States may specify unconditional or conditional outputs. Unconditional outputs assigned to a state are active when the system is in that state, without regard to the input conditions. Conditional outputs assigned to a state are active when the system is in that state *and* when input conditions are as specified for the output (see *Branch* above).

Each state can specify unconditional and conditional outputs. The same output can be unconditional in one state and conditional in another state.

An asynchronous machine moves to the next state when an input change occurs.

Asynchronous machine: The present state is the state of the q_j at the time before inputs change. The next state is the state of the q_j after the circuit has reacted to input changes. (Restrictions on input changes are discussed in Chapter 11.)

Synchronous machine: The present state is the state of the q_j at the time before a clock edge occurs when edge triggered flip-flops are used. The next state is the state of the q_j after the clock edge occurs and after the circuit has reacted to the clock edge. The present state and the input values immediately before the clock edge determine *which* state is next. The clock edge determines *when* the state machine goes to the next state.

Algorithmic state machine chart: This documents the controller's algorithm. The chart is a network of states with unconditional outputs, branches, and conditional outputs. When we do not have the algorithm clearly in mind, the process of building the chart helps us refine the algorithm.

A synchronous machine moves to the next state when an active clock edge occurs.

To summarize, then: A synchronous sequential logic machine includes a sequence-of-events controller that moves the machine from state to state when clock edges occur. Each clock edge asks the question "Which next state do we move to now?" The controller uses the present state and the present values of the input variables to decide which next state to move to, and clock edges decide when to move. Constructing an algorithm requires gathering information on the desired sequence of output events. This information is translated into sequences of states linked by paths between states. A path may or may not include branch decision points. If a path does *not* include branch decision points, the next clock edge orders the machine to follow the path from its present state to the unique next state. When a path includes a branch decision point(s), the selected path is the path activated by the inputs. (These ideas are illustrated in Figure 7.19 on page 372). The set of states linked by paths defines the possible sequences of states that can be specified by inputs. When a logic machine is in a state, none, one, two, or more output events can be specified to occur. The outputs may be conditional or unconditional.

7.2 State Diagram Method

Prior to development of the algorithm state machine method, designers used the *state diagram method*. This was during the era of the Mealy and the Moore machine structures. In state diagram terms, an ASM machine with only unconditional outputs is a Moore machine, and an ASM machine with only conditional outputs is a Mealy machine. The ASM method recognizes that the truly general machine is a combination of both structures: a Mealy-Moore machine structure. The basic goals of any state machine synthesis method is determination of the present-state output function and the next-state function. We start with an analysis of a state machine using the state diagram method.

7.2.1 State Machine Diagram Notation

Circles represent states and arrows point to next states.

State   The circle is state s_j's symbol (Figure 7.1a). The state identifier s_j is placed inside the circle.

Unconditional outputs   One or more unconditional outputs active in state s_j are listed beneath the state identifier inside the circle.

Arrow   If n input variables are associated with a state, then there are 2^n next states. This means there are 2^n arrows connecting the present-state circle to the next-state circles. When $n = 2$, then four arrows connect the present state to four possible next states. The associated function of input variables is written above each arrow (Figure 7.1b).

FIGURE 7.1
State diagram notation

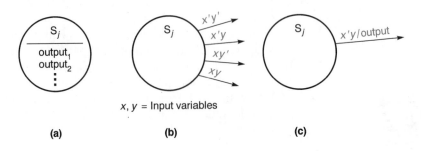

x, y = Input variables

(a)   (b)   (c)

Conditional outputs One or more conditional outputs active in state s_j are listed as expressions above the arrow (Figure 7.1c).

7.2.2 State Machine Analysis

Analysis of state machine circuits allows us to introduce a number of state machine design tools. The analysis method goes as follows.

STEP 1   Derive the flip-flop next-state equations from the circuit.

STEP 2   Derive the output equations from the circuit.

STEP 3   Make a state transition table with columns for the present state (PS), each input, the next state (NS), and each output.

STEP 4   Fill the input side of the state transition table. List all combinations of values for the present-state and machine inputs.

STEP 5   Fill the output side of the state transition table. Evaluate the next-state equations and the output equations for each combination of values in the input side.

STEP 6   Use the information in the state table to make the state diagram.

Specific state machine analysis The goal is to determine the state diagram that represents the state machine in Figure 7.2. The next-state equations q_1^+, q_0^+ and the output z are readily derived from the circuit.

FIGURE 7.2
111 sequence detector

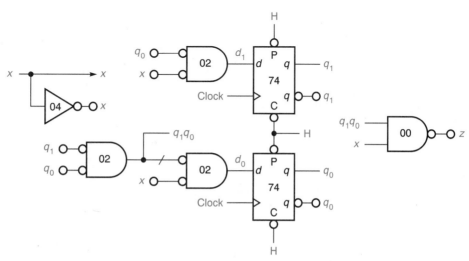

Use flip-flop defining
equations.

$$q_1^+ = d_1 = xq_0$$

$$q_0^+ = d_0 = x(q_iq_0)'$$

$$z = xq_1q_0$$

Two flip-flop outputs q_1, q_0 represent two bits, which have four present states that are entered into the PS column of the state transition table in Figure 7.3a. The next-state equations show that input variable x is active in every state. This means there are two exit paths from each state: one when $x = 0$, the other when $x = 1$. Each exit path is represented by a row in the truth table. So at every state we enter 0 and 1 in the input column. (When the number of states is s, the number of inputs is i, and every input is active in every state, then the number of truth tables rows is 2^{s+i}.)

The NS and output columns are filled by evaluating the equations for each present state with x equal to 1 or 0. Finally, we replace the binary state numbers with state identifiers (s_0, s_1, s_2, s_3) and rewrite the table as shown in Figure 7.3b. This table is referred to as a *state table*. The columns of the state table in Figure 7.3b provide us with the information we need to draw the state diagram in Figure 7.4.

In a transition table, states are represented by binary state numbers assigned to the states.

In a state table, states are represented by state identifiers.

FIGURE 7.3
Tables for 111 sequence detector

(a) Transition Table					(b) State Table			
PS[1]	PI[2]	NS[3]	PO[4]		PS	Inputs	NS	Outputs
q_1q_0	x	$q_1^+q_0^+$	z		s_j	x	s_j	z
0 0	0	0 0	0		s_0	x'	s_0	0
0 0	1	0 1	0		s_0	x	s_1	0
0 1	0	0 0	0		s_1	x'	s_0	0
0 1	1	1 1	0		s_1	x	s_3	0
1 0	0	0 0	0		s_2	x'	s_0	0
1 0	1	0 1	0		s_2	x	s_1	0
1 1	0	0 0	0		s_3	x'	s_0	0
1 1	1	1 0	1		s_3	x	s_2	1

[1] Present state.
[2] Present inputs.
[3] Next state.
[4] Present outputs.

FIGURE 7.4
State diagram for 111 sequence detector—Mealy type

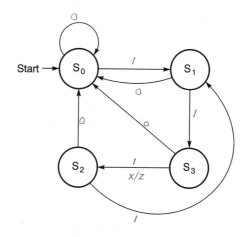

EXERCISE 7.1

The next-state equation for a circuit is $q_0^+ = q_0$ xor x. Derive the state diagram. *Hint:* derive the state transition table.

Answer:

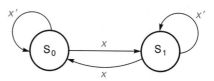

EXERCISE 7.2

The next-state equations for a circuit are $q_1^+ = q_0$, $q_0^+ = x$. Derive the state diagram. *Hint:* derive the state transition table.

Answer:

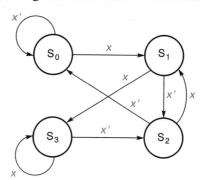

7.2.3   State Machine Synthesis

State machine synthesis begins with the development of a state diagram. This difficult task has been replaced by the ASM chart method making the transition to the ASM synthesis method we will take up next.

Given a state diagram, the synthesis process starts by taking information off the state diagram to construct the state transition table. This step includes selection of the type of flip-flop to be used in the state register. Next, the flip-flop input equations and the output equations are derived from the table. Finally, the state machine and output circuits are synthesized from the equations. We use the state machine in Figure 7.5 to explain the process.

A state machine with eight states requires three state variables. Three variables x, y, and z generate eight (2^3) minterms, and each minterm is true for one combination of the constants 0 and 1 (which are not numbers). For example, the minterm m_3 ($= x'yz$) is true for the combination of constants 011. If we interpret 011 as a *number,* we can say a state machine is in state s_3, or state three, when the minterm $x'yz$ is True.

The state diagram representing the state machine has eight nodes (the circles in Figure 7.5), and arrows connect the nodes to each other. That this diagram represents a synchronous state machine is not shown on the diagram. This is an example of what we call *hidden logic*. At each clock edge the machine will move to the next state, as dictated by the arrows. State s_0 differs from the other states because the next state depends on the input variable p. Input variables such as p establish conditions for state transitions. An arrow without an input associated with it, such as the connection from state s_2 to s_3, has no conditions placed on it. The name END beneath state s_7 in the state-7 node is an unconditional output. When in state s_7, output END is True without conditions. We say END is unconditionally True when in state s_7. Let us synthesize a circuit implementing this state diagram.

Next state conditions are entered above the arrows.

FIGURE 7.5

**State
diagram—Moore
structure**

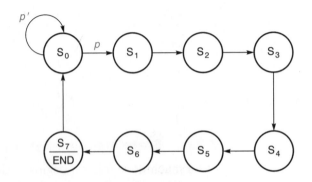

Three-state variables $q_2 q_1 q_0$ are implemented with three flip-flops. The three flip-flops constitute the state register. Each flip-flop q output is a state variable. Concatenation of the q_j is the state register state: e.g., $s_5 = q_2 q_1' q_0$. The state identifier is assigned a binary number equal to its subscript. These convenient assignments may not be the best set of assignments.

STEP 1   List the present states (PS) in a column, as shown in **Figure 7.6**. In effect this is a list of the states found in the state diagram (Figure 7.5).

STEP 2   Make one column per input variable (Figure 7.6). There is one row in the table for each combination of variables associated with each state. In this case, input variable p is associated only with state 000. Therefore state 000 has two rows corresponding to the p' and p combinations of one variable. We mark the column with 0 or 1 in the rows. Observe that p is a "don't care" input variable for all of the other states.

FIGURE 7.6

State transition table with state identifiers

Step: 1		2	3		4	5
PS		PI	NS		Flip-Flop Inputs	PO
s_j	$q_2 q_1 q_0$	p	s_j^+	$q_2^+ q_1^+ q_0^+$	$t_2 t_1 t_0$	END
s_0	000	0	s_0	000	000	0
s_0	000	1	s_1	001	001	0
s_1	001	–	s_2	010	011	0
s_2	010	–	s_3	011	001	0
s_3	011	–	s_4	100	111	0
s_4	100	–	s_5	101	001	0
s_5	101	–	s_6	110	011	0
s_6	110	–	s_7	111	001	0
s_7	111	–	s_0	000	111	1

[1] Present state.
[2] Present inputs.
[3] Next state.
[4] Present outputs.

STEP 3   List the next states (NS) in a column, as shown in Figure 7.6. The state diagram arrows' destinations with or without conditions yield the NS list.

STEP 4   Make one column for each flip-flop input. Select a flip-flop type (select T for this example). (*Reminder:* if JKs are used, each JK flip-flop requires two columns.) We calculate from the flip-flop defining equation the input values setting up the next state.

$$q^+ = tq' + t'q$$

Briefly stated, $t = 1$ toggles the flip-flop, and $t = 0$ holds the present value. Therefore if NS bit q_j^+ does not equal PS bit q_j, we let $t = 1$; and if NS bit q_j^+ equals PS bit q_j, we let $t = 0$.

STEP 5   Make one column per output variable. Mark the END column with 1 on the PS lines the output(s) is true.

STEP 6   Derive the t flip-flop input equations for $t_2 t_1 t_0$ from the state transition table. We use the present-state and associated inputs, but not the next state.

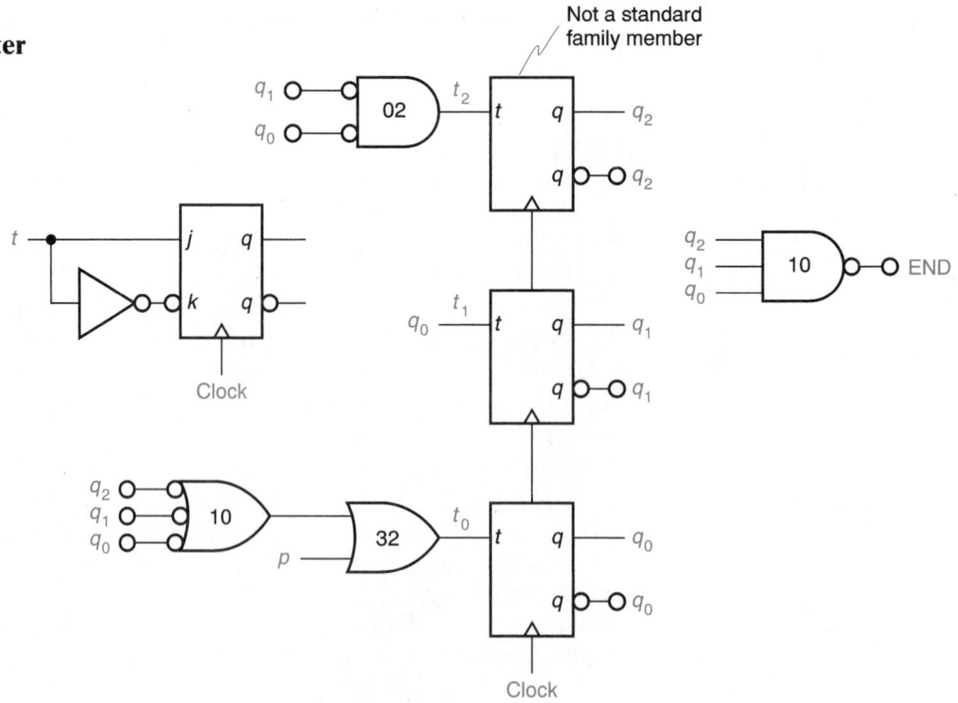

FIGURE 7.7

Three-bit counter

In the t_0 column there is only one 0, so we find the equation for $t_0{}'$.

$$t_0{}' = m_0 p'$$

$$t_0 = m_0{}' + p = (q_2{}' q_1{}' q_0{}')' + p = q_2 + q_1 + q_0 + p$$

By inspection, $t_1 = q_0$, $t_2 = q_1 q_0$, and END $= q_2 q_1 q_0$.

STEP 7   Synthesize the state machine circuit from the flip-flop input equations. The circuit implementing the state diagram of Figure 7.5 is shown in Figure 7.7. A notation for parallel connection of the clock to flip-flop clock inputs is also illustrated in Figure 7.7. The clock wire is routed under each flip-flop.

EXAMPLE 7.1   Converting State Diagram to State Table

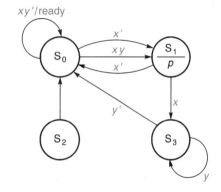

Four states require two state variables q_1, q_0. We enter four state identifiers in the PS column.

State s_0 has variables x and y associated with it. We enter four combinations of xy values in the x and y input columns.

State s_1 has variable x associated with it. We enter two combinations of x values in the x column and two "don't care" dashes in the y column.

State s_2 has no variables associated with it. We enter dashes in the x and y columns.

State s_3 has variable y associated with it. We enter two combinations of y values in the y column and two "don't care" dashes in the x column.

We enter the next-state numbers in the NS column. The arrows point to the next states.

The conditional ready output is asserted when the machine is in s_0 and the minterm xy' is asserted. The unconditional p output is asserted when the machine is in state s_1.

PS	PI		NS	PO	
s_j	x	y	s_j	ready	p
s_0	0	0	s_1	0	0
	0	1	s_1	0	0
	1	0	s_0	1	0
	1	1	s_1	0	0
s_1	0	–	s_0	0	1
	1	–	s_3	0	1
s_2	–	–	s_0	0	0
s_3	–	0	s_0	0	0
	–	1	s_3	0	0

EXAMPLE 7.2   Converting State Table to State Diagram

PS	PI		NS	PO	
s_j	x	y	s_j	c	u
s_0	0	–	s_2	1	0
	1	–	s_1	0	0
s_1	–	–	s_0	0	0
s_2	–	0	s_2	0	1
	–	1	s_0	0	1
s_3	–	–	s_0	0	0

Four states require four state symbols. We draw four circles to start the diagram.

State s_0 has only variable x associated with it because y is a "don't care." We draw two arrows from s_0 to the next states s_2 and s_1. Output c is conditional, so we mark the s_0-to-s_2 arrow with x'/c and the s_0-to-s_1 arrow with x.

State s_1 has no variables associated with it. We draw one arrow from s_1 to the next state s_0.

State s_2 has only variable y associated with it because x is a "don't care." We draw two arrows from s_2 to the next states s_2 and s_0. We mark the s_2-to-s_2 arrow with y' and the s_2-to-s_0 arrow with y. Output u is unconditional, so enter u in the s_2 circle.

State s_3 has no variables associated with it. We draw one arrow from s_3 to the next state s_0.

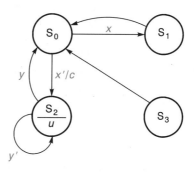

EXERCISE 7.3

Find the next-state equation for the following state diagram. Use the state assignment:

(a) $s_0 = 0, s_1 = 1$ (b) $s_0 = 1, s_1 = 0$

Answer: (a) $q_0^+ = q_0'x' + q_0y'$ (b) $q_0^+ = q_0'x + q_0y$

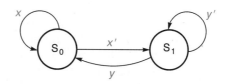

EXERCISE 7.4

Find the next-state equation for the following state diagram. Use the state assignments $s_0 = 00$, $s_1 = 01$, $s_2 = 10$, and $s_3 = 11$.

Answer: $q_1^+ = q_1$ xor pq_0, $q_0^+ = q_0$ xor p

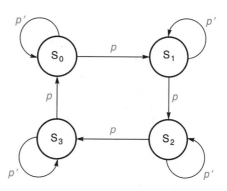

7.3

ASM charts use three symbols: state box, branch diamond, and conditional output oval.

The algorithmic state machine method uses straightforward procedures to create reliable sequential state machines. Via the ASM method we can develop designs with style and elegance. However, the ASM method does not eliminate the need for creativity. Creativity is required to construct the basic algorithm that solves the current problem. At the same time, the ASM method is an excellent assistant during the creative phase. Furthermore, the ASM chart notation consists of only three symbols: state, branch, and conditional output.

State   The rectangle is state s_j's symbol (Figure 7.8a). The symbol has only one entrance and one exit.

For a synchronous state machine each active clock transition causes a change of state from the present state to the next state. Given the present state, the next state must be determined without ambiguity for any values of state and input variables. Having arrived at the next state, this next state becomes the present state.

Whereas any number of paths may lead to a state rectangle's single entry point, only one path may lead away from the state rectangle's one and only exit point.

Branch   The diamond is the branch symbol (Figure 7.8b). The extended diamond is a compact form representing a *tree* of diamonds (Figures 7.8b and 7.9b). *Decision box* is an alias for *branch diamond*.

The diamond symbol with True and False exit paths represents a decision. The condition placed in the box may be any Boolean function of input variables. The True (False) exit path is taken when the condition is True (False). The extended diamond has one exit path per minterm of the input variables involved in the decision. (These are *not* minterms of all the state machine input variables.) In Figure 7.8b, a decision box with two variables x, y has one exit for each minterm ($x'y'$, $x'y$, xy', and xy). Boolean variables such as x in a branch function (Figure 7.8d) usually represent inputs from the outside world. In addition, the inputs can originate from within the same logic machine.

Suppose y is a "don't care" when x is True. This means the two exits corresponding to minterms xy' and xy merge into one exit (Figure 7.8e). This is why the y diamond on the $x = T$ side of the tree can be removed.

Each branch diamond has only one entry point. A branch exit path leads either to only one other branch diamond or to only one (next) state so that the next state is determined without ambiguity for any values of state and input variables. Only one of the parallel exit paths can be active if the next state is to be uniquely determined.

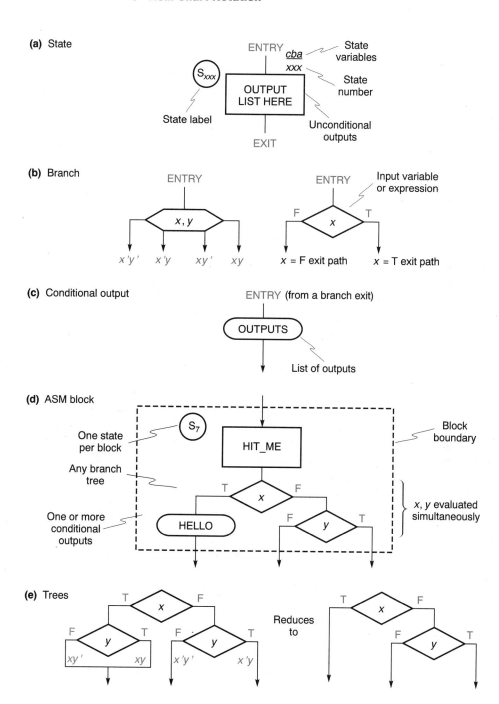

FIGURE 7.8
ASM chart notation

FIGURE 7.9
**Incorrect and
correct versions of
an ASM binary tree**

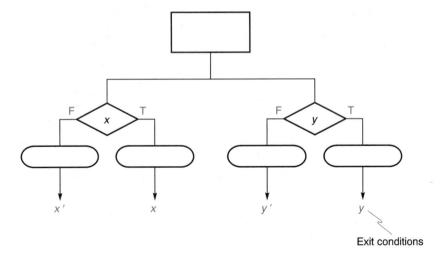

(a) Incorrect: always two simultaneous exit paths (*x* and *y* independent)

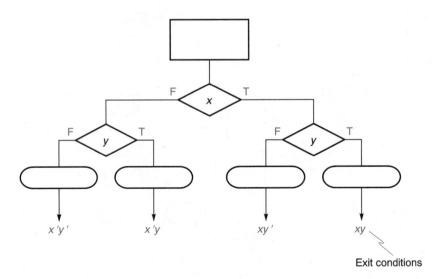

(b) Correct: always only one exit path at at time

An input variable does not have to be associated with every decision. Input variables not involved in a decision are treated as "don't cares" in that decision. For example, there may be five input variables, but only x, y are associated with some state.

A branch must be associated with a state to have meaning. There must be only one possible next state for each set of input conditions.

> **Fundamental Rule** Every path must lead to only one state. Therefore we activate only one path at a time.

When this is not the case, the logic machine cannot select a unique next state, which is a design error (Figure 7.9a).

The branch is activated during the entire state time. *State time* is the clock period prior to the next clock edge that executes the decision. The exit path selected by the input values immediately before the clock edge occurs determines the next state.

Unconditional output One or more unconditional outputs are specified by entering the outputs' names inside the desired state rectangle (Figure 7.8a). There is no special symbol for an unconditional output; the only symbol is the name.

Unconditional outputs depend solely on the state. They do not depend on inputs in any way. Unconditional outputs are active during the associated state time. (In state diagram terms, an ASM machine with only unconditional outputs is a Moore machine.)

Note: When there are no unconditional outputs, a state is an "empty box." We do not delete the empty box from the ASM chart simply because it is empty. There may be other reasons why the state exists.

Conditional output The oval is the symbol of conditional output (Figure 7.8c). The oval is always associated with a branch diamond. Conditional output names are entered in the oval. A conditional output is active during a state time if the associated branch condition is True during that state time. The output is conditioned by the state and branch that feeds it. (In state diagram terms, an ASM machine with only conditional outputs is a Mealy machine.) In Figure 7.8d output HELLO is active when the system is in state s_7 and when x is True.

Note: When there are no conditional outputs, the oval is "empty." The empty oval is deleted because there are no other possible reasons for its presence.

All events in an ASM block execute simultaneously.

ASM block   An ASM block consists of one entry line, one state rectangle, and any number of associated branch diamonds, conditional output ovals, and mutually exclusive exit lines (Figure 7.8d). Any ASM chart is an assembly of ASM blocks connected by paths (see Example 7.3).

An ASM block representing state$_j$ describes the state machine's operation while the state machine is in state$_j$. The time spent in state$_j$ may be an indefinite number of clock periods when a branch diamond is associated with the state. The time spent in state$_j$ ends when the input variables close the path that returns to the present state$_j$ and open a path to a next state$_k$.

In an ASM chart the operations described in an ASM block are executed simultaneously, i.e., in parallel. In the ASM block of Figure 7.8d the following events occur during time spent in state s_7: Unconditional output HIT_ME is activated, and, in parallel, input variables x and y are evaluated simultaneously, activating the path corresponding to the $f(x, y)$ minterm that is True. When either minterm m_2 or m_3 is True, x is True. This activates conditional output HELLO. When m_0 or m_1 is true, HELLO is not activated and the $x'y'$ or $x'y$ exit is taken. Until this ASM block is embedded in an ASM chart, and the characteristics of input x are known, we do not know how many clock periods are spent in state s_7. This differs from a flow chart, where the events occur sequentially and event duration is not specified.

EXAMPLE 7.3   **From Statement to ASM Chart**

A state machine is needed to cycle through states s_0, s_1, s_2, and s_3 when input variable x is asserted, and to cycle through states s_0 and s_3 when input variable x is not asserted. The essential deduction is that s_1 and s_2 are skipped when x is not asserted. That is, from present state s_0 the next state is s_1 when $x = T$, and when $x = F$ the next state is s_3. Because x is evaluated in state s_0, input variable x is active in, associated with, s_0. Furthermore, the machine steps unconditionally from s_1 to s_2, s_2 to s_3 and s_3 to s_0. The ASM chart follows.

Two ASM charts are shown [with and without clock (ck) as an input]. When clock is omitted as an input associated with every state, the implicit assumption is that the state machine is a synchronous machine. The ASM chart with clock omitted as an input to every state is not only simplified significantly, it is clear.

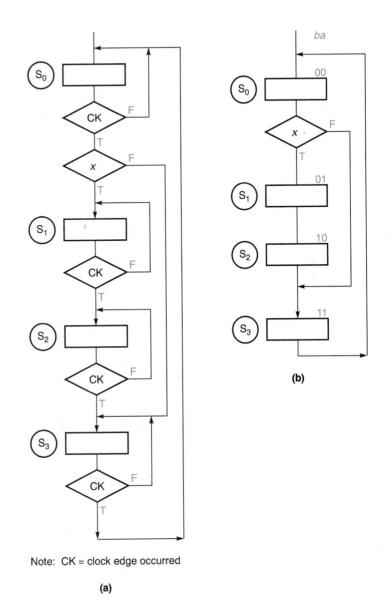

Note: CK = clock edge occurred

(a)

(b)

EXAMPLE 7.4   From Sequencer Statement to ASM Chart

Suppose a sequential circuit has one input x and one output g. Output g is asserted whenever the most recent inputs are 111, where the most recent input is the last digit in the string. Overlapping of 111 sequences is not allowed, but "adjacent" 111 sequences can occur (e.g., . . . 01111110 . . . produces . . . 00010010 . . .).

The number of states required is not known at the outset. Let state s_0 be the rest state. The circuit remains in the s_0 state when zeros are received ($x = 0$). Let the machine advance to state s_1 when a one is received ($x = 1$). Therefore s_1 represents one 1 received. We draw s_0 and s_1 state boxes and a branch diamond with input x at the s_0 exit. We complete the path from s_0 to s_0 by drawing an arrow from the diamond x-equals-0 output to the s_0 state box entrance. We draw an arrow from the diamond x-equals-1 output to the s_1 state box entrance.

If a zero is received while in s_1, a return to s_0 restarts the sequence detection process. At the s_1 state box exit we draw a branch diamond with input x. We complete the path from s_1 to s_0 by drawing an arrow from the diamond x-equals-0 output to the s_0 state box entrance. If a one is received, we advance to s_2 so that s_2 represents the fact that a sequence of two 1s has been received. We draw an arrow from the s_1 diamond x-equals-1 output to the s_2 state box entrance.

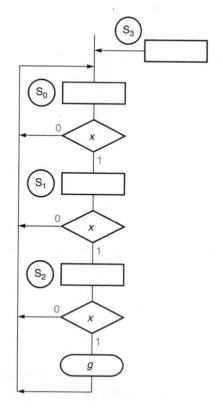

If a zero is received while in s_2, a return to s_0 restarts the sequence detection process. At the s_2 state box exit, we draw a branch diamond with input x. We complete the path from s_2 to s_0 by drawing an arrow from the diamond x-equals-0 output to the s_0 state box entrance. If a third one is received, we output a one ($g = 1$) on return to s_0, which then represents that a sequence of three 1s has been received. We draw an arrow from the s_2 diamond x-equals-1 output to an oval's entrance. We enter g in the oval. This will output the one report. We draw an arrow from the oval's exit to the s_0 state box entrance.

When three state numbers are encoded with two binary digits, a fourth state (s_3) is possible. Good practice* dictates that all unused states have a next state by design. The "rest" state s_0 is a practical next state for unused states. So we add a state box for unused state s_3. We draw an arrow from the s_3 box exit to the s_0 state box entrance.

EXERCISE 7.5   Derive the ASM chart for Figure P6.2 on page 326.

Answer:

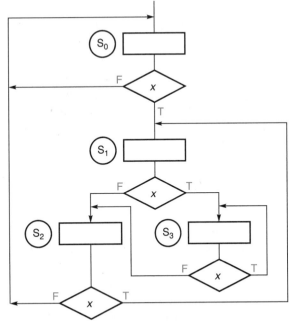

---

* s_3 can be entered when the power is turned on, in which case the next clock edge moves the machine to s_0 so that it is ready to execute the algorithm correctly.

EXERCISE 7.6   Derive the ASM chart for Figure P6.6 on page 327.

Answer:

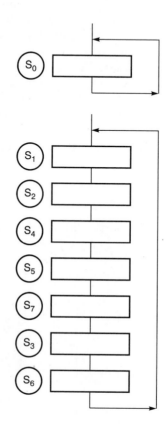

EXAMPLE 7.5 From ASM Chart to State Machine

A state machine is needed to cycle through states s_0, s_1, s_2, and s_3 when input variable x is asserted, and to cycle through states s_0 and s_3 when input variable x is not asserted. The ASM chart for this application was developed in Example 7.3. The PS-Input-NS truth table is derived from the ASM chart via a process similar to the state machine synthesis process described in Section 7.2.3. The complete synthesis process, from ASM to circuit, is described in Section 7.5.

PS		PI	NS		
s_j	$q_1 q_0$	x	s_j	$d_1 d_0$	$= q_1^+ q_0^+$
s_0	00	0	s_3	11	
		1	s_1	01	
s_1	01	–	s_2	10	
s_2	10	–	s_3	11	
s_3	11	–	s_0	00	

$$d_1 = x's_0 + s_1 + s_2 = x'q_1'q_0' + q_1 \text{ xor } q_0$$
$$d_0 = s_0 + s_2 = q_1'q_0' + q_1 q_0' = q_0'$$

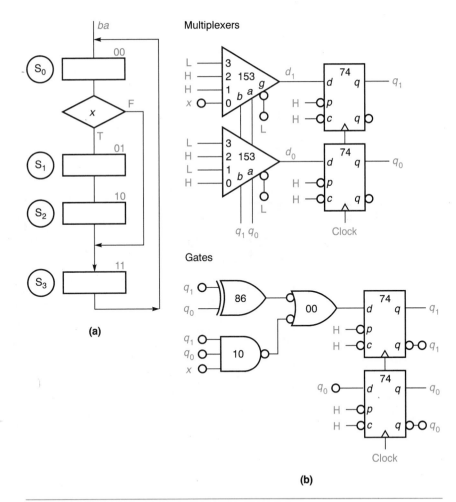

(a)

(b)

EXAMPLE 7.6   **From ASM Chart to Sequencer State Machine**

A sequential circuit has one input x and one output g. Output g is asserted whenever the most recent inputs are 111, where the most recent input is the last digit in the string. Overlapping of 111 sequences is not allowed. This sequencer ASM chart was developed in Example 7.4.

s_j	PS $q_1 q_0$	PI x	s_j	NS $d_1 d_0 = q_1^+ q_0^+$	PO g
s_0	00	0	s_0	00	0
		1	s_1	01	0

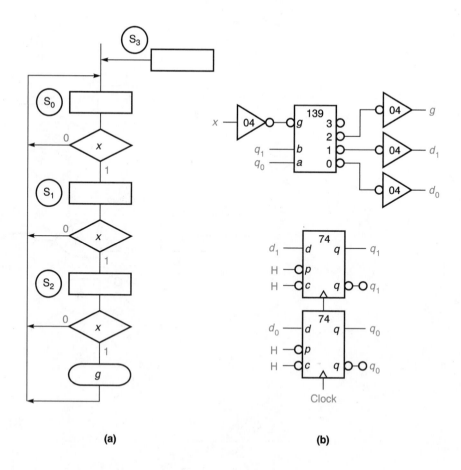

(a)   (b)

PS		PI	NS			PO
s_j	$q_1 q_0$	x	s_j	$d_1 d_0$	$= q_1^+ q_0^+$	g
s_1	01	0	s_0	00		0
		1	s_2	10		0
s_2	10	0	s_0	00		0
		1	s_0	00		1
s_3	11	–	s_0	00		0

$$d_1 = xs_1 = xq_1'q_0$$
$$d_0 = xs_0 = xq_1'q_0'$$
$$g = xs_2 = xq_1q_0'$$

EXERCISE 7.7

Derive the next-state equation and the output equation for the following ASM chart (for state assignments $s_0 = 0$, $s_1 = 1$).

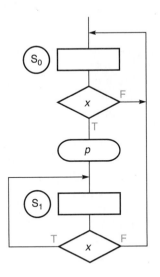

Answer: $q_0^+ = x$, $p = xq_0'$ ∎

EXERCISE 7.8

Derive the next-state equations and the output equation for the following ASM chart ($s_0 = 00$, $s_1 = 01$, $s_2 = 10$, $s_3 = 11$).

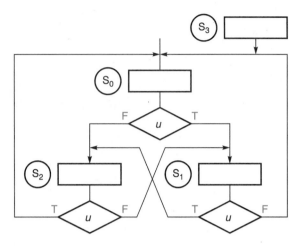

$$\text{Answer: } q_1{}^+ = q_1{}'(q_0 \text{ xor } u'), \qquad q_0{}^+ = q_0{}'(q_1 \text{ xor } u) \qquad \blacksquare$$

7.4   ASM Sequences of States

Sequences of states are determined by input minterms.

The purpose of any state machine is execution of one or more sequences of states in response to various combinations of input variable values. Each combination of values sets up paths in the chart; paths determine a sequence of states.[1] (The minterms of a function of the input variables represent combinations of values.) In principle there is one sequence of states for each constant combination of values (minterm) that executes in, say, n clock periods[2] ($n\tau$). For a fixed set of input values there corresponds a sequence of states determined by the paths set up by those input values.

When the input minterm is held constant, the state machine executes the corresponding sequence of states either once or with period $n\tau$, depending on the ASM algorithm. The input minterm is held constant when there are modes of operation. For example, one minterm sets up the machine to generate a periodic clock waveform; another minterm sets up the machine to generate a single pulse whenever the single-pulse input is activated. Then there is the infinite variety of sequences of states generated by an infinite number of sequences of input minterms. For example, the sequence of minterms changes the

---

[1] In Example 7.3 on page 352 the sequence of states is $s_0 s_1 s_2 s_3$ when input x is True, and $s_0 s_3$ when x is False.

[2] In Example 7.3, $n = 4$ when input x is True, and $n = 2$ when x is False.

operating mode from single-pulse to periodic clock generator, and vice versa.

In many cases a state machine is designed to perform some set of tasks, where each task is selected by an input minterm. Then various sequences of input minterms are used to create various combinations of tasks in time sequence. The clock generator is one example: switch to periodic clock, switch to single pulse for testing, switch back to periodic clock, and so forth. This switching back and forth can be done manually. On the other hand, a memory module controller is one example where tasks are selected by electronic means in rapid-fire order: read, read, write, nop ("no operation"), read, write, write, and so forth.

Input minterms held constant Two variables x and y are inputs to the ASM chart in Figure 7.10. A function of two variables has four minterms. This implies that the chart generates four sequences of states when input minterms are held constant. In this case two of the sequences are identical for the paths set up by the minterms $x'y$ and $x'y'$. Starting from s_0 we trace the paths when $x = F$ to verify this statement; notice how the branch diamonds with input y are bypassed. Tracing a path raises the *when* and *which* questions regarding state-to-state moves.

Input minterms select which state is the next state.

FIGURE 7.10

ASM chart

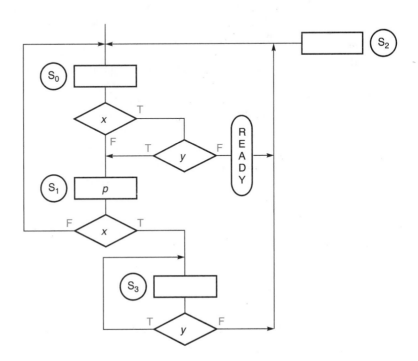

Active clock edges
determine when a
move to the next state
occurs.

A synchronous state machine moves from any present state to the next state when a clock edge occurs. Clock edges determine *when* moves occur. This is why sequences of states are intimately related to timing diagrams. Input variables determine *which* state is the next state to move to from the present state. A path from state s_j to s_k may have no branch diamonds in it, in which case the next-state decision is independent of the input variables. (A path from state s_j to s_k with no branch diamonds in it may be thought of as a path with a full binary tree of x and y diamonds, as in Figure 7.9b, with all exits leading to the same state.)

Producing timing diagrams from ASM charts with input minterms held constant   Here is a straightforward process leading to the sequences of states for input terms $x'(= x'y' + x'y)$, xy', and xy.

$x = \mathbf{F}\ (x' = \mathbf{T})$ Let us assume the machine is resting in state zero (s_0) of the ASM chart in Figure 7.10. Observe that the input variable x is associated with s_0. With $x = \mathrm{F}$, the path from s_0 to s_1 is active; therefore the clock edge ending cycle 1 (Figure 7.11) moves the machine from s_0 to s_1. We note that x is also associated with s_1. Input x is still False; thus the clock edge ending cycle 2 moves the machine from s_1 to s_0. Clearly, input variable y has no role in this particular repetitive sequence consisting of $s_0 s_1$ cycles. Furthermore, each time the ASM is in s_1, the output variable p is True (Figure 7.11). This is the periodic mode.

$x = \mathbf{T},\ y = \mathbf{T}\ (xy = \mathbf{T})$ Again, let us assume the machine is resting in s_0 of the ASM chart in Figure 7.10. With $x = \mathrm{T}$ and $y = \mathrm{T}$, another path from s_0 to s_1 is active, so that the clock edge ending cycle 1 moves the machine from s_0 to s_1. The clock edge ending cycle 2 moves the machine from s_1 to s_3 because $x = \mathrm{T}$. Input variable $y = \mathrm{T}$ activates the path from the s_3 exit to the s_3 entrance. This is how the state machine finds itself stuck in s_3. Output variable p is True for only one clock period in this sequence of states (Figure 7.12).

FIGURE 7.11

**Timing diagram
when $x = F(x' = T)$**

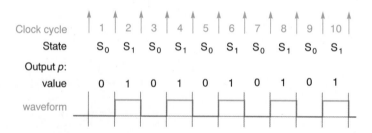

Clock cycle	1	2	3	4	5	6	7	8	9	10
State	S_0	S_1	S_0	S_1	S_0	S_1	S_0	S_1	S_0	S_1
Output p: value	0	1	0	1	0	1	0	1	0	1
waveform										

FIGURE 7.12
**Timing diagram
when $x = $ T, $y = $ T
($xy = $ T)**

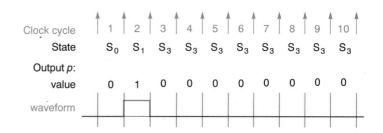

$x = $ **T, $y = $ F ($xy' = $ T)**   Let's assume the machine is resting in s_3 of
the ASM in Figure 7.10 with $y = $ T. When y switches to F in cycle 2,
the machine moves from s_3 to s_0 at the next clock edge. Now, with
$x = $ T and $y = $ F, yet another path from s_0 is active; this time the path
leads back to s_0, so that the clock edge ending cycle 3 moves the
machine from s_0 to s_0. This is how the state machine finds itself stuck
in s_0 with the ready output active. The machine does not pass through
s_1, so output p is never True in this sequence of states (Figure 7.13).
   The state machine is actually standing by reporting it is "ready"
to perform when $xy' = $ T. If y switches from F to T, output p is active
for one clock period (Figure 7.12) and the machine dwells in s_3 until
y switches back to F. A merger of the two sequences gives them
purpose.

$x = $ **T, $y = $ F, T, F**   Now let's assume the machine is resting in s_0 of
the ASM in Figure 7.10 with $x = $ T. With $y = $ F the machine repeatedly cycles from s_0 to s_0 at each clock edge. This is how we found the
machine apparently stuck in s_0. We say "apparently" because there is

FIGURE 7.13
**Timing diagram
when $x = $ T, $y = $ F
($xy' = $ T)**

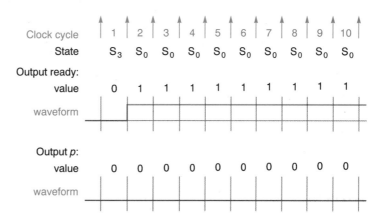

FIGURE 7.14
Timing diagram for single-pulse sequence of states

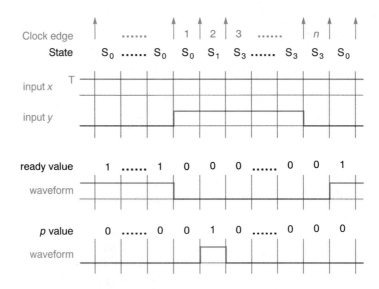

Indefinite time intervals marked as ......

another interpretation: the machine is in the ready-to-perform state when $x = T$ and $y = F$. In Figure 7.12, when $x = T$ and $y = T$ we see that the output p is what is called a *single pulse*.

If we make $y = T$ for some (indefinite) amount of time, then the F-T pair generates the $s_0 s_1 s_3$ sequence in cycles 1, 2, 3, followed by the dwell-in-s_3 sequence. When y switches back to the final F in cycle n, this allows the machine to return to s_0 to wait for the next $y = T$ input. This is the *single-pulse mode*.

EXERCISE 7.9   What is the sequence of states in Figure P7.2 on page 400 when input minterm $x = 1$?

Answer: $s_0 s_1 s_1 \ldots$ ■

EXERCISE 7.10   What is the sequence of states in Figure P7.6 on page 406 when input minterm $x'y = 1$?

Answer: $s_0 s_1 s_2 s_3 s_0 \ldots$ ■

EXERCISE 7.11   What is the sequence of states in Figure 7.15 when input minterm $wr't = 1$?

Answer: $s_0 s_2 s_0 s_2 s_0 \ldots$ ■

FIGURE 7.15

Memory module ASM chart

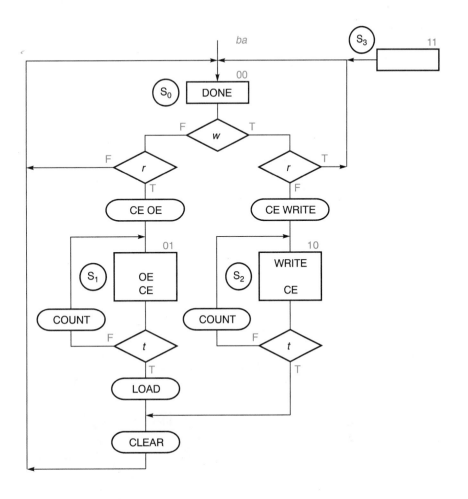

Input minterms implement branch decisions.

Sequence of input minterms   To illustrate how we can determine the sequence of states corresponding to a sequence of input minterms, let us consider the ASM chart in Figure 7.15. The ASM chart represents a controller for a dynamic RAM system (discussed in detail in Chapter 9). For now we consider it a black box with three inputs (w, r, t) and a number of outputs (DONE, CE, OE, . . .). We assume the physical implementation is a synchronous sequential machine. Then each positive clock edge is a time when change to the next state takes place.

The timing diagram in Figure 7.16 shows inputs w, r, and t as functions of time. This is a time sequence of input minterms. With values for inputs r and w held constant, some path from the s_0 exit to the entrance of s_0 or s_1 or s_2 is active. On the ASM chart when min-

FIGURE 7.16

**ASM chart
sequence of states
to read a memory**

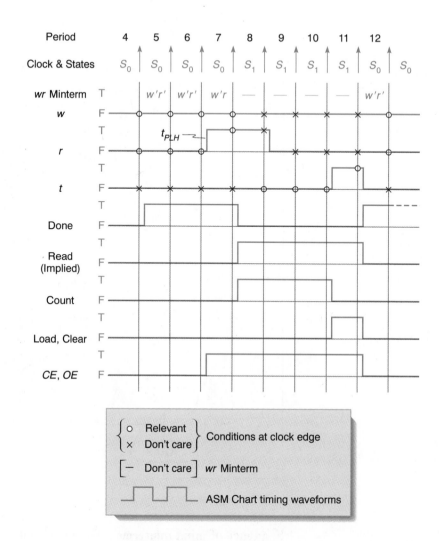

FIGURE 7.16

**ASM chart
sequence of states
to read a memory**

term $w'r'$ is True ($w = r = $ F), the active path is now from s_0 through
the branch diamond tree to the s_0 entrance (Figure 7.15). Thus s_0 is the
next state. Clock edge 6 ends period 6 as it sets the system to the next
state s_0.

In some other part of the system we assume clock edge 6 also
switches r to True after a small time delay t_{PLH} (see period 7). Now
minterm $w'r$ is True ($w = $ F, $r = $ T).

At clock edge 7, the ASM is still in s_0, so that input t is a "don't
care." With minterm $w'r = $ True, the ASM moves to state s_1 on clock
edge 7 (Figure 7.16). In state s_1 input variable t is active and inputs w, r
are "don't cares." At the end of periods 8, 9, and 10 the t input is

False and w, r are "don't cares." With $t = F$, the path from s_1 exit to s_1 entrance is now active (Figure 7.15). The ASM dwells in s_1 until t goes True in period 11, and during this period the path from s_1 to s_0 is open (Figure 7.15). Clock edge 11 moves the ASM from state s_1 to state s_0.

On the waveforms, the values of "don't care" variables are marked with an x and the values of "care" variables are marked with a small circle (○). For example, input w is associated with s_0. When the state machine is in s_0, w is active (we care, so we use ○). When the state machine is in other states, w is inactive (we don't care, so we use x).

EXERCISE 7.12   Starting from state s_2 in Figure 7.15, what is the sequence of states when the input minterm sequence is $w'r't$, $w'r't$, $wr't'$, $wr't'$, $wr't'$, $w'r't'$, $w'r't$, $w'r't'$, . . . ?

Answer: $s_2 s_0 s_0 s_2 s_2 s_2 s_2 s_0 s_0$ . . .   ■

7.5 Synthesis: From ASM Chart to Circuit

We now return to our main topic of ASM-based design to discuss the synthesis of the state machine. Recall that synthesis begins with selection of the algorithm being implemented, continues with design of the data path, and proceeds to the design of the state machine controlling the data path. Once the state machine design step is reached, the designer will have developed an intimate understanding of the data path and of the signals needed to implement the algorithm. We have included at the end of Chapter 8 (Section 8.4) an example of this process to guide the neophyte designer. Here we assume the data path is defined or that the ASM method is being used to design a state machine that does not control a data path, such as a counter.

The multiplexer solution clearly shows the next state for any present state.

For large state machines ROM data words can represent the next state and present outputs.

The logic circuit synthesis method to be presented shortly is direct in the sense that we provide what is called the standard solution (Figure 7.17). Our intuition tells us that this solution is valid for any number of states. To us the multiplexer solution is understood in an instant. Nevertheless, sixty-four-to-one multiplexers, for example, are not necessarily a practical solution for a state machine with 64 states. There are a number of approaches to solutions, such as attempts at state reduction. Our preferred solution for large state machines is to have a read-only memory (ROM) (see Chapter 9) replace the multiplexers, as discussed in Section 10.5.3, and to use the ROM in combination with linked state machines, as discussed in Section 7.8.2. These alternate solutions of complex problems are straightforward applications of basic concepts we now present.

FIGURE 7.17
Standard ASM circuit

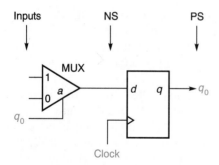

(a) Two states, one state variable
a ($2 = 2^1$)

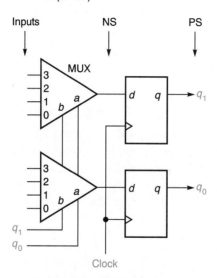

(b) Four states, two state variables
ba ($4 = 2^2$)

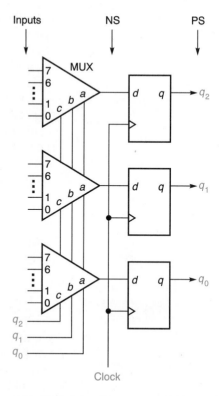

(c) Eight states, three state variables
cba ($8 = 2^3$)

Logic circuit design requires equations. In turn, equations create a need for a truth table. So our method starts by extracting a truth table from the ASM chart. Logic equations are derived from the truth table after adding D flip-flop excitation columns to the truth table. Then the equations are recast into multiplexer equation format and directly applied to the standard multiplexer circuit solution. There is no one-and-only right answer to state machine design because there are a multitude of possible circuit formats. Nevertheless, we offer this one very practical method that always results in one very practical answer. The importance of the method lies in the ideas behind it. To illustrate the method, let us consider a specific example.

We design only the controller of the memory module in Figure 7.18. The memory module ASM chart is shown in Figure 7.15. Any memory module consists of a controller (sequential state machine and output circuits) and a data path, as shown in Figure 7.18. The data path is controlled by the state machine. (Data path design is covered in later chapters).

7.5.1 State Machine Synthesis

STEP 1   *Create the ASM chart.* The ASM chart (Figure 7.15) is repeated here for convenience.

STEP 2   *Extract the state machine truth table from the ASM chart* (Figure 7.15). The design of a state machine centers on calculating the next-

FIGURE 7.18
Memory module block diagram

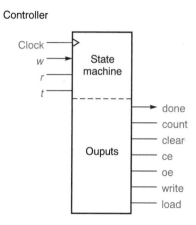

Controller

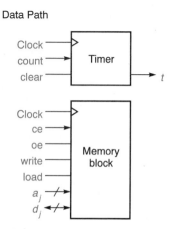

Data Path

FIGURE 7.15

**Memory module
ASM chart
(repeated)**

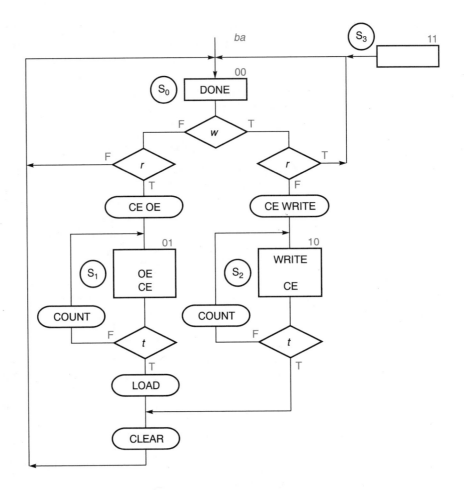

state number from the present-state number and the present-state input variables. The chart has four states and three input variables, w, r,
and t. Therefore there are four present states, which we list in the
Present State column as s_0, s_1, s_2, and s_3 (see Table 7.1). Three
columns (w, r, and t) are needed for the three input variables; one
column is needed for the next-state number. Here is how to analyze
the ASM chart to fill the columns.

State s_0   Variables w, r are in the s_0 ASM block. Two variables have
four combinations of values, resulting in four truth table rows, as
shown in Table 7.1. Variable t is not associated with s_0, so we mark
the t column with four dashes, signifying "don't care."
  The ASM chart tells us the next state for any path. In Figure 7.10,
starting from the s_0 rectangle the $w'r'$ path returns to s_0, the $w'r$ path

goes to s_1, the wr' path goes to s_2, and the wr path returns to s_0. Figures 7.19a and 7.19b illustrate two of these paths.

State s_1 Variable t is in the s_1 ASM block. One variable has two combinations of values, resulting in two truth table rows, as shown in Table 7.1. Variables w and r are not associated with s_1, so we mark the w and r columns with four dashes. Starting from the s_1 rectangle, the t' path returns to s_1 and the t path goes to s_0. Figures 7.19c and 7.19d illustrate these paths.

State s_2 Variable t is in the s_2 ASM block. One variable has two combinations of values, resulting in two truth table rows, as shown in Table 7.1. Variables w and r are not associated with s_2, so we mark the w and r columns with four dashes. Starting from the s_2 rectangle, the t' path returns to s_2 and the t path goes to s_0.

State s_3 There are no variables assigned to this ASM block. Starting from the s_3 rectangle the path goes to s_0.

STEP 3

Flip-flop defining equations bridge truth tables to circuits.

Add D flip-flop excitation columns (one column for each flip-flop). Two D flip-flops constitute the state register. They hold the present-state number. The number at the d inputs will be stored in the D flip-flops at the next clock edge. When s_0 is to be the next state, the number at the d inputs should be 00, which in voltage format is LL

Truth table rows represent paths in ASM charts.

TABLE 7.1 Memory Module Controller's Truth Table

PS[1]	PI[2]			NS[3]
	w	r	t	
s_0	F	F	–	s_0
	F	T	–	s_1
	T	F	–	s_2
	T	T	–	s_0
s_1	–	–	F	s_1
	–	–	T	s_0
s_2	–	–	F	s_2
	–	–	T	s_0
s_3	–	–	–	s_0

[1] Present state.
[2] Present inputs.
[3] Next state.

FIGURE 7.19
**ASM chart paths to
next state**

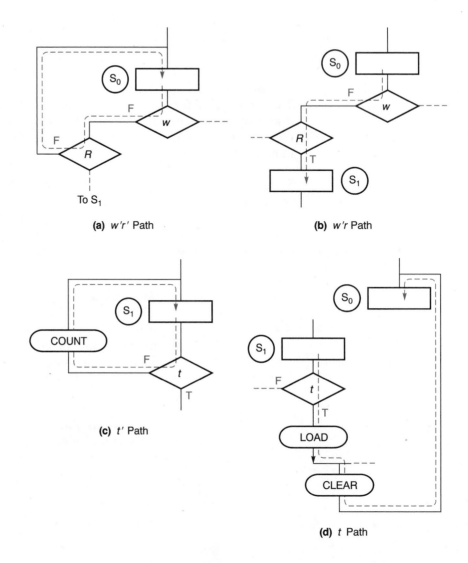

(a) $w'r'$ Path

(b) $w'r$ Path

(c) t' Path

(d) t Path

because we use the active high assignment T := H. The simplicity of
the D flip-flop next-state equation $q^+ = d$ allows us to enter numbers
in the d columns that are equal to the next-state number. We enter the
binary number corresponding to the next state (Table 7.2); we enter 00
for s_0, 01 for s_1, and so forth to define the D flip-flop functions d_1 and
d_0.

STEP 4   ***Derive logic equations.*** We write D flip-flop excitation input equations
using information from the truth table in Table 7.2. To make a state

TABLE 7.2 **Memory Module Controller's Truth Table (continued)**

| PS | PI | | | NS | | |
	w	r	t	s_j	d_1	d_0
s_0	F	F	–	s_0	0	0
	F	T	–	s_1	0	1
	T	F	–	s_2	1	0
	T	T	–	s_0	0	0
s_1	–	–	F	s_1	0	1
	–	–	T	s_0	0	0
s_2	–	–	F	s_2	1	0
	–	–	T	s_0	0	0
s_3	–	–	–	s_0	0	0

assignment, we replace the state identifiers s_j with minterms m_j of variables $q_1 q_0$.

$$d_1 = s_0 w r' + s_2 t' = m_0 w r' + m_2 t'$$

$$d_0 = s_0 w' r + s_1 t' = m_0 w' r + m_1 t'$$

The next-state number delivered to the D flip-flop inputs is $d_1 d_0$. The d_1 and d_0 equations are implemented by a combinational-logic network. The state variables q_1, q_0 and the input variables w, r, t are inputs to the combinational-logic network that we call the *next-state-number generator*.

Anticipating the use of multiplexers, we cast the equations into the form of multiplexer equations. First we recognize that all multiplexer inputs must be connected to some source. We therefore include all missing states by adding zeros in the form $s_j \times 0$. Adding zeros does not invalidate the equations. Next we replace state numbers with minterms. This is permissible because our state assignments make minterms equal the states ($m_j = s_j$).

$$d_1 = s_0 w r' + s_2 t'$$

$$d_1 = s_0 w r' + s_1 \times 0 + s_2 t' + s_3 \times 0$$

$$d_1 = w r' m_0 + 0 \times m_1 + t' m_2 + 0 \times m_3$$

$$d_0 = s_0 w' r + s_1 t'$$

$$d_0 = s_0 w' r + s_1 t' + s_2 \times 0 + s_3 \times 0$$

$$d_0 = w' r m_0 + t' m_1 + 0 \times m_2 + 0 \times m_3$$

STEP 5   *Use the standard state machine multiplexer circuit to implement the logic equations.* The minterm coefficients are the multiplexer inputs. This means we can write down the minterm coefficients as inputs to the multiplexers in the standard circuit with two flip-flops (see Figure 7.17b). The result is the circuit in Figure 7.20, which implements the memory module ASM chart state machine. This completes the state machine design. Next we design the output circuits.

EXERCISE 7.13   Find the circuit for the ASM chart.

Answer: Figure 7.2 on page 339.

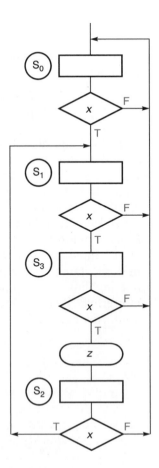

FIGURE 7.20

**Memory module
state machine**

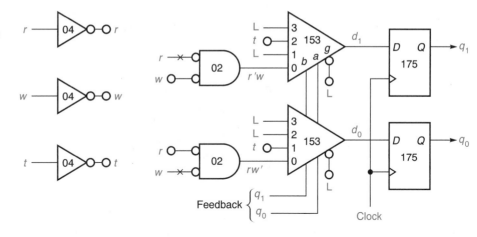

7.5.2   Output Circuit Synthesis

The memory module state machine we have designed reports the present state PS. Unconditional outputs depend only on the PS, whereas conditional outputs depend on the PS *and* the PI (present inputs). We start with a copy of the memory module truth table PS and PI columns from Table 7.2.

STEP 1

Add Output Columns to the Truth Table. For each unconditional and conditional output we add one column to the truth table (Table 7.3). There is no need to note the type of output.

Unconditional outputs are present outputs asserted in a present state.

Unconditional outputs are found inside ASM chart state rectangles. An output listed in state s_j is asserted when the ASM is in s_j. Since the assertion is not dependent on the input variables, for each output listed in state s_j we enter a T on each row corresponding to state s_j. For example, DONE is active when the ASM is in state s_0. There are four s_0 rows in the truth table because two input variables are active in s_0. We make four T entries in the DONE column, one per s_0 row. We repeat this process for all other unconditional outputs in all states.

Conditional outputs are asserted in a present state if the associated input minterm is true.

Conditional outputs are found inside ovals. Ovals are part of each ASM block's binary tree. An output listed in state s_j is asserted when the ASM is in s_j *and* the path to the oval is active. Since the assertion is dependent on the input variables, for each output listed in the state s_j binary tree we enter a T on each row corresponding to the active path in state s_j. For example, the ASM chart specifies that CLEAR is active when the ASM is in state s_1 *and* input t is True. We make one T entry in the CLEAR column in the s_1 row where $t = $ T. We repeat this process for all other conditional outputs in all states. Table 7.3 is constructed by this method. Observe that an output, such as OE, is conditional in s_0 and unconditional in s_1.

STEP 2 *Derive Output Logic Equations.* In the usual manner we obtain equations for the outputs from the truth table. There is one equation per column and one term per T entry. Notice how we write down the equations. The format anticipates using multiplexers as well as minimizing the potential for errors. This time we do not add states with zero coefficients to the equations, because we now know the missing states in effect specify L sources to corresponding multiplexer inputs.

$$\text{DONE} = s_0 \qquad\qquad\qquad\qquad = m_0$$
$$\text{CE} = s_0(rw' + r'w) + s_1 \ \ + s_2 = m_0(rw' + r'w) + m_1 \ \ + m_2$$
$$\text{OE} = s_0(rw' \qquad) + s_1 \qquad\quad = m_0(rw' \qquad) + m_1$$
$$\text{WRITE} = s_0(\qquad + r'w) \qquad + s_2 = m_0(\qquad + r'w) + m_2$$
$$\text{COUNT} = \qquad\qquad s_1 t' + s_2 t' = \qquad\qquad m_1 t' + m_2 t'$$
$$\text{CLEAR} = \qquad\qquad s_1 t \ + s_2 t \ = \qquad\qquad m_1 t \ + m_2 t$$
$$\text{LOAD} = \qquad\qquad\quad s_1 t \qquad\quad = \qquad\qquad\quad m_1 t$$

STEP 5 *Use the standard output multiplexer circuit to implement the logic equations* (Figure 7.21). The minterm coefficients are the multiplexer

TABLE 7.3 **Truth Table for Memory Module Outputs**

PS	PI			PO						
s_j	w	r	t	DONE	CE	OE	WRITE	COUNT	CLEAR	LOAD
s_0	F	F	–	T	.	.	.	.	.	.
	F	T	–	T	T	T	.	.	.	.
	T	F	–	T	T	.	T	.	.	.
	T	T	–	T	.	.	T	.	.	.
s_1	–	–	F	.	T	T	.	T	.	.
	–	–	T	.	T	T	.	.	T	T
s_2	–	–	F	.	T	.	T	T	.	.
	–	–	T	.	T	.	T	.	T	.
s_3	–	–	–	.	.	.	.	.	.	.

FIGURE 7.21

**Memory module
standard output
circuit**

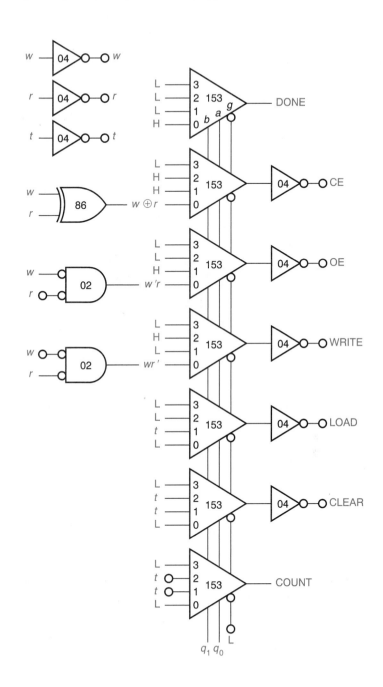

FIGURE 7.22
Memory module traditional output circuits

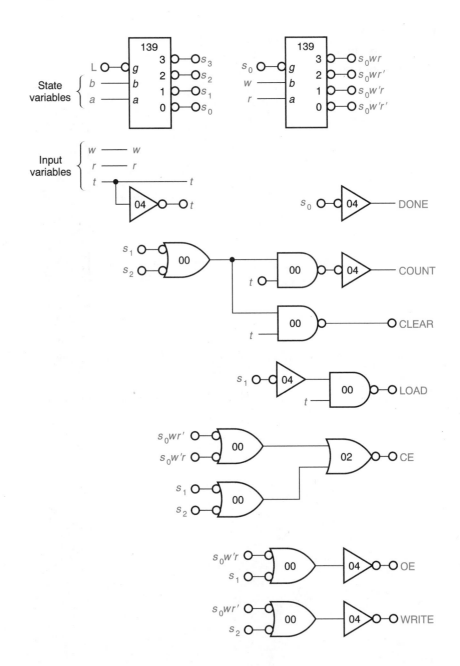

inputs. This means we can write down the minterm coefficients as inputs to the multiplexers in the standard output circuit. The result is the circuit in Figure 7.21, which implements the memory module ASM chart outputs. Compare this circuit to a more traditional circuit (Figure 7.22). In practice we do not know how to predict which circuit form uses fewer gates and combinational building blocks.

7.6 Analysis: From Circuit to ASM Chart

If we did not know the ASM chart for the circuits in Figure 7.21 and Figure 7.23, the analysis questions would be as follows. Given the circuits, what is the ASM chart (that is, what is the next state for each present state)? Second, what are the active outputs in each present state? The analysis starts by extracting flip-flop input equations from the state machine circuit for the purpose of using them to create the present-state/next-state truth table. Then ASM blocks are drawn from information in the truth table. The assembly of ASM blocks results in the ASM chart. The second part of the analysis starts by extracting output equations from the output circuits. The equations determine where outputs are inserted into the ASM chart.

STEP 1   ***Derive the d_1, d_0 flip-flop input equations.*** The multiplexer outputs are the flip-flop d_1, d_0 inputs (Figure 7.23). Minterm notation simplifies the equations and provides a one-to-one correspondence to state numbers s_j (i.e., the state assignment). The zero coefficients correspond to the L inputs.

$$d_1 = wr'm_0 + 0 \times m_1 + t'm_2 \quad + 0 \times m_3$$
$$= wr's_0 \; + 0 \times s_1 \; + t's_2 \quad + 0 \times s_3$$
$$d_0 = w'rm_0 + t'm_1 \quad + 0 \times m_2 + 0 \times m_3$$
$$= w'rs_0 \; + t's_1 \quad + 0 \times s_2 \; + 0 \times s_3$$

STEP 2   ***Extract a truth table from the d_j equations.*** The given state machine circuit (Figure 7.23) has a D flip-flop state register. This two-bit register represents two variables, q_1 and q_0. The variables q_1 and q_0 represent four possible states s_j numbered, in binary, 00, 01, 10, and 11.

We build a truth table in Table 7.4 by assigning $q_1 q_0$ the four state numbers 0, 1, 2, and 3, in sequence, in the Present State PS column. Then we add columns for d_1 and d_0 and enter the minterm coefficients. Next we add PI columns, one for each input variable used in the coefficients. Adding rows as needed, we fill in the PI columns with all input variable 0, 1 combinations for variables active in a state. Then we add "don't care" dashes for (inactive) variables not associated with a state.

Minterms' coefficients are next-state conditions.

FIGURE 7.23
State machine

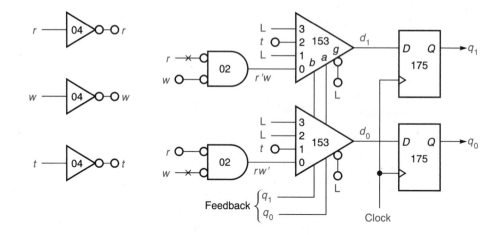

We calculate the next-state numbers NS by evaluating the terms in the d_1, d_0 columns using the input variable values in each row. Then we use the assignments $m_j = s_j$ to convert next-state numbers to state identifiers s_j. *Note:* each row of the truth table represents a path in the ASM chart that ends at the next state.

STEP 3   ***Draw an ASM block for each state.***

State 0   The two variables w and r are active in state zero. This means the two input variables w, r are in the state-zero (s_0) ASM block. Input variables w and r form a binary tree of branch diamonds at the state-

Truth tables show the input minterms associated with a present state.

TABLE 7.4   Next-State Truth Table

PS			PI			NS		
s_j	d_1	d_0	w	r	t	q_1^+	q_0^+	NS
s_0	wr'	$w'r$	0	0	–	0	0	s_0
			0	1	–	0	1	s_1
			1	0	–	1	0	s_2
			1	1	–	0	0	s_0
s_1	0	t'	–	–	0	0	1	s_1
			–	–	1	0	0	s_0
s_2	t'	0	–	–	0	1	0	s_2
			–	–	1	0	0	s_0
s_3	0	0	–	–	–	0	0	s_0

FIGURE 7.24
ASM block for s_0

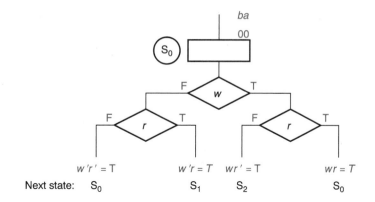

FIGURE 7.24
ASM block for s_0

zero output line. The w, r binary tree has four exits to next states (Figure 7.24).

State 1 One variable t is active in state one. Therefore only input variable t is in the state-one (s_1) ASM block. The t binary tree has two exits to next states (Figure 7.25).

State 2 One variable t is active in state two. Therefore only input variable t is in the state-two (s_2) ASM block. The t binary tree has two exits to next states (Figure 7.26).

State 3 No variables are active in state three. Therefore the state rectangle constitutes the state-three (s_3) ASM block (Figure 7.27).

STEP 4 **Form the ASM chart from the ASM blocks.** We interconnect the blocks in Figures 7.24 to 7.27 so that the states shown at the end of each exit path are the next states listed in the truth table. The result is the ASM chart without outputs in Figure 7.28.

FIGURE 7.25
ASM block for s_1

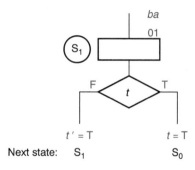

FIGURE 7.26
ASM block for s_2

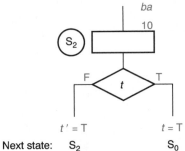

FIGURE 7.27
ASM block for s_3

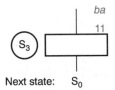

FIGURE 7.28
ASM chart with no outputs

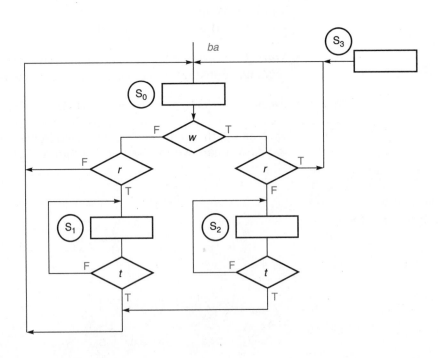

STEP 5   ***Extract equations from the output circuits.*** Equations are derived from the output circuits (see Figure 7.21 or Figure 7.22) in the usual manner.

$$\text{DONE} = s_0$$

$$\text{COUNT} = t's_1 + t's_2 \quad \textit{Moore}$$
$$\quad \textit{mealy}$$

$$\text{CLEAR} = ts_1 + ts_2$$

$$\text{LOAD} = ts_1$$

An output equation term with 1 as the state coefficient is an unconditional output.

$$\text{CE} = (w'r + wr')s_0 + s_1 + \quad s_2 \quad \textit{Both}$$

$$\text{OE} = (w'r \qquad)s_0 + s_1$$

$$\text{WRITE} = (\qquad wr')s_0 \qquad + s_2$$

FIGURE 7.29
ASM chart with outputs

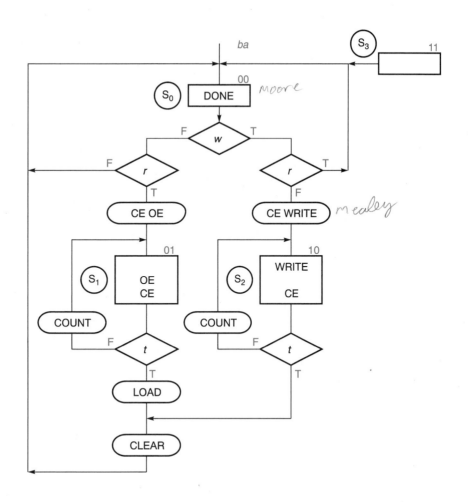

An output equation term with an expression as the state coefficient is a conditional output.

Equation terms with a coefficient of one are unconditional. Unconditional outputs are entered into state rectangles (Figure 7.29) according to the equations. Equation terms with variables as coefficients are conditional. Conditional outputs are entered into ovals placed at the end of appropriate binary tree paths (Figure 7.29) defined by the equations.

7.7   Asynchronous and Synchronous Inputs

The timing diagrams in Figure 7.30 show the effect on outputs AF and AT when input variable X is synchronous or asynchronous.

Asynchronous X   Logical errors may occur if X changes state during a required setup or hold time. Also, the AT waveform may end up being too narrow (causing its own errors) when X goes high, as shown. The AF waveform goes low during the clock period. AF being low may also be "too narrow" before the next clock edge, in the sense of violating a setup time requirement.

Synchronous X   Now the AT waveform duration is always high a full clock period and AF goes low at the beginning of the clock period.

Synchronize asynchronous sources before connecting them to synchronous machine inputs

A suggested synchronizer circuit is discussed in Section 6.4.4.

FIGURE 7.30

Synchronous and asynchronous inputs

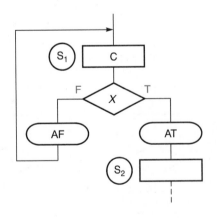

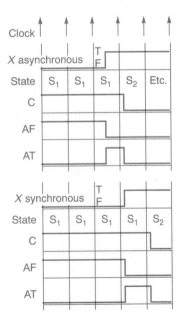

Let us consider the ASM fragment and timing diagram in Figure 7.30 when X is synchronous. If we assume the logic machine is in state s_1, then:

Unconditional output C is active
and
Conditional output AF is active if input variable X is False. The state machine moves to state s_1 on the next clock edge. In this case the next state is the same state. The system dwells in present state s_1.

or

Unconditional output C is active
and
Conditional output AT is active if input variable X is True. The state machine moves to state s_2 on the next clock edge.

The next state is loaded at each active clock edge. When X is False, the next state is s_1, which may give the impression that nothing happens. However, something always happens; the active clock edge moves the machine from s_1 to s_1 or from s_1 to s_2.

7.8 Practical ASM Topics

The ASM executes all events in any ASM block at the same time; the ASM executes events in parallel. This advantage leads to complications with loops, which are discussed first. Then the concept of the linked ASM is introduced.

7.8.1 Control Loops

An ASM dwelling in one state is one form of control loop (similar to a programming language loop). This simple form also illustrates one trap a designer can fall into. The problem arises from the fact that equations such as $c = c + a$ are executed by the clock edge at the *end* of clock periods. The possible negative consequences are not necessarily clear until a timing diagram is drawn. Once understood, the negative consequences are avoided by judicious choice of the initial how-many-times-to-loop value or the type of loop.

The ASM outputs "add_a_to_c" and "sub_1_from_b" are inputs to the data path controlled by the state machine (Figure 7.31). The data path that executes the ASM outputs and evaluates $c = c + a$ and

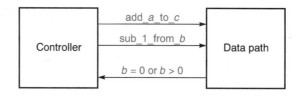

$b = b - 1$ uses the same clock as the ASM state machine. Therefore the controller and the controlled are synchronous. The actions corresponding to ASM outputs "add_a_to_c" and "sub_1_from_b" are data path actions in the *next state* resulting from the present-state inputs to the data path.

The begin-until loop uses unconditional outputs. The begin-until loop in Figure 7.32a always spends one or more clock cycle(s) in its state, because the exit condition is evaluated at the end of a loop cycle. Unlike the begin-until loop, the while-do-repeat loop uses conditional outputs. The while-do-repeat loop in Figure 7.32b also spends one or more clock cycles in its state, because the exit condition is still evaluated at the end of a loop cycle.

The sequence of events for the begin-until loop when $b = 0$ is as follows (Figure 7.33). At the end of clock cycle$_2$ the state machine moves from state s_{n-1} to s_n to enter the loop. Variable values do not change during cycle$_3$ because the equations are not executed until the end of cycle$_3$.

Unconditional output "equations" are executed by the active clock edge at the end of a state time.

At the end of cycle$_3$, b is decremented and a is added to c. Since $b = 0$ during cycle$_3$, the state machine moves from s_n to s_{n+1} in cycle$_4$. In begin-until loop ASM fragments, unconditional equations execute

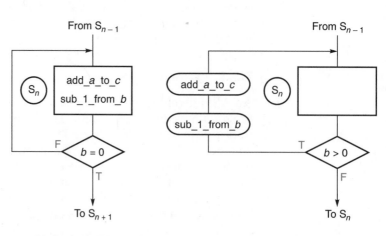

(a) Begin-Until Loop ASM   **(b)** While-Do-Repeat Loop ASM

FIGURE 7.33
Begin-until loop timing ($b = 0$)

Clock cycle	↑ 1	↑ 2	↑ 3*	↑ 4	↑ 5	↑ ...
State	s_{n-2}	s_{n-1}	s_n	s_{n+1}	...	
Loop cycle	—	—	1	—	—	
Values: a	5	5	5	5	5	...
b	0	0	0	−1	−1	...
c	0	0	0	5	5	...

* loop entered

$b + 1$ times. The case when $b = 3$ is illustrated in **Figure 7.34**. This is typical, and comparable to what can happen in a computer program.

The value of b determines the next state in Figure 7.34. At the end of cycle$_3$, cycle$_4$, and cycle$_5$ the next-state decision is to go to s_n, because b is greater than zero. At the end of cycle$_6$ the next-state decision is to go to s_{n+1}, because $b = 0$. The machine dwells in s_n for four cycles. This is why $c = 4a = 20$ in cycle$_7$.

The sequence of events for the while-do-repeat loop when $b = 0$ is as follows (Figure 7.35). At the end of clock cycle$_2$ the state machine moves from state s_{n-1} to s_n to enter the loop. Variable values do not change during cycle$_3$, because the equations are not executed until the end of cycle$_3$. At the end of cycle$_3$ the equations are not executed, because $b = 0$. And since $b = 0$ during cycle$_3$, the state machine moves from s_n to s_{n+1} in cycle$_4$.

Conditional output "equations" are also executed by the active clock edge at the end of a state time.

In while-do-repeat loop ASM fragments, conditional equations execute b times. The case when $b = 3$ is illustrated in **Figure 7.36**.

Observe that conditional outputs may be added to **begin-until** loops, and unconditional outputs may be added to **while-do-repeat** loops.

FIGURE 7.34
Begin-until loop timing ($b = 3$)

Clock cycle	↑ 1	↑ 2	↑ 3*	↑ 4	↑ 5	↑ 6	↑ 7	↑ 8	↑ 9	↑ 10	↑
State	s_{n-2}	s_{n-1}	s_n	s_n	s_n	s_n	s_{n+1}	...			
Loop cycle	—	—	1	2	3	4					
Values: a	5	5	5	5	5	5	5	...			
b	3	3	3	2	1	0	−1	...			
c	0	0	0	5	10	15	20	...			

* loop entered

FIGURE 7.35
**While-do-repeat
loop timing ($b = 0$)**

Clock cycle		↑ 1	↑ 2	↑ 3*	↑ 4	↑ 5	↑ ...
State		s_{n-2}	s_{n-1}	s_n	s_{n+1}	...	
Loop cycle		—	—	1	—	—	
Values:	a	5	5	5	5	5	...
	b	0	0	0	0	0	...
	c	0	0	0	0	0	...

* loop entered

7.8.2 Linked ASMs

Linked ASMs reduce the number of states required to implement the system, thereby simplifying the hardware circuit. Any digital function that is used repeatedly by an algorithm may be implemented as a stand-alone ASM linked to the mainline ASM implementing the algorithm. Linking allows repeated use of a function's circuit. This repeated use represents states not added to the mainline ASM. (Readers with computer programming experience will recognize that the linked ASM is analogous to the subroutine.) Links are established when an output from ASM_j is an input to ASM_k, and vice versa. A mainline ASM linked to ASMs that are in turn linked to other ASMs and so on is an example of top-down design at its best. Perhaps several examples can make the point.

A link is one ASM's output connected to another ASM's input.

Let us suppose that the fragment of the mainline ASM in Figure 7.37 is linked to an ASM implementing function R, where R is whatever function we please. There may be many links to the R-ASM from various mainline states. Further, let's suppose function R is at rest until called by the mainline ASM.

FIGURE 7.36
**While-do-repeat
loop timing ($b = 3$)**

Clock cycle		↑ 1	↑ 2	↑ 3*	↑ 4	↑ 5	↑ 6	↑ 7	↑ 8	↑ 9	↑ 10 ↑
State		s_{n-2}	s_{n-1}	s_n	s_n	s_n	s_n	s_{n+1}	...		
Loop cycle		—	—	1	2	3	4				
Values:	a	5	5	5	5	5	5	5	...		
	b	3	3	3	2	1	0	0	...		
	c	0	0	0	5	10	15	15	...		

* loop entered

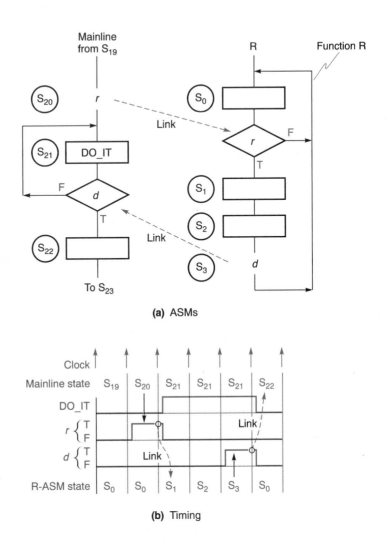

FIGURE 7.37
Linked ASM

(a) ASMs

(b) Timing

Links are established by outputs r from mainline states to input r in the R-ASM (Figure 7.37). Mainline output r is input r to the R-ASM. With input r asserted the R-ASM starts and executes its task. Meanwhile, the mainline dwells in waiting state s_{21} until input d is asserted (Figure 7.37). Mainline input d is output d from the R-ASM. For example, when the mainline needs to output the DO_IT command (from mainline state s_{21} in Figure 7.37) for a specific time, R-ASM provides the time delay function. This situation is shown in Figure 7.37a, where states s_1, s_2, and s_3 define a three-clock-period delay. The wise designer draws a timing diagram to verify correct performance (Figure 7.37b).

FIGURE 7.38
Linked ASM—
Multiple calls

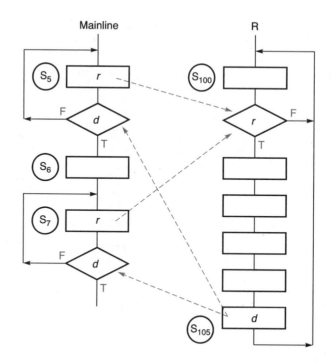

FIGURE 7.39
Linked ASM—
Multiple calls—
No feedback

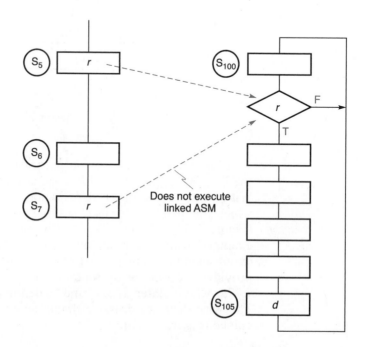

Multiple calls are readily implemented (Figure 7.38). Observe how the waiting state, s_5 or s_7, emits output r while waiting. The R-ASM does not care about r once the R-ASM has moved from its resting state, s_{100}. Also note that output d must be in s_{105} and not in s_{100}.

Linked ASMs usually have fewer states than one ASM performing the same algorithm.

The linked ASM can execute in parallel with the mainline when so desired. The mainline calls the link ASM to start it executing. Then the mainline continues on with its task as the link ASM executes in parallel. In this case the branch back to the same state (when d is False) is not used and the d feedback link is omitted (Figure 7.39). When the d feedback link is omitted, execution of R depends on how many mainline states intervene between calls (Figure 7.39). R is not executed when the intervening number of mainline states is less than the number of executing states in R (Figure 7.39). A programmer would say subsequent calls to R are not executed unless the current call has completed. Caution is needed with this high-risk practice.

EXERCISE 7.14

Find the parallel sequences of states for the linked ASM in Figure P7.8 on page 407 (s_0 to s_0). Start from s_0 and s_{10}.

Answer:

ASM_0	s_0	s_1	s_1	s_1	s_1	s_2	s_3	s_0
ASM_1	s_{10}	s_{10}	s_{11}	s_{12}	s_{13}	s_{10}	s_{10}	s_{10}

■

EXERCISE 7.15

Find the parallel sequences of states for the linked ASM in Figure P7.9 on page 408 when R is False (s_0 to s_0). Start from s_0 and s_{10}.

Answer:

ASM_0	s_0	s_1	s_2	s_3	s_0
ASM_1	s_{10}	s_{10}	s_{10}	s_{10}	s_{10}

■

EXERCISE 7.16

Find the parallel sequences of states for the linked ASM in Figure P7.9 on page 408 when R is True (s_0 to s_0 to s_0). Start from s_0 and s_{10}.

Answer:

ASM_0	s_0	s_1	s_2	s_3	s_0	s_1	s_2	s_3	s_0
ASM_1	s_{10}	s_{10}	s_{10}	s_{10}	s_{11}	s_{11}	s_{11}	s_{11}	s_{10}

■

7.9   Design Example: Computer Controller

At this point we are in a position to design any state machine we please. However, the straightforward process we have just learned could burden us unduly unless we gain further insight, by studying more complex examples such as the ASM representing a simple pipelined computer controller shown in Figure 7.40. (The details of the names and the control process are left to a computer design course.)

FIGURE 7.40

**ASM for a simple
pipelined computer
controller**

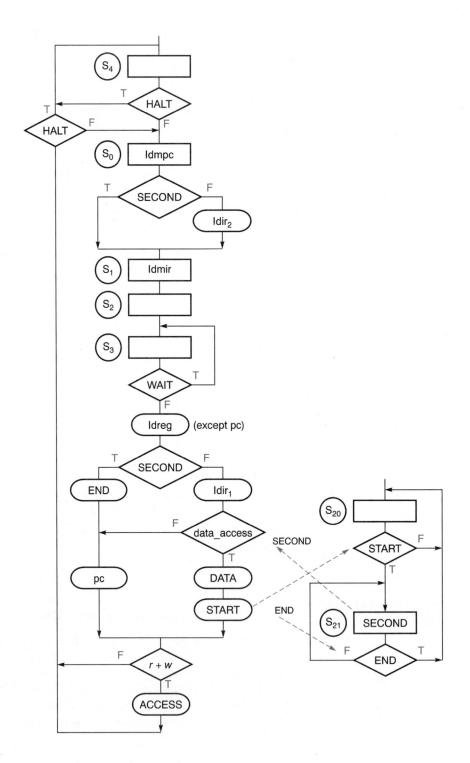

State machine truth table The conventional procedure would be to make a (large and complex) state machine truth table with one column for every input variable. This is not necessary if we make two observations:

1. s_1 is the state following s_0 regardless of the value of the input Second: in s_0, Second is a "don't care." The next state is always s_1.
2. The input tree from the WAIT diamond to the $r + w$ diamond has only one entrance and one exit, and goes nowhere else. The same is true of the $r + w$ diamond.

In effect the only inputs in s_3 that influence the state machine's next-state decisions are HALT and WAIT, which is why the truth table in Table 7.5 is simple. (Try building it using every input variable.)

Insight: look for simplifying equivalent paths.

A very different consideration is the weight given to inputs by implication when they are listed in the truth table. In Table 7.5 WAIT has been given greater weight than HALT by virtue of its position, because in s_3 HALT is below WAIT in the tree. To appreciate this fine point, make a table with HALT in the first position.

Insight: select input weight carefully.

TABLE 7.5 **Computer Controller Next-State Truth Table**

PS	PI		NS
s_j	WAIT	HALT	s_j^+
s_0	–	–	s_1
s_1	–	–	s_2
s_2	–	–	s_3
s_3	F	F	s_0
	F	T	s_4
	T	–	s_3
s_4	–	F	s_0
	–	T	s_4

TABLE 7.6 **Computer Controller Output Truth Table**

Let w = WAIT, s = SECOND, d = data_access, $r = r + w$

PS	PI				OUTPUTS									
s_j	w	s	d	r	ldmpc	$ldir_2$	ldmir	ldreg	$ldir_1$	data	st	end	pc	access
s_0	–	F	–	–	T	T	·	·	·	·	·	·	·	·
	–	T	–	–	T	·	·	·	·	·	·	·	·	·
s_1	–	–	–	–	·	·	T	·	·	·	·	·	·	·
s_2	–	–	–	–	·	·	·	·	·	·	·	·	·	·
s_3	T	–	–	–	·	·	·	·	·	·	·	·	·	·
	F	F	F	–	·	·	·	T	T	·	·	·	·	·
	F	F	T	–	·	·	·	T	T	T	T	·	T	·
	F	T	–	–	·	·	·	T	·	·	·	T	T	·
	–	–	–	T	·	·	·	·	·	·	·	·	·	T
s_4	–	–	–	–	·	·	·	·	·	·	·	·	·	·

Output truth table Here, too, input weight selection is important. Tree hierarchy determines the weight. In s_3 the hierarchy is WAIT, SECOND, data_access, and $r + w$.

Insight (repeated): select input weight carefully.

Input variables in independent trees are "don't cares" in foreign trees. Variable $r + w$ is foreign to the tree starting with WAIT in the input diamond. The variables WAIT, SECOND, and data_access are foreign to the $r + w$ tree. See s_3 in Table 7.6.

Insight: foreign variables are "don't cares."

SUMMARY

State variable: A quantity capable of assuming the value 0 or 1.

State: The condition of the sequential digital circuit, as specified by a set of state variables q_j, reflecting the history of the state machine.

Branch: Associated with each state are none, one, two, or more input variables. The present state of these input variables in some combination selects the next state. The next state is determined without conditions when there are no input variables.

Outputs: States may specify unconditional or conditional outputs. Unconditional outputs assigned to a state are active when the system is in that state. Conditional outputs assigned to a state are active when the system is in that state *and* when input conditions are as specified (see *Branch* above).

Asynchronous machine: The present state is the state of the q_j at the time before inputs change. The next state is the state of the q_j after the circuit has reacted to input changes.

Synchronous machine: The present state is the state of the q_j at the time before clock edge occurs. The next state is the state of the q_j after clock edge occurs and after the circuit has reacted to the clock edge. The inputs and the present state determine which state is next; the clock edge determines when the state machine goes to the next state.

There is always a next state: The next state is loaded at each active clock edge. When the next state is the same as the present state, the impression that nothing happens may be created. However, something always happens; the active clock edge moves the machine from the present state to the next state.

Digital Design Method

Step 1.   Create the ASM chart from the algorithm. Use the three symbols for state, branch, and conditional output to build ASM blocks, which are then connected together to make up the chart. Do not be surprised to find that this step takes up most of the design time for any reasonably complex digital design. You will find yourself constantly reworking the ASM chart. This process is a source of ideas and insight.

Step 2.   Make a state machine truth table. Draw up a truth table relating input variables and the present state to the present (conditional and unconditional) outputs and the next state. This information is taken from the ASM chart.

Step 3.   Add columns for flip-flop inputs and outputs to the truth table.

Step 4.   Derive logic equations. Draw up the Boolean equations for flip-flop excitation inputs, the unconditional outputs, and the conditional outputs. Recast the equations into multiplexer format.

Step 5.   Synthesize the hardware from the equations.

Our standard circuit for the state machine and output circuits uses a multiplexer format (see Figures 7.17 and 7.21).

REFERENCES

Breeding, K. J. 1989. *Digital Design Fundamentals*. Englewood Cliffs, N.J.: Prentice-Hall.

Clare, C. R. 1971. *Designing Logic Systems Using State Machines*. New York: McGraw-Hill.

Roth, C. H. 1992. *Fundamentals of Logic Design*, 4th ed. St. Paul, Minn.: West.

PROJECT 09 Demonstration Report

Name ______________________________ SID# ________________ Section _____

(last) (initials)

Approved by ______________________________ Date ______________________________

Grade on report ______________________

System Clock Design

1. Design the single-pulse/periodic clock circuit.

_____ truth table from ASM chart with D flip-flops
_____ equations for d_j inputs to flip-flops and outputs
_____ state machine circuit
_____ output circuits
_____ debounced switch input circuits

2. Build the circuit.

_____ circuit documents

3. Use the truth table signal generator to test the circuit.

_____ input/output timing diagram

Background: A single pulse circuit emits a signal that is true for one complete clock period for each press of a push button. The signal is synchronous with the clock edges.

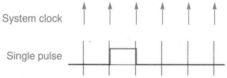

On the other hand, a periodic clock with 50% duty cycle emits a sequence of complete clock periods, where each is two-system-clock-periods in duration. By 50% duty cycle we mean that the signal is at level H 50% of the time and at level L the other 50% of the time.

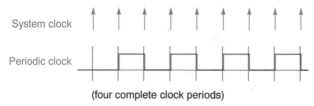

(four complete clock periods)

So-called system clocks usually have a mode switch for selecting single-pulse or periodic mode of operation.

The important design task is to emit only *complete* events when the mode switch is transferred from single to periodic and vice versa.

Project Tasks

1. Design a single-pulse/periodic clock circuit with debounced switch inputs using the ASM chart in Figure 7.10. Let q_1 and q_0 be the ASM state variables.

2. Build the circuit. Use 74LS00 gates to debounce switches. Use 74LS00, 74LS153, and 74LS175. Use an LED for the Ready report. Save it on your breadboard for future projects.

3. Test the circuit. Connect signal generator output clk to the clock input, and let $x = q_3$, $y = q_2 + q_1'$. Plot q_1, q_2, q_3, y, p, and Ready.

PROBLEMS

Note: In all problems requiring timing diagrams show "real" waveforms with finite propagation delays after clock edges.

7.1 The equations for a two-flip-flop circuit using a 74LS74 are: $d_0 = q_1'$, $d_1 = q_0$. Make the PS, NS table. Derive the state diagram, and draw the circuit. Start with all q_j low (state zero), and plot the q_j outputs on a timing diagram.

7.2 The equations for a two-flip-flop circuit using a 74LS109 are: $j_0 = q_1'$, $k_0 = q_0$, $j_1 = q_0$, $k_1 = q_1$. Make the PS, NS table. Derive the state diagram, and draw the circuit. Start with all q_j low (state zero), and plot the q_j outputs on a timing diagram.

7.3 The equations for a two-flip-flop circuit using a 74LS74 are: $d_0 = q_1'$, $d_1 = q_1'q_0 + q_1q_0'$. Make the PS, NS table. Derive the state diagram, and draw the circuit. Start with all q_j low (state zero), and plot the q_j outputs on a timing diagram.

7.4 The equations for a three-flip-flop circuit using a 74LS74 are: $d_0 = q_2$, $d_1 = q_2$ xor q_0, $d_2 = q_2$ xor q_1. Make the PS, NS table. Derive the state diagram, and draw the circuit. Start with all q_j high (state 7), and plot the q_j outputs on a timing diagram.

7.5 The equations for a three-flip-flop circuit using a 74LS74 are: $d_0 = q_2$, $d_1 = q_2$ xor q_0, $d_2 = q_1$. Make the PS, NS table. Derive the state diagram, and draw the circuit. Start with all q_j high (state 7), and plot the q_j outputs on a timing diagram.

7.6 The equations for a two-flip-flop circuit using a 74LS109 are: $j_0 = 1$, $k_0 = 1$, $j_1 = q_0$, $k_1 = q_0$. Make the PS, NS table. Derive the state diagram, and draw the circuit. Start with all q_j low (state zero), and plot the q_j outputs on a timing diagram.

7.7 The equations for a two-flip-flop circuit using a 74LS109 are: $j_0 = q_0'$, $k_0 = q_0$, $j_1 = q_1$ xor q_0, $k_1 = q_1$ xor q_0'. Make the PS, NS table. Derive the state diagram, and draw the circuit. Start with all q_j low (state zero), and plot the q_j outputs on a timing diagram.

7.8 The equations for a two-flip-flop circuit using a 74LS109 are: $j_0 = 1$, $k_0 = 1$, $j_1 = q_0'$, $k_1 = q_0'$. Make the PS, NS table. Derive the state diagram, and draw the circuit. Start with state 2, and plot the q_j outputs on a timing diagram.

7.9 The equations for a two-flip-flop circuit using a 74LS109 are: $j_0 = q_0'$, $k_0 = q_0$, $j_1 = q_1$ xor q_0', $k_1 = q_1$ xor q_0. Make the PS, NS table.

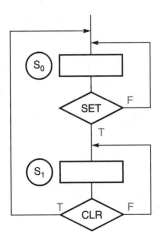

Derive the state diagram, and draw the circuit. Start with all q_j low (state zero), and plot the q_j outputs on a timing diagram.

7.10 Design the ASM specified by the ASM chart in Figure P7.1.

(a) Use D flip-flops and multiplexers.
(b) Use JK flip-flops, gates, and decoders.

7.11 Design the ASM specified by the ASM chart in Figure P7.2.

(a) Use D flip-flops and multiplexers.
(b) Use JK flip-flops, gates, and decoders.

FIGURE P7.2

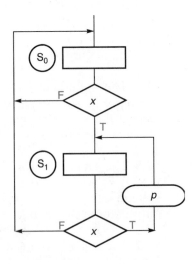

FIGURE P7.3

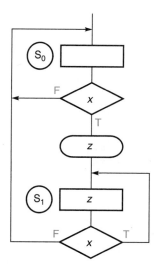

FIGURE P7.4

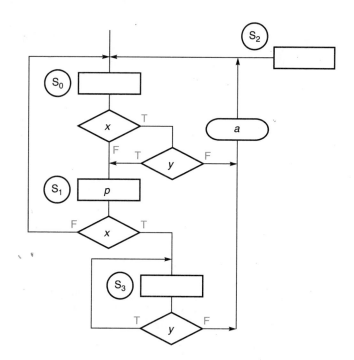

7.12 Design the ASM specified by the ASM chart in Figure P7.3.

(a) Use D flip-flops and multiplexers.
(b) Use JK flip-flops, gates, and decoders.

7.13 Design the ASM specified by the ASM chart in Figure P7.4.

(a) Use D flip-flops and multiplexers.
(b) Use JK flip-flops, gates, and decoders.

7.14 Given the x, y waveforms, plot p and a for the ASM chart in Figure P7.4. Show the sequence of states.

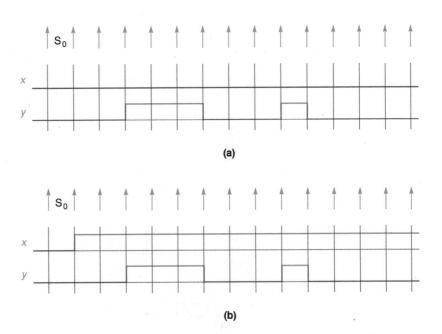

(a)

(b)

7.15 Design the ASM specified by the ASM chart in Figure P7.5.

(a) Use D flip-flops and multiplexers.
(b) Use JK flip-flops, gates, and decoders.
(c) Use D flip-flops and one-hot format.

7.16 Given the x waveforms, plot $p_1 p_2 r_1 r_1$ for the ASM chart in Figure P7.5. Show the sequence of states.

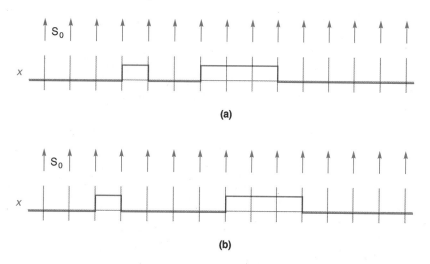

(a)

(b)

7.17 Design an ASM that generates the sequence of states 0132013 etc. Use $q_1 q_0$ as state variables. Draw the ASM chart.

(a) Use D flip-flops and multiplexers.
(b) Use JK flip-flops, gates, and decoders.

7.18 Design an ASM that generates the sequence of states 243675124 etc. Use $q_2 q_1 q_0$ as state variables. Draw the ASM chart.

(a) Use D flip-flops and multiplexers.
(b) Use JK flip-flops, gates, and decoders.

7.19 The sequence of states is 320132 etc. when input u is True. When u is False the sequence of states is 1023102 etc. Draw the ASM chart. Design the circuit.

(a) Use D flip-flops and multiplexers.
(b) Use JK flip-flops, gates, and decoders.

7.20 Convert the timing diagram ($0 = L$, $1 = H$) to an ASM chart. Design the circuit. q is a conditional output.

Input x 0 1 1 1 0 0 1 0 1 1 1 0 1 0

Output q 0 0 1 0 1 0 1 0 0 1 0 1 1 0

Use D flip-flops in a one-hot format.

FIGURE P7.5

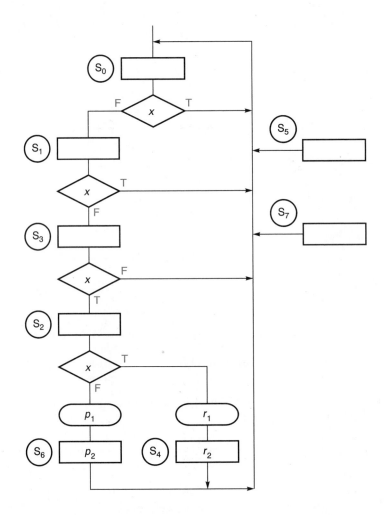

7.21 Design the ASM specified by the ASM chart in Figure P7.6.

(a) Use D flip-flops and multiplexers.
(b) Use JK flip-flops, gates, and decoders.

7.22 Design the ASM specified by the ASM chart in Figure P7.7. Use D flip-flops in a one-hot format.

7.23 Given the x waveform, enter the sequence of states for the state machine in Example 7.5.

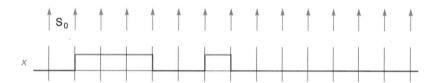

7.24 Given the x waveform, enter the sequence of states for the state machine in Example 7.6, and plot the g output waveform.

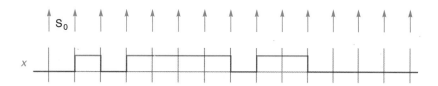

7.25 In Figure 7.32a, replace the exit condition $b = 0$ with $b = 1$, and repeat the analysis of Figure 7.34.

7.26 In Figure 7.32b, replace the exit condition $b > 0$ with $b = 0$, and repeat the analysis of Figure 7.36.

7.27 Design the circuit specified by the linked ASM chart in Figure P7.8.

7.28 Design the circuit specified by the linked ASM chart in Figure P7.9.

7.29 Refer to the ASM chart linked to a 163 chip in Figure P7.10. Design the state machine specified by the ASM chart. Make a diagram showing the sequence of states, and waveforms for count, end, w, and load_6. Show the count number in each clock period.

7.30 Modify the ASM in Figure P7.10. Replace load_6 with load_n, where $n = 0$ to F hex. How would you set n? Draw three timing diagrams for three different n. Let $n = 7$, E, and F.

7.31 Refer to the ASM chart linked to a 163 chip in Figure P7.11. Design the state machine specified by the ASM chart. Make a diagram showing the sequence of states, and waveforms for count, end, load_n, and w. Show the count number in each clock period.

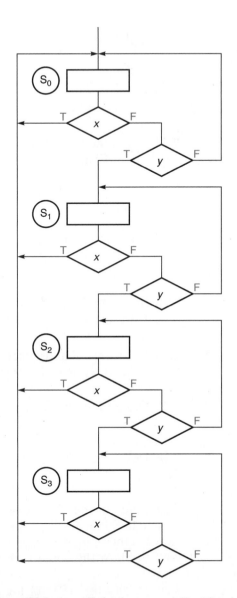

7.32 Refer to the ASM chart linked to two 163 chips in Figure P7.12. Design the state machine specified by the ASM chart. Make a diagram showing the sequence of states, and waveforms for count, =FF, =4F, load_EF, and w. Show the count number in each clock period.

7.33 Refer to Figure 7.40. Design the state machine.

7.34 Refer to Figure 7.40. Design the output circuits.

FIGURE P7.7

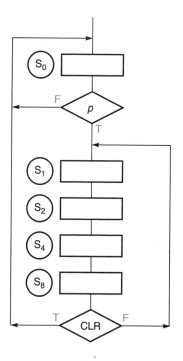

FIGURE P7.8

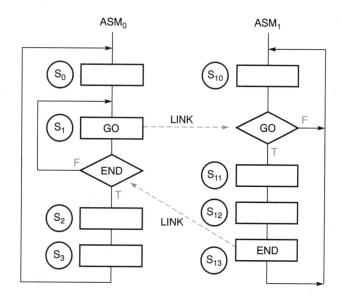

FIGURE P7.9

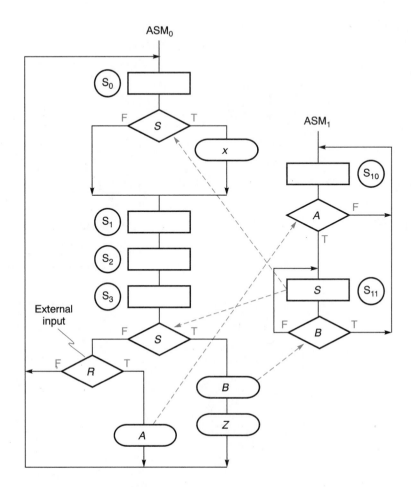

FIGURE P7.10

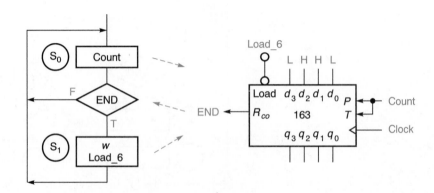

FIGURE P7.11

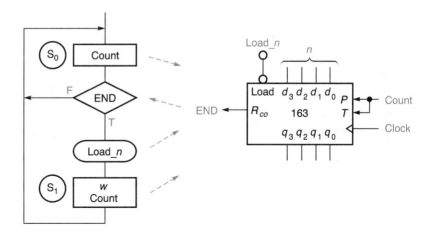

FIGURE P7.12

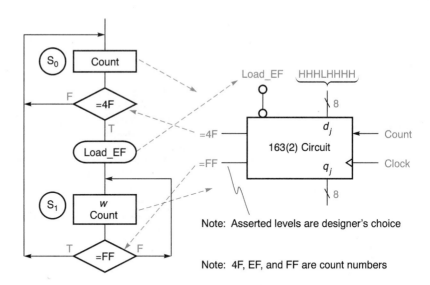

Note: Asserted levels are designer's choice

Note: 4F, EF, and FF are count numbers

8 SEQUENTIAL BUILDING BLOCKS

We need to move up the ladder of complexity if we are to be more productive as designers. Most controllers and data paths are too complex to be implemented with a mass of gates and flip-flops. We need sequential building blocks, which happen to take the form of registers. Registers implement complex operators to store data, count events, and perform lateral one-bit moves to transform data by factors of 2.

Sequential building blocks allow us to move further up the ladder of complexity. A flip-flop can change state once per clock period at every rising (or falling) active clock edge. This property is unsuitable for storing data. A multiplexer is added at the D flip-flop input so that the flip-flop changes the stored data only when a load command is asserted. This is the enabled D flip-flop, which is the basic element of any storage register.

Next we study counters, which are storage registers to which gates and wiring have been added. Our new knowledge about the ASM chart and ASM circuit analysis and synthesis is applied to the design of up and down synchronous counters. In the process, experience is gained with combinational mixed-logic circuit analysis and more is learned about the power of XOR.

Shift registers are storage registers with additional gates and wiring. The power and simplicity of the ASM method simplifies the process of designing circuits for a variety of shift operators.

INTRODUCTION

Storage register, counter, and shift register functions are basic to many applications. Before the specific assemblies of gates and flip-flops implementing these functions became available as standard parts, many engineers found themselves designing these logic functions over and over again on an as-needed basis. This is why the world was ready for sequential building blocks when fabrication of what are known as medium-scale integrated (MSI) circuits became feasible.

8.1 Storage Register

A *register* is a group of two or more flip-flops whose bits are related. For example, an eight-bit register stores one byte of data.

D flip-flop stores input data at every active clock edge.

The D flip-flop defining equation, $q^+ = d$, implies that, at one rising or one falling clock edge per period, a D flip-flop stores the present value of d, which becomes the next value of q (Figure 8.1). We will refer to the edge where storage takes place as the active edge.

FIGURE 8.1
**D flip-flop
input/output
relationship**

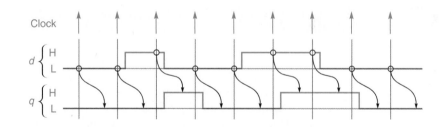

o = active input and active clock edge implement the defining equation.

Data is stored, however, with the intention of saving the data for future use. When a register is part of a controlled data path, data is modified as required by the controlling state machine. New values are saved until a change is once again required. Hence the act of storing data, manifested as the changing of values in a register, takes place at specified times and certainly not at each active clock edge. Clearly, a register must include more than just a group of D flip-flops.

8.1.1   Load Command

Asserted load input AND a clock edge store input data in a register.

The act of storing data is called *loading*. We say that at specific active clock edges we want to load data synchronously into a register. This implies the need for an input control line named *load*. The load function transforms a D flip-flop into a register flip-flop, also known as an *enabled D flip-flop*. When a load input line is asserted, the data at the register input is loaded at the next active clock edge. When the load line is not asserted, the stored data does not change at the next active clock edge. The present data is on hold until the next time the load line is asserted (Figure 8.2). This is why a register is an assembly of enabled D flip-flops.

FIGURE 8.2
Load command

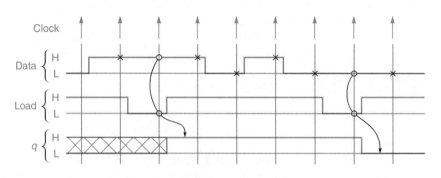

o = active   × = Inactive (ignore)

Given this information we can write the following defining equation for a data storage flip-flop. The corresponding truth table is shown in Figure 8.3.

$$q^+ = \text{load } d \text{ or hold } q$$

$$q^+ = gd + g'q$$

where

q = present value stored

d = signal at input, which is the data to be stored

g = load command (We choose g because g enables data loads; we do not use L because L stands for "low voltage.")

We need to design a circuit for the one-bit enabled D flip-flop. First we let Boolean variable q represent the state of one D flip-flop, which has two states, 0 and 1. Then, when g in the defining equation $d = gd + g'q$ represents a multiplexing variable, the circuit in upcoming Figure 8.6 results. Next we learn that the "standard" ASM method results in a more complex circuit.

The truth table based on the defining equation is put into ASM format (Figure 8.4). Next, we synthesize from the truth table a "standard" multiplexer circuit for the enabled D flip-flop. The equation for the function d_0 defined in the truth table is

$$d_0 = q_0'gd + q_0g' + q_0gd$$

$$= q_0'gd + q_0(g' + gd)$$

$$= q_0'gd + q_0(g' + d)$$

This equation is in a form suitable for a standard ASM mux circuit, shown in Figure 8.4, where q_0 is the multiplexing variable. The equation is converted into the defining equation when the roles of g and q_0 are exchanged so that g becomes the multiplexing variable.

FIGURE 8.3
Enabled D flip-flop truth table

Defining equation: $q^+ = gd + g'q$
Excitation table: q is a table-entered variable.

Row	g	d	q^+	Property
0	0	0	q	hold q, $q^+ = q$
1	0	1	q	hold q, $q^+ = q$
2	1	0	0	load d, $q^+ = d$
3	1	1	1	load d, $q^+ = d$

$$d_0 = q_0'gd + q_0g' + q_0gd$$
$$= gd(q_0' + q_0) + g'q_0$$
$$= gd + g'q_0$$

The point is that, while a "standard" method may always yield a solution, seeking a simplified circuit can be rewarding.

Another useful exercise is to derive the ASM chart from the truth table in Figure 8.4. We start the chart by drawing two state rectangles for states when $q_0 = 0$ and $q_0 = 1$ (Figure 8.5a). From the truth table in Figure 8.4 we see that the two input variables g and d are associated with each state. This means a two-variable binary tree is associated with each state rectangle. The truth table's next-state column specifies the next state at the exit of each path through the binary trees (Figure 8.5b). We assemble the enabled D flip-flop ASM chart by connecting the two ASM blocks together (Figure 8.5c). The chart is simplified by recognizing that d is a "don't care" when $g = 0$ (Figure 8.5d).

Defining equation: $q^+ = gd + g'q$

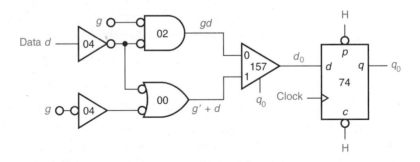

PS	PI		NS
q_0	g	d	$q_0^+\text{-}d_0$
0	F	F	0
	F	T	0
	T	F	0
	T	T	1
1	F	F	1
	F	T	1
	T	F	0
	T	T	1

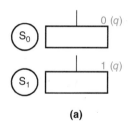

(a)

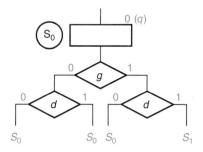 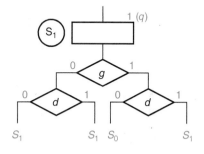

(b)

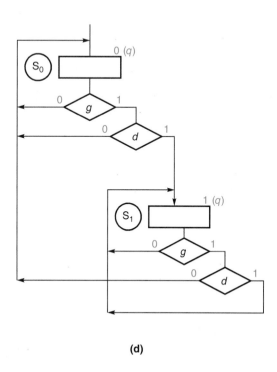

(c) (d)

FIGURE 8.5
**Enabled D flip-flop
ASM chart
evolution**

FIGURE 8.6
FIGURE 8.6
Enabled D flip-flop circuit

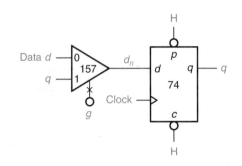

FIGURE 8.7
173 Four-bit register

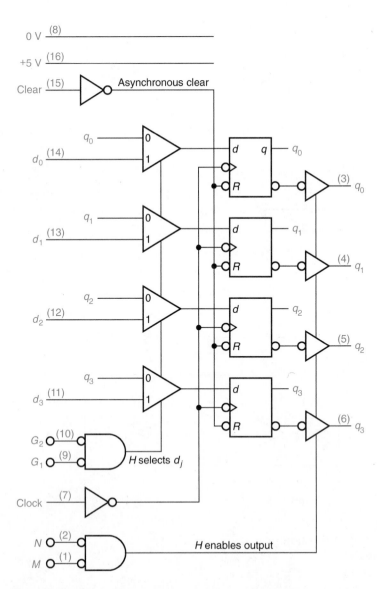

8.1.2   173 Four-Bit Enabled D Flip-Flop Register

The enabled D flip-flop circuit with g as the multiplexing variable in Figure 8.6 is one bit of standard family member 173 (Figure 8.7). The load command $g = G_1 G_2$ (both inputs are asserted). When g is not asserted, the next rising clock edge loads the present state $q_3 q_2 q_1 q_0$ into the four-bit 173 register, holding the data in storage. When g is asserted, the next rising clock edge loads $d_3 d_2 d_1 d_0$ into the register. The three-state output circuit facilitates connect and disconnect to a bus according to $s = MN$. An asynchronous clear input is provided. In a synchronous system, asynchronous clear is used only when power is turned on, to prevent the flip-flops from entering the metastable state.

8.2

Counting Register

Synchronous look-ahead carry up and down counters are now introduced. We start with two-bit counters with no controls, so that we can focus on basic principles. Then controls are added: count, up/down, load, and clear.

8.2.1   Two-Bit Synchronous Up Counter

Use deductive reasoning or standard method converting ASM chart to circuit.

A two-bit binary up counter can count from 0 to 3 on a recycling basis: 0, 1, 2, 3, 0, 1, 2, 3, ... The ASM chart is assembled from four state rectangles directly connected from one state to the next (Figure 8.8). The truth table, D flip-flop excitation equations, and circuit (Figure 8.9) are derived from the ASM chart via the ASM synthesis procedure in Section 7.5.

FIGURE 8.8

Up counter ASM chart

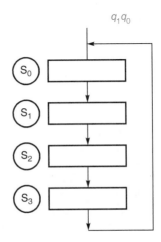

FIGURE 8.9
Up counter design using D flip-flops

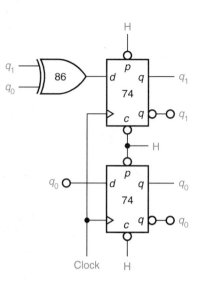

PS	NS		
s_j		q_1^+	$q_0^+ = d_1 d_0$
s_0	s_1	0	1
s_1	s_2	1	0
s_2	s_3	1	1
s_3	s_0	0	0

$$d_0 = s_0 + s_2 = q_1'q_0' + q_1q_0' = q_0' = q_0 \text{ xor } 1$$
$$d_1 = s_1 + s_2 = q_1'q_0 + q_1q_0' = q_1 \text{ xor } q_0$$

When $q_0^+ = d_0 = q_0'$, the q_0 bit toggles on every clock pulse to produce the q_0 string 01010101 . . . . Bit q_0 is always in the toggle mode. The truth table with q_0 as a table-entered variable is:

q_0^+	q_0 Bit Property
q_0'	toggle

The more complex equation for q_1^+ produces the q_1 string 00110011 . . . . To understand this result, we use q_1 as a table-entered variable in the truth table for $q_1^+ = d_1 = q_1'q_0 + q_1q_0'$:

q_0	q_1^+	q_1 Bit Property
0	q_1	hold
1	q_1'	toggle

D flip-flops count with input q' or hold present count with input q.

When $q_0 = 0$, the q_1 value is held over to the next period to produce a 00 or 11 pair while bit q_0 toggles to 1. When $q_0 = 1$, the q_1 bit toggles its value for the next period while q_0 toggles to 0.

Our understanding is perhaps enhanced if we emphasize that the XOR gates are used as programmable inverters, inverting or not inverting to toggle or to hold the count as required.

FIGURE 8.10

Up counter q_1q_0 waveforms

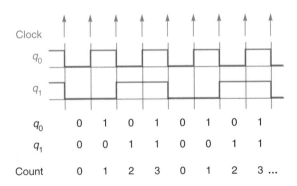

q_0	0	1	0	1	0	1	0	1
q_1	0	0	1	1	0	0	1	1
Count	0	1	2	3	0	1	2	3 ...

XOR TRUTH TABLE

q_0	q_1	d_1	Function
0	0	0	not inverting q_1 (hold)
0	1	1	
1	0	1	inverting q_1 (toggle)
1	1	0	

Plotting the values of q_1 and q_0 over time, we get voltage waveforms and equivalent binary strings as a function of time. Observe that the q_1, q_0 waveforms in Figure 8.10 are a q_1q_0 truth table row generator (00, 01, 10, 11). Also note that when $q_0 = 0$, the present value is held by q_1 for one more clock period. And when $q_0 = 1$, the q_1 bit toggles at the next active clock edge.

EXAMPLE 8.1 Up Counter Design Using JK Flip-Flops

A two-bit up counter design is based on the ASM chart of Figure 8.8. The JK defining equation, $q^+ = jq' + k'q$, specifies that the modes are

$$jk = 00 \text{ hold,}$$

$$jk = 11 \text{ toggle,}$$

$$jk = 01 \text{ clear, and}$$

$$jk = 10 \text{ set.}$$

These various jk modes offer opportunities for simplifying the circuits driving the jk inputs, such as the two choices shown in the truth table. The hold/toggle choice results in the no-gates-required jk input equations. The clear/set choice converts the JK to D flip-flops.

PS	$q_1 q_0$	NS	$q_1^+ q_0^+$	HOLD/TOGGLE		CLEAR/SET	
				$j_1 k_1$	$j_0 k_0$	$j_1 k_1$	$j_0 k_0$
s_0	00	s_1	01	00	11	01	10
s_1	01	s_2	10	11	11	10	01
s_2	10	s_3	11	00	11	10	10
s_3	11	s_0	00	11	11	01	01

$$j_1 = k_1 = q_0 \qquad\qquad j_1 = k_1' = q_1 \text{ xor } q_0$$

$$j_0 = k_0 = 1 \qquad\qquad j_0 = k_0' = q_0'$$

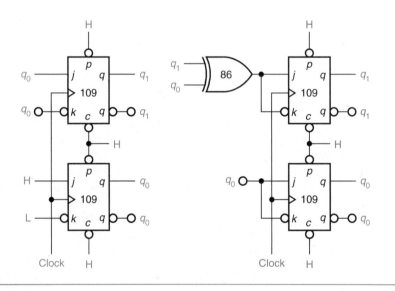

EXERCISE 8.1   Use the JK defining equation, $q^+ = jq' + k'q$, to derive the q_1^+, q_0^+ next-state equations for the JK hold/toggle mode in Example 8.1. What is the equation relating the j and k inputs?

Answer: $q_1^+ = q_1 \text{ xor } q_0, \qquad q_0^+ = q_0 \text{ xor } 1, \qquad j_1 = k_1, \qquad j_0 = k_0$ ■

EXERCISE 8.2   Use the JK defining equation, $q^+ = jq' + k'q$, to derive the q_1^+, q_0^+ next-state equations for the JK clear/set mode in Example 8.1. What is the equation relating the j and k inputs?

Answer: $q_1^+ = q_1 \text{ xor } q_0, \qquad q_0^+ = q_0 \text{ xor } 1, \qquad j_1 = k_1', \qquad j_0 = k_0'$ ■

FIGURE 8.11
Up counter count control ASM block for state$_j$

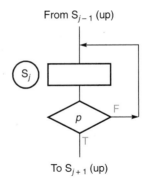

Adding a count control input p   A synchronous count control input p is associated with every counter state (Figure 8.11) because, if p goes to the not-asserted level when the counter is in state$_j$, the counter stops in state$_j$. The counter dwells in state$_j$ as long as p is not asserted. When p is asserted, the counter moves from state$_j$ to state$_{j+1}$ at the next active clock edge. When it reaches the maximum count, the counter resets to zero, as discussed previously. An N-state counter is represented by an assembly of N one-state ASM blocks, as shown in Figure 8.11.

Count control p is not asserted when a counter holds a count, and p is asserted when the count increments.

FIGURE 8.12
Up counter design with count control

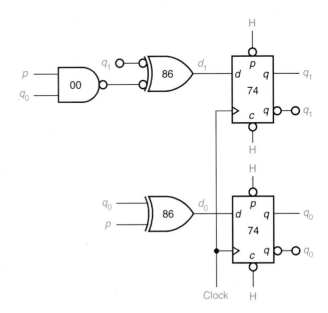

PS	PI p	NS	$q_1^+ q_0^+ = d_1\ d_0$
s_0	0	s_0	0 0
	1	s_1	0 1
s_1	0	s_1	0 1
	1	s_2	1 0
s_2	0	s_2	1 0
	1	s_3	1 1
s_3	0	s_3	1 1
	1	s_0	0 0

$$d_0 = s_0 p + s_1 p' + s_2 p + s_3 p' = q_0 \text{ xor } p$$

$$d_1 = s_1 p + s_2 + s_3 p' = q_1 \text{ xor } q_0 p$$

The truth table, D flip-flop excitation equations, and circuit (Figure 8.12) are derived from the ASM chart as shown in Section 7.5. Observe that one 86 XOR gate is shown with two bubbles at the input. We believe this is more clear than a circuit with the bubbles omitted. See Figure 4.23 on page 191.

EXAMPLE 8.2   Up Counter K Maps Without Count Control p

The next-state equations for a two-bit up counter are:

$$q_0^+ = q_0 \text{ xor } 1 = q_0' \qquad\qquad q_1^+ = q_1 \text{ xor } q_0 = q_1 q_0' + q_1' q_0$$

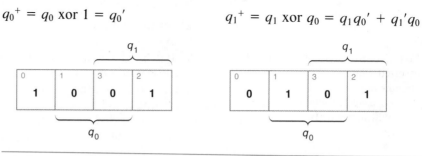

EXAMPLE 8.3   Up Counter K Maps with Count Control p

The three-variable K map and the two-variable K map with map-entered variable p are shown for both bits q_1 and q_0. The equations are put in sum-of-terms format for mapping purposes. The next-state equations for a two-bit up counter with count control are as follows:

$$q_0^+ = q_0 \text{ xor } p = q_0 p' + q_0' p = p'(q_0) + p(q_0')$$

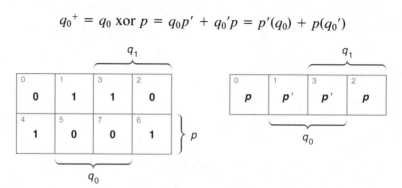

$$q_1^+ = q_1 \text{ xor } pq_0 = q_1(pq_0') + q_1'(pq_0) = q_1q_0' + p'(q_1) + p(q_1'q_0)$$

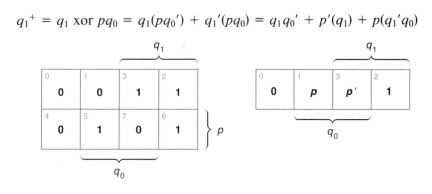

EXERCISE 8.3
Use JK flip-flops in an up counter design with p control. Find the equations for the j and k inputs from the next-state equations.

$$q_1^+ = q_1 \text{ xor } pq_0 \qquad q_0^+ = q_0 \text{ xor } p$$

Answer: $j_1 = k_1 = pq_0 \qquad j_0 = k_0 = p$ ∎

8.2.2 Two-Bit Synchronous Down Counter

The ASM chart for a two-bit down counter is the same as for an up counter except for a renumbering of the states (Figure 8.13). Again, the truth table, D flip-flop excitation equations, and circuit (Figure 8.14) are derived from the ASM chart (Section 7.5).

Since $d_1 = q_1 \text{ xor } q_0'$, the only circuit change from an up to a down counter is to change q_0 to q_0' (compare the circuits in Figures 8.9 and 8.14). With two-bit counters, an alternative is to use the equation $d_1 =$

FIGURE 8.13
Down counter ASM chart

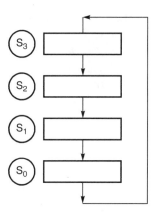

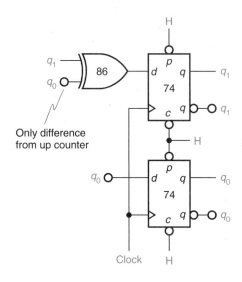

PS q_1q_0	NS	$q_1^+q_0^+ = d_1$	d_0
s_3 s_2	1	0
s_2 s_1	0	1
s_1 s_0	0	0
s_0 s_3	1	1

$$d_0 = s_2 + s_0 = q_1q_0' + q_1'q_0' = q_0' = q_0 \text{ xor } 1$$

$$d_1 = s_3 + s_0 = q_1q_0 + q_1'q_0' = q_1 \text{ xor } q_0'$$

FIGURE 8.14

Down counter design using D flip-flops

q_1' xor q_0. When more bits are involved, only the q_0' form is used, as we will see shortly.

When $q_0^+ = q_0'$, bit q_0 toggles on every clock pulse to produce the string 01010101 . . . . Or is it 101010 . . . ? We cannot distinguish until we relate it to the q_1 string. This applies to the up counter as well. In any event, bit q_0 is always in the toggle mode. The truth table with q_0 as a table-entered variable is once again the following:

q_0^+	q_0 Bit Property
q_0'	toggle

Down counter toggles on 0's.

Observe that this time the equation for q_1^+ produces the string 1100110011 . . . . To obtain an insight into this result, we use q_1 as a table-entered variable in the truth table for $q_1^+ = q_1q_0 + q_1'q_0'$. To emphasize that q_0 in the equation has been replaced by q_0', we write the table in reverse row order, as follows:

q_0	q_1^+	b Bit Property
1	q_1	hold
0	q_1'	toggle

This result is the complement of the up counter q_1^+ result, which is consistent with q_0 changing to q_0' in the schematic (Figure 8.14). The values of q_1 and q_0 are plotted in Figure 8.15 to obtain their voltage waveforms and equivalent binary strings as a function of time. The q_1,

FIGURE 8.15

Down counter q_1q_0 waveforms

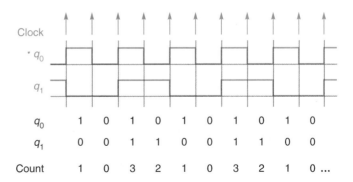

q_0 waveforms generate the q_1q_0 truth table rows in reverse order, consistent with a down count. This time we start the waveforms in the middle of the 3210 sequence with $state_1$ in order to show that the counter works correctly no matter where the waveforms start.

EXAMPLE 8.4 **Down counter design using JK flip-flops**

A two-bit down counter design is based on the ASM chart of Figure 8.13. The JK defining equation, $q^+ = jq' + k'q$, specifies that the modes are

$$jk = 00 \text{ hold,}$$

$$jk = 11 \text{ toggle,}$$

$$jk = 01 \text{ clear, and}$$

$$jk = 10 \text{ set.}$$

These various jk modes offer opportunities for simplifying the circuits driving the jk inputs, such as the two choices shown in the truth table. The hold/toggle choice results in the no-gates-required jk input equations.

PS	q_1q_0	NS	$q_1^+q_0^+$	HOLD/TOGGLE j_1k_1	j_0k_0	CLEAR/SET j_1k_1	j_0k_0
s_0	00	s_3	11	11	11	10	10
s_1	01	s_0	00	00	11	01	01
s_2	10	s_1	01	11	11	01	10
s_3	11	s_2	10	00	11	10	01

$$j_1 = k_1 = q_0' \quad j_1 = k_1' = q_1 \text{ xor } q_0'$$

$$j_0 = k_0 = 1 \quad j_0 = k_0' = q_0'$$

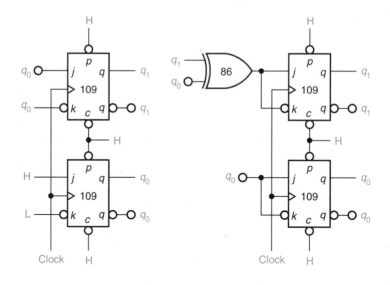

EXERCISE 8.4   Use the JK defining equation, $q^+ = jq' + k'q$, to derive the $q_1{}^+$, $q_0{}^+$ next-state equations for the JK hold/toggle mode in Example 8.4. What is the equation relating the j and k inputs?

Answer: $q_1{}^+ = q_1$ xor q_0', $q_0{}^+ = q_0$ xor 1, $j_1 = k_1$, $j_0 = k_0$ ∎

EXERCISE 8.5   Use the JK defining equation, $q^+ = jq' + k'q$, to derive the $q_1{}^+$, $q_0{}^+$ next-state equations for the JK clear/set mode in Example 8.4. What is the equation relating the j and k inputs?

Answer: $q_1{}^+ = q_1$ xor q_0', $q_0{}^+ = q_0$ xor 1, $j_1 = k_1'$, $j_0 = k_0'$ ∎

Adding a count control input p   A synchronous count input p is associated with every counter state (Figure 8.16), because, if p goes to the not-asserted level when the counter is in state$_j$, the counter stops in state$_j$. The counter dwells in state$_j$ as long as p is not asserted. When p is asserted, the counter moves from state$_j$ to state$_{j-1}$ at the next active clock edge. When it reaches zero, the counter resets to the maximum count, as discussed previously. An N-state counter is represented by an assembly of N one-state ASM blocks, as shown in Figure 8.16. The design (Figure 8.17) is the same as the up counter circuit except for q_0' replacing q_0.

FIGURE 8.16
Down counter count control ASM block for state_j

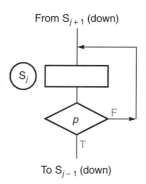

From S_{j+1} (down)

S_j

p F

T

To S_{j-1} (down)

8.2.3 Two-Bit Synchronous Up/Down Counter

The up/down counter is a merger of up and down counter circuits to which is added an input control line u specifying the up (u) or the down (u') condition. This intuitive merging of up and down counter circuits is confirmed by merging the up and down counter equations while including the up/down control variable u.

FIGURE 8.17
Down counter design with count control

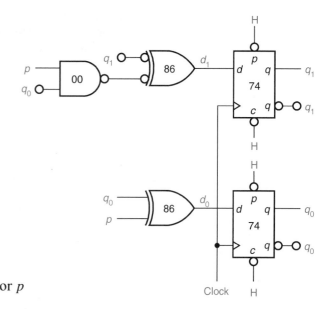

PS	PI p	NS	$q_1^+q_0^+ = d_1\ d_0$
s_0	0	s_0	0 0
	1	s_3	1 1
s_1	0	s_1	0 1
	1	s_0	0 0
s_2	0	s_2	1 0
	1	s_1	0 1
s_3	0	s_3	1 1
	1	s_2	1 0

$$d_0 = s_0p + s_1p' + s_2p + s_3p' = q_0 \text{ xor } p$$

$$d_1 = s_0p + s_2p' + s_3 = q_1 \text{ xor } q_0'p$$

Up

$d_0 = q_0'$ $\qquad$ $\rightarrow q_0^+ = q_0'$

$d_1 = q_1 \text{ xor } q_0$ $\quad \rightarrow q_1^+ = q_1 q_0' + q_1' q_0$

Down

$d_0 = q_0'$ $\qquad$ $\rightarrow q_0^+ = q_0'$

$d_1 = q_1 \text{ xor } q_0'$ $\quad \rightarrow q_1^+ = q_1 q_0 + q_1' q_0'$

Up/down

$d_0 = q_0'$

$d_1 = $ up condition or down condition

$\quad = (q_1 \text{ xor } q_0)u + (q_1 \text{ xor } q_0')u'$ $\quad$ Where $u = $ up, $u' = $ down

$\quad = (q_1 \text{ xor } q_0)u + (q_1 \text{ xor } q_0)'u'$

$\quad = (q_1 \text{ xor } q_0)' \text{ xor } u$

$\quad = (q_1 \text{ xor } q_0') \text{ xor } u$

$\quad = q_1 \text{ xor } (q_0' \text{ xor } u)$

$\quad = q_1 \text{ xor } (u \text{ xor } q_0')$ $\quad$ Figure 8.18

or

$\qquad d_1 = q_1 \text{ xor } (uq_0 + u'q_0')$ $\quad$ Figure 8.19

There is another kind of connection between the waveforms and the count (the number in the counter register). The connection can be made by noting the $q_1 q_0$ waveforms (Figure 8.20). If the numbers

Figure 8.18
Up/down counter circuit with XOR-XOR format

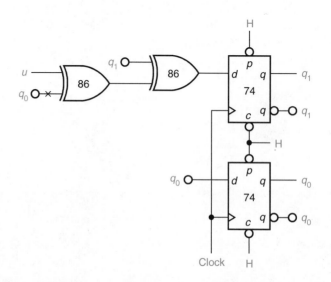

FIGURE 8.19

Up/down counter circuit with AND-OR-XOR format

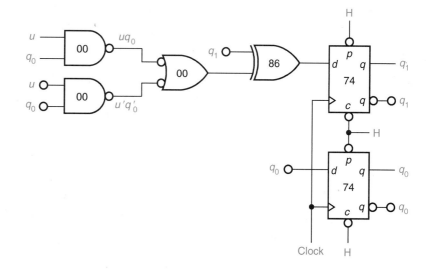

represented by the $q_1 q_0$ waveforms are studied while counting ones, the count is up, whereas the count is down for the $q_1 q_0'$ waveforms.

EXERCISE 8.6

Plot the $q_1 q_0$ waveforms shown in Figure 8.20. Mark each clock period with 1 or 0. Mark each clock period with the state number.

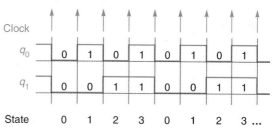

FIGURE 8.20

$q_1 q_0$ waveforms count up and $q_1 q_0'$ waveforms count down

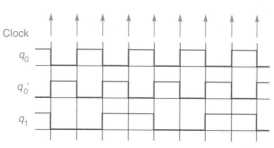

Counting 1's in both cases:

$q_1 q_0$	0	1	2	3	0	1	2	3 ...
$q_1 q_0'$	1	0	3	2	1	0	3	2 ...

EXERCISE 8.7   Plot the $q_1 q_0'$ waveforms shown in Figure 8.20. Mark each clock period with 1 or 0. Mark each clock period with the state number.

Answer:

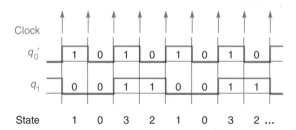

8.2.4   Counter Control Lines and Their Function

Counter registers are controlled by one to four variables. In this discussion all operations are synchronous, which is not necessarily true of all commercial counters. For example, the 160 and 161 have an asynchronous clear control. Any glitch with sufficient energy will clear these counters, causing functional crashes. We prefer the 162 and 163 counters, which have synchronous clear lines. This reduces their vulnerability significantly because they are vulnerable only to glitches that straddle the active clock edge—one reason why we do not use asynchronous controls.

Multiple control lines require a priority scheme.

Clear c: When clear is asserted, the counter is reset to state$_0$ at the next active clock edge. All bits are reset to zero.

Load g: When load is asserted, the counter is set to state$_n$ at the next active clock edge (n equals the number represented by the input data).

Count p: When count is asserted, the number in the counter register is incremented by one (up counter) or decremented by one (down counter) at the next active clock edge.

Up u: When up is asserted, the number in the register increments (see Count p). When up is *not* asserted, the number in the register decrements (see Count p).

The question of priority arises when more than one control line is asserted in the same clock period. (Guaranteeing assertion of one line at a time is extremely difficult and definitely not desirable. In fact, the converse is desirable because the resulting system is invariably simplified.) Priority, which determines what actually happens in that clock period, is implemented by logically relating the inputs with a binary tree associated with each state (Figure 8.21). Observe how the binary tree establishes priority automatically. The load function is active only when clear is False; and when the load function is active, the count function is disabled. In descending order, the priority hierarchy used in standard logic families is as follows:

Function	Symbol
clear	c
load	g
count	p
up/down	u

FIGURE 8.21
ASM block with counter control line logic

c	g	p	u	Function
1	–	–	–	clear state register to state$_0$
0	1	–	–	load data d_n into state register
0	0	0	–	hold current state (count)
0	0	1	0	count down
0	0	1	1	count up

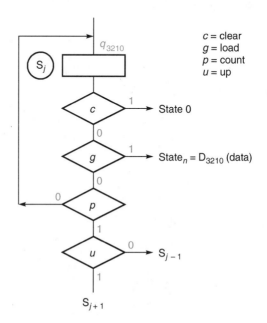

EXERCISE 8.8   Refer to Figure 8.21. Derive the input minterms that define the next state to be s_0, s_n, s_j, s_{j-1}, s_{j+1}.

Answer:

s_0	c
s_n	$c'g$
s_j	$c'g'p'$
s_{j-1}	$c'g'pu'$
s_{j+1}	$c'g'pu$

■

8.2.5   163 Four-Bit Synchronous Up Counter

Our objective in this section is to transform the 163 specification into an ASM chart and to design the 163 state machine. Perhaps this will give you some idea of the creativity required. Once you know how to design blocks such as the 163, you are in a position to use such blocks creatively because you understand the circuit. Contrast this capability to use based on memorization of input line functions. A by-product is presentation of the commercial 163 IC.

The 163 is a four-bit register that loads input data or increments the count on each active clock edge.

The 163 is like a four-bit 173 register plus gates and wiring connecting the four bits as a binary up counter (Figure 8.22). The next-state circuits use an XNOR gate driven by active low q_j and an active high "pq_j" term. Consider f_2, for example, where the "pq_j" term equals pq_1q_0. There are two mixed-logic points of view that yield the same result:

1. Assign $--$ to f_2:

$$f_2 = q_2' \oplus pq_1q_0$$
$$d_2 = f_2' = (q_2' \oplus pq_1q_0)' = q_2 \oplus pq_1q_0$$

2. Assign $++$ to f_2:

$$f_2 = q_2' \odot pq_1q_0 = q_2 \oplus pq_1q_0$$
$$d_2 = f_2 = q_2 \oplus pq_1q_0$$

At each active clock edge the 163 is cleared, or the 163 is loaded with data, or the 163 adds one to the count, or the 163 holds the current count. Control line priority is as defined in Figure 8.21.

The 163 is cleared by asserting the synchronous clear line. The asserted clear line disables the load and count circuits, forcing zeros into the flip-flops on the next active clock edge. If clear is not asserted, then the 163 can be synchronously loaded with data by asserting the

FIGURE 8.22
163 four-bit up counter

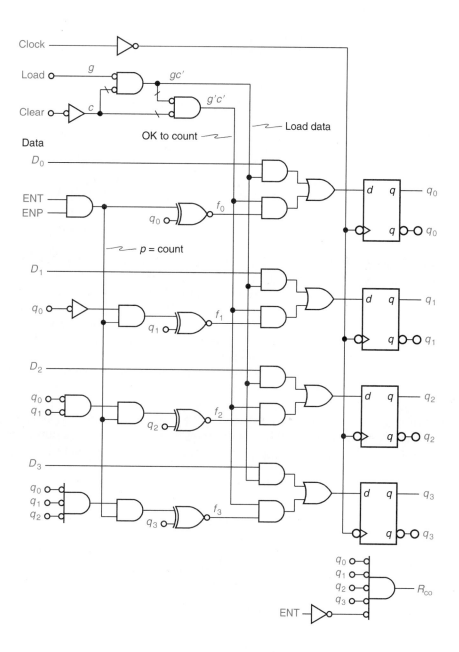

load line so that $gc' = 1$ (Figure 8.21). When clear is not asserted and load is asserted, the term gc' is asserted. Then data inputs D_j determine the next state$_n$. This process is also known as *presetting* the counter. [There is a potential symbol problem because data and D flip-flop start with the letter d. The D_j are the data input bits, and the d_j are the D flip-flop inputs ($j = 0, 1, 2, 3$).] If clear and load are not asserted ($g'c' = 1$), then the 163 synchronously counts active clock edges when p, represented in the 163 by ENP × ENT, is asserted. Then when p is asserted, current count plus one is the next-state number ($n \leftarrow n + 1$).

The R_{co} (ripple carry out) line is the carry output to the next 163 in a 163 chain that makes up a counter with more than four bits. The equation for R_{co} is ENT × $q_3 q_2 q_1 q_0$. Carry output is asserted when ENT is asserted *and* the 163 is in state$_{1111}$ ($1111_2 = F_{16}$).

In addition to count control, the EN*able*P and the EN*able*T inputs allow a designer to implement chip-to-chip ripple carry or look-ahead carry in multichip counters, as shown in data books.

Ripple carry processes bits serially.	***Ripple carry connection*** P is connected (Figure 8.23a) to the first chip's ENP and ENT inputs, and to the other chips' ENP inputs. The chip-to-chip connection is now from R_{co} to ENT. We assume the first chip is in state 1110_2 and all other chips are in state 1111_2. This is ripple carry because the next active clock edge advances 1110_2 to 1111_2, asserting the first R_{co}. This signal has to propagate through all output carry gates before a carry is forwarded to the last chip so that downstream chips can advance from 1111 to 0000.
Look-ahead carry is faster than ripple carry because it processes bits in parallel.	***Look-ahead carry connection*** Count control p is connected (Figure 8.23b) to the first chip's ENP and ENT inputs and to the ENT input of the second chip. The first chip's R_{co} output is connected to all other chips' ENP inputs. R_{co} is connected to ENT for the remaining chip-to-chip carries. With *all* chips in state 1111_2, the first R_{co} asserts all other chips' ENP inputs for a full clock period minus t_{PHL} or t_{PLH}. All chips

FIGURE 8.23
Ripple and look-ahead carry connections

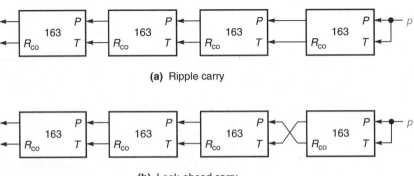

(a) Ripple carry

(b) Look-ahead carry

P = ENP; T = ENT

are able to count. The next active clock edge advances the count to 0000_2 in all chips at the same time.

EXERCISE 8.9

Refer to Figure 8.23a. Derive the equation for R_{co3} out of, and P_3 into, the most significant 163 chip.

Answer: $P_3 = P$

$$R_{co3} = q_F q_E q_D q_C R_{co2} = q_F q_E q_D q_C (q_B q_A q_9 q_8 R_{co1}) \ldots$$

so that

$$R_{co3} = q_F q_E q_D q_C (q_B \ldots (q_7 \ldots (q_3 \cdot q_1 q_0 P)))$$

where each parenthesis implies one time delay, creating a ripple carry. ∎

EXERCISE 8.10

Refer to Figure 8.23b. Derive the equation for R_{co3} out of, and P_3 into, the most significant 163 chip.

Answer: $P_3 = q_3 q_2 q_1 q_0 P$

$$R_{co3} = q_F q_E q_D q_C R_{co2} = q_F q_E q_D q_C (q_B q_A q_9 q_8 R_{co1}) \ldots$$

so that

$$R_{co3} = q_F q_E q_D q_C (q_B \ldots (q_7 \ldots (P)))$$

where each parenthesis implies one time delay that does not matter, because P and all $q_j = 1$. ∎

Four-bit up counter synthesis procedure The synthesis procedure is decomposed into a manageable series of smaller procedures. The words ''four-bit up counter with control lines'' represent the basic specification, which implies the sequence of states from which follows the truth table in upcoming Figure 8.25 without provision for control lines. (Note that this procedure can be applied to any sequence of states, not just 0, 1, 2, 3, ...) This table is useful because the next-state counting equations are independent of the control lines. The ASM fragment in Figure 8.21 is the basis for the 163 counter equations, including control lines. A key creative insight is recognition that the ASM block is the same for each counter state. This is why there is no need to draw the n-state ASM chart. The circuit follows from the equations. The procedure has four major steps.

1. Up counter → sequence of states 0, 1, 2, 3, . . . (Figure 8.24).
2. Sequence of states → next-state truth table (Figure 8.25).
3. Next-state truth table → d_j equations. Equations are derived in two steps:

  a. Equations without control variables (Figure 8.25).
  b. Equations with control variables (Figure 8.21).
4. d_j equations → counter circuit (Figure 8.22)

STEP 1

Next state does not have to be present state plus one.

A four-bit up counter's maximum counting range is from 0 to F_{16} = 1111_2. The four bit outputs q_3, q_2, q_1, and q_0 represent the state number. The four state variables generate a 16-row truth table, which in Figure 8.24 is displayed horizontally instead of vertically. The truth table rows are also displayed in waveform format.

STEP 2

K-maps reveal their value in this step.

The four-bit counter ASM chart is assembled from 16 state rectangles directly connected from one state to the next. This chart, which is not shown, is analogous to the four-state two-bit counter chart in Figure 8.8. The truth table and D flip-flop excitation equations in Figure 8.25 are derived from the ASM chart using the ASM synthesis procedure of Section 7.5.

STEP 3

Equations are derived in two steps. We leave Step 3a, deriving the d_j equations, as exercises for the reader. The d_j equations are shown in Figure 8.25. We execute Step 3b as follows. The ASM chart fragment in Figure 8.21 shows control line impact on the next-state equations

FIGURE 8.24

Four-bit up counter truth table rows and q_j waveforms

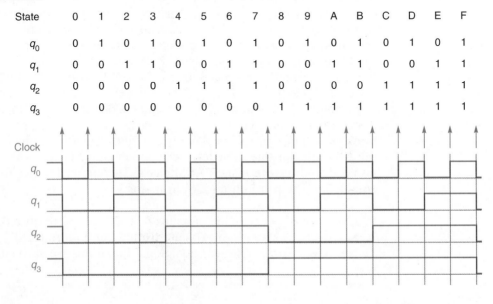

FIGURE 8.25

Four-bit up counter design with no control lines

PS	NS	d_3	d_2	d_1	d_0
s_0	s_1	0	0	0	1
s_1	s_2	0	0	1	0
s_2	s_3	0	0	1	1
s_3	s_4	0	1	0	0
s_4	s_5	0	1	0	1
s_5	s_6	0	1	1	0
s_6	s_7	0	1	1	1
s_7	s_8	1	0	0	0
s_8	s_9	1	0	0	1
s_9	s_A	1	0	1	0
s_A	s_B	1	0	1	1
s_B	s_C	1	1	0	0
s_C	s_D	1	1	0	1
s_D	s_E	1	1	1	0
s_E	s_F	1	1	1	1
s_F	s_0	0	0	0	0

We write each $q_j^+ = d_j$ column to a K map. We read each map to find the minimized d_j equations:

$$d_0 = q_0 \text{ xor } 1$$
$$d_1 = q_1 \text{ xor } q_0$$
$$d_2 = q_2 \text{ xor } q_1 q_0$$
$$d_3 = q_3 \text{ xor } q_2 q_1 q_0$$

q_j^+. The next-state s_j^+ equation is readily taken from this ASM chart fragment.

The next-state equation is derived from the control line input tree.

$$s_j^+ = cs_0 + gc's_n + g'c'p's_j + g'c'ps_{j+1}$$
$$= cs_0 + gc's_n + g'c'(p's_j + ps_{j+1})$$

Dropping down to the bit level:

$s_j^+ \rightarrow q_j^+$ The next-state variable q_j^+ replaces the next-state number s_j^+ at the bit level.

$s_0 \rightarrow 0$ The four bits representing s_0 are all zero, so the first term drops out.

$s_n \rightarrow D_j$ The state number s_n is replaced by the data-bit D_j.

$s_j \rightarrow q_j$ The state number s_j is the concatenation of $q_3 q_2 q_1 q_0$. This is why s_j reduces to bit q_j.

$s_{j+1} \rightarrow d_j$ The next state s_{j+1} is determined by the present state d_j because $q_j^+ = d_j$.

Since q_j^+ is the next state of q_j, the q_j^+ equation replaces the s_j^+ equation at the bit level:

$$s_j^+ = cs_0 + gc's_n + g'c'(p's_j + ps_{j+1})$$
$$q_j^+ = c \times 0 + gc'D_j + g'c' (p'q_j + pd_j)$$
$$q_j^+ = gc'D_j + g'c'w \quad \text{where } w = p'q_j + pd_j$$

This may not be obvious, but the last two terms need reduction. Applying the d_j equations from Figure 8.25, we substitute in the *last* term of w the expression for d_j. (We set $g'c'$ aside for the moment.)

Define a_j:

$$pd_j = p(q_j \text{ xor } a_j) \quad \text{where } a_j = q_{j-1} \ldots q_0, \text{ and } a_0 = 1$$

Start with w:

$$w = p'q_j + pd_j$$

Substitute for d_j:

$$w = p'q_j + p(q_j \text{ xor } a_j)$$

Expand the XOR functions:

$$w = p'q_j + p(q_j a_j' + q_j' a_j)$$

Rearrange terms so that q_j and q_j' are factors:

$$w = q_j(p' + pa_j') + q_j' pa_j$$

Use the theorem $p' + pr' = p' + r'$:

$$w = q_j(p' + a_j') + q_j' pa_j$$

Invoke DeMorgan's Theorem in the first term:

$$w = q_j(pa_j)' + q_j'(pa_j)$$

$$= q_j \text{ xor } pa_j$$

Substitute reduced w into q_j^+:

$$q_j^+ = gc'D_j + g'c'w$$

$$= gc'D_j + g'c' (q_j \text{ xor } pa_j)$$

Introduce the count-with-look-ahead-carry functions f_j:

$$q_j^+ = gc'D_j + g'c'f_j \quad \text{where } f_j = q_j \text{ xor } pa_j$$
$$f_0 = q_0 \text{ xor } p$$
$$f_1 = q_1 \text{ xor } pq_0$$
$$f_2 = q_2 \text{ xor } pq_0q_1$$
$$f_3 = q_3 \text{ xor } pq_0q_1q_2$$

R_{co} detects state 15.

The f_j are the count-with-look-ahead-carry functions (see Figures 8.22 and 8.12). The carry output equation has one term:

$$R_{co} = q_3 q_2 q_1 q_0 \text{ENT}$$

Finally, $p = \text{ENT} \times \text{ENP}$. The 163 four-bit counter circuit is synthesized from these equations (Figure 8.22). Bubbles are added in Figure 8.22 to make the 163 data book schematic more clear.

EXAMPLE 8.5 **Up Counter with Count Control Using JK Flip-Flops**

When load is False, $g' = 1$; when clear is False, $c' = 1$. We let $g'c' = 1$, so that

$$q_m^+ = gc'D_m + g'c'f_m = f_m$$

where

$$f_m = q_m \text{ xor } pa_m$$

so that

$$
\begin{aligned}
q_{0+} &= f_0 = q_0 \text{ xor } p && = (p)q_0' && + (p)'q_0 \\
q_{1+} &= f_1 = q_1 \text{ xor } pq_0 && = (pq_0)q_1' && + (pq_0)'q_1 \\
q_{2+} &= f_2 = q_2 \text{ xor } pq_0q_1 && = (pq_0q_1)q_2' && + (pq_0q_1)'q_2 \\
q_{3+} &= f_3 = q_3 \text{ xor } pq_0q_1q_2 && = (pq_0q_1q_2)q_3' && + (pq_0q_1q_2)'q_3
\end{aligned}
$$

For the JK,

$$q_m^+ = (j_m)q_m' + (k_m)'q_m$$

Equating coefficients of the four q_m^+ and q_m^+, we get

$$
\begin{aligned}
j_0 &= k_0 = p \\
j_1 &= k_1 = pq_0 \\
j_2 &= k_2 = pq_0q_1 \\
j_3 &= k_3 = pq_0q_1q_2
\end{aligned}
$$

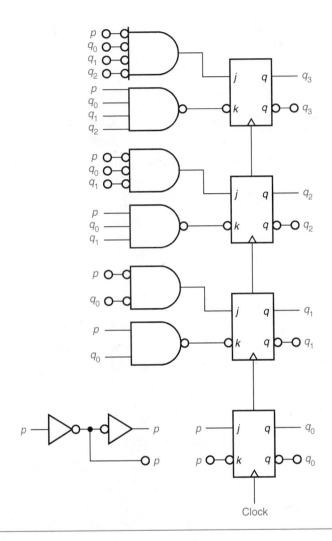

EXAMPLE 8.6   **Decade Up Counter D Flip-Flop Design Using K Maps**

As explained in Section 1.5.2, the BCD digit range is 0 to 9. This is why a BCD counter counts modulo-10, recycling through states 0000 to 1001. A four-bit decade counter is required for each BCD digit. The truth table for one BCD digit is straightforward. We start by designating the last six next states as "don't cares."

PS		NS	d_3	d_2	d_1	d_0
s_0	0000	s_1	0	0	0	1
s_1	0001	s_2	0	0	1	0
s_2	0010	s_3	0	0	1	1
s_3	0011	s_4	0	1	0	0
s_4	0100	s_5	0	1	0	1
s_5	0101	s_6	0	1	1	0
s_6	0110	s_7	0	1	1	1
s_7	0111	s_8	1	0	0	0
s_8	1000	s_9	1	0	0	1
s_9	1001	s_0	0	0	0	0
s_A	1010	–	–	–	–	–
s_B	1011	–	–	–	–	–
s_C	1100	–	–	–	–	–
s_D	1101	–	–	–	–	–
s_E	1110	–	–	–	–	–
s_F	1111	–	–	–	–	–

The design steps are:

1.1 Write the d_j truth table columns onto K maps.
1.2 Read the K maps to derive the d_j equations. (This step assigns 0 or 1 to the "don't care" dashes.)
1.3 Use the equations to design a circuit.

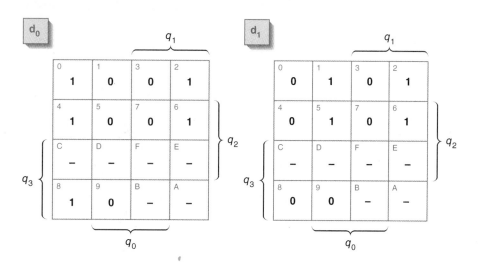

d₂

q_1

0	1	3	2
0	0	1	0

4	5	7	6
1	1	0	1

C	D	F	E
–	–	–	–

8	9	B	A
0	0	–	–

q_2 · q_3 · q_0

d₃

q_1

0	1	3	2
0	0	0	0

4	5	7	6
0	0	1	0

C	D	F	E
–	–	–	–

8	9	B	A
1	0	–	–

q_2 · q_3 · q_0

The following steps make the "don't care" assignments explicit.

1.4 Redraw the K maps with 0 or 1 replacing the "don't care" dashes, using the assignments from Step 1.2.
1.5 Rewrite the truth table with 0 or 1 replacing the "don't care" dashes, as shown in the K maps from Step 1.4

Steps 2 through 5 are left as exercises.

EXAMPLE 8.7 **Modulo-13 Up Counter—0-to-12 Sequence**

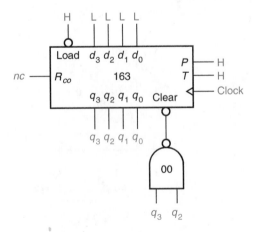

Note: this and the next example use intuitive circuit design based on waveform analysis. The normal truth table, K map, circuit design process is more straightforward. (see upcoming Section 8.2.7.)

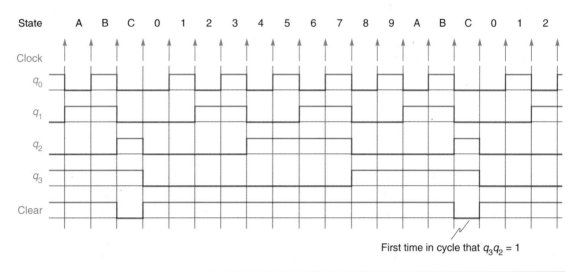

First time in cycle that $q_3 q_2 = 1$

EXAMPLE 8.8 **Modulo-9 Up Counter—3-to-11 Sequence**

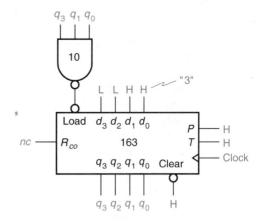

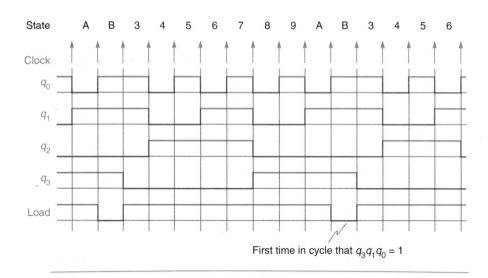

First time in cycle that $q_3 q_1 q_0 = 1$

8.2.6   169 Four-Bit Synchronous Up/Down Counter

This is another example of transformation of a specification into an ASM chart and design of the (169) state machine. A by-product is presentation of the commercial 169 IC.

The 169 is like a four-bit 173 register plus gates and wiring connecting the four bits as a binary up/down counter (Figure 8.26). At each active clock edge the 169 is loaded with data, or the 169 adds/subtracts one to the count, or the 169 holds the current count.

The 169 is a four-bit register that loads input data, and increments or decrements the count on each active clock edge.

The 169 does not have a clear line because the 169 package pin, analogous to the 163 pin used for clear, is used for the 169 up/down counter input u/d. The counter is cleared by loading 0000. The 169 is synchronously loaded with data by asserting the load line ($g = 1$). Loading presets the counter to the state number n represented by the data input number. If load is not asserted ($g' = 1$), then the 169 counts active clock edges up or down according to the u (for 169 u/d) input line ($u = 1$ to count up) when ENP and ENT are both asserted.

The R_{co} line is the carry output to the next 169 in a 169 chain that makes up a counter with more than four bits. R_{co} is asserted when ENT is asserted *and* the 169 is in state$_{1111}$ while counting up or when in state$_{0000}$ while counting down. Analysis and synthesis of the 169 is left to the exercises.

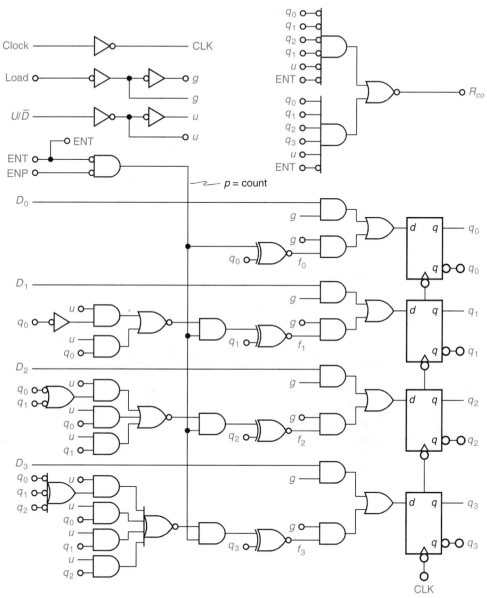

FIGURE 8.26
169 four-bit
up/down counter

8.2.7   Counters as Waveform Generators

Synchronous counters are used to synthesize precision waveforms for a variety of applications. Here is the rationale. Waveforms such as the one in Figure 8.27 are either at the H level or at the L level. A waveform transition from H to L or L to H is an event that occurs one propagation time after an active clock edge in a synchronous system. Therefore the times between waveform events are some multiple n_j of the system clock period τ. Waveform transitions from H to L or L to H are precisely timed events a counter can generate. Counters are readily adapted to produce the multipliers n_j.

One key idea is to decode counter state numbers in any clock period *prior* to an output event's occurrence. The number decoders produce "pulses" one clock period in duration. The "pulses" are converted to levels by JK flip-flops, which produce waveforms without glitches. Another key idea is to have the counters recycle the pattern of "pulses" so they have the same period as the desired waveform. The waveforms in Figure 8.27 show that this happens automatically. (The circuit is shown in upcoming Figure 8.30.)

One-hex-digit counter application   One hexadecimal digit represents all the states in a four-bit counter. A 163 is a one-hex-digit counter that generates the state numbers 0 to F_{16}. Suppose the output waveform w with period 10τ in Figure 8.27 is to be generated. Two events define this waveform. The events occur at the *end* of cycles 9 and 2. The cycle numbers may or may not be coincident with state numbers. Two possible sets of state numbers are shown in Figure 8.27. The significance of the various sets of state numbers that can be assigned is that one of the sets is usually implemented with a smaller number of gates than the others.

The first set of state numbers, 0 to 9, requires number decoders for 9 and 2 with outputs n_9 and n_2 (Figure 8.28). Output n_9 sets output w to the H level and loads state 0000_2 into the 163 counter. Output n_2 resets

Precision waveforms are readily derived from counters.

Detect two numbers to define a pulse duration.

Can use clear to recycle the waveform.

FIGURE 8.27
Periodic waveform

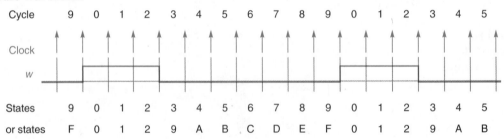

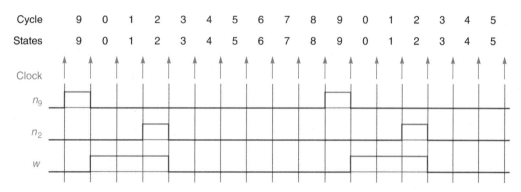

FIGURE 8.28
Number decoder n_9 and n_2 output waveforms

w to the L level. We leave as exercises for the reader the design of circuits for various sets of state numbers.

The number decoder outputs shown in Figure 8.29 are required for state set 6 through F. The 163 R_{co} output is a number decoder for state$_F$; therefore this output can be used as the n_F output. The n_F

Use load to recycle the waveform.

output sets w to the H level and loads state 0110_2 into the 163. The n_8 output resets w to the L level (Figure 8.30). The design is left as an exercise for the reader.

Please note that signals flow from right to left in Figures 8.30. (The reason we prefer this format is more clear in Figure 8.33).

EXERCISE 8.11   Refer to Figure 8.30. Derive the equation for n_8.

$$Answer: n_8 = q_3 m_0 = q_3 q_2' q_1' q_0'$$ ∎

FIGURE 8.29
Number decoder n_F and n_8 output waveforms

Cycle	9	0	1	2	3	4	5	6	7	8	9	0	1	2	3	4	5
States	F	6	7	8	9	A	B	C	D	E	F	6	7	8	9	A	B

Clock

n_F

n_8

w

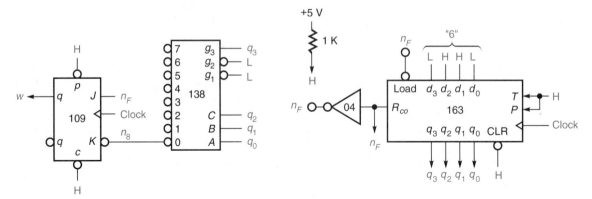

FIGURE 8.30
**Waveform
generator using
state set 6 to F**

Two-hex-digit counter application Two or more four-bit binary counters can be cascaded when we need more than 16 states. For example, suppose the waveform in Figure 8.31 is specified for a cathode ray tube screen raster scan display application. (The waveform is not drawn to scale.) Two hex digits represent eight bits, labeled h_j ($j = 0$ to 7). H means horizontal. The set of h_j address 80_{10} columns from 00_{16} to $4F_{16}$ to display their contents as the electron beam sweeps from left to right across the screen. The numbers 00 to 4F are used to fetch the graphics bits to be displayed. The sweep is blanked and retraces from right to left as the h_j scan from 50_{16} to 60_{16}. (When the h_j are scanning from 50 to 60, nothing is displayed: the screen is blank.) This means we do not care what numbers are scanned by the counter during the blank interval.

The straightforward solution uses number decoders with outputs n_{4F} and n_{60}. Output n_{4F} sets w to H and loads 50 into the counter at the next active clock edge. Output n_{60} resets w to L and loads 00 into the counter at the next active clock edge. But remember R_{co}. A simpler solution results if we ask what number precedes 00. FF does. The

FIGURE 8.31
**Horizontal sweep
raster scan periodic
waveform**

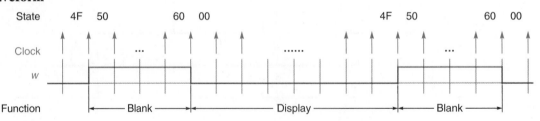

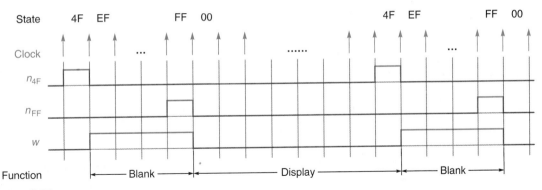

FIGURE 8.32
Horizontal periodic waveform using state FF

You do not have to use consecutive numbers.

range 50 to 60 hex uses 17 clock periods. If FF is the last number, then EF is the first of the 17-number range. So we use a new set of states (Figure 8.32). The waveform generator is shown in Figure 8.33. The design is left as an exercise for the reader.

Please note that signals flow from right to left in Figure 8.33. The reason why we prefer this format is that this is more clear when the signal lines representing the 8 bit number $h_7 h_6 h_5 h_4 h_3 h_2 h_1 h_0$ appear on

FIGURE 8.33
Horizontal waveform generator

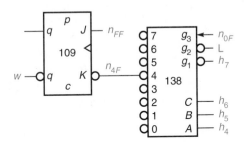

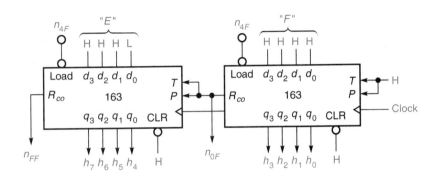

the schematic with the line for the most significant bit h_7 at the left consistent with h_7's position in a written number.

Shift Register

A register is called a *shift register* when all bits stored in the register can move one or more positions to the left or to the right of their present position by asserting control lines. A shift register holds the bit values until a shift command is received. This explains why the register flip-flops must be enabled D flip-flops. A shift register is like a 173 with gates and wiring added to allow lateral one-bit moves per clock edge.

8.3.1   Left- and Right-Shift Operations

Commonly used one-bit shift operations are as follows.

Operator	Definition*	Function
ROTL	Shift all bits left one position. Put original MSB in LSB position.	Rotate left
ROTR	Shift all bits right one position. Put original LSB in MSB position.	Rotate right
SRL	Shift all bits right one position. Put a zero in MSB position.	Shift right logical
SRA	Shift all bits right one position. Put a copy of original MSB in MSB position	Shift right arithmetic
SLL	Shift all bits left one position. Put a zero in LSB position.	Shift left logical
SLA**	Shift all bits left one position. Put a zero in LSB position.	Shift left arithmetic

* MSB is *most significant bit*; LSB is *least significant bit*.
** Same function as SLL; term is used in industry.

One-bit shift operations vacate the LSB or MSB. Therefore rules are needed for filling the LSB or MSB. The necessary rules are included in the operator definitions. Let the bits in the shift register represent a binary number. The bit in position 0 at the far right is the least significant bit. The LSB has weight $2^0 = 1$, the next bit has weight 2^1, the bit in position n has weight 2^n, and so forth. The most

Shift operators can multiply and divide by 2.

significant bit is at the far left position. In a 16-bit register the MSB weight is 2^{15}. Moving a bit to the left in the register doubles the value of a bit. Moving a bit to the right halves the value of a bit.

SRA and SLA are considered arithmetic operators because the shift rules are consistent with signed binary arithmetic.

A one-bit left shift doubles the value of each bit. If, at the same time, a zero is shifted into the LSB position, the original *unsigned* number is multiplied by two. For example, the SLA operation converts 0111 into 1110 ($+7 \rightarrow +14$).

A one-bit right shift halves the value of each bit. If, at the same time, a copy of the original MSB is shifted into the MSB position, the original number is divided by two, leaving the quotient in the register. For *all* shifted numbers the remainder of 0 or 1 is discarded. Consequently, positive numbers are rounded down in magnitude and negative numbers are rounded up in magnitude. We illustrate the SRA operation on all four-digit signed binary numbers in Table 8.1. Shifting in a copy of the MSB into the MSB position leaves the sign of the number unchanged, referred to as *sign extension*.

Shift operator definitions are also presented in the form of displays of before–after bit patterns. Table 8.2 illustrates the displays using an eight-bit register.

TABLE 8.1 **SRA Operations on Binary Numbers**

Before	Operator	After	n	q	r	$\dfrac{n}{2} = q + \dfrac{r}{2}$
0111	SRA	0011	($+7 \rightarrow +3$)		1	
0110		0011	($+6 \rightarrow +3$)		0	
0101		0010	($+5 \rightarrow +2$)		1	
0100		0010	($+4 \rightarrow +2$)		0	
0011		0001	($+3 \rightarrow +1$)		1	
0010		0001	($+2 \rightarrow +1$)		0	
0001		0000	($+1 \rightarrow +0$)		1	
0000		0000	($+0 \rightarrow +0$)		0	
1111		1111	($-1 \rightarrow -1$)		1	
1110		1111	($-2 \rightarrow -1$)		0	
1101		1110	($-3 \rightarrow -2$)		1	
1100		1110	($-4 \rightarrow -2$)		0	
1011		1101	($-5 \rightarrow -3$)		1	
1010		1101	($-6 \rightarrow -3$)		0	
1001		1100	($-7 \rightarrow -4$)		1	
1000		1100	($-8 \rightarrow -4$)		0	

TABLE 8.2   **Shift Operators and Bit Patterns**

	BIT PATTERN (STATE) BEFORE EACH OPERATION							
	q_7	q_6	q_5	q_4	q_3	q_2	q_1	q_0
OPERATOR	BIT PATTERN (STATE) AFTER EACH OPERATION							
ROTR	q_0	q_7	q_6	q_5	q_4	q_3	q_2	q_1
ROTL	q_6	q_5	q_4	q_3	q_2	q_1	q_0	q_7
SRL	0	q_7	q_6	q_5	q_4	q_3	q_2	q_1
SRA	q_7	q_7	q_6	q_5	q_4	q_3	q_2	q_1
SLL (= SLA)	q_6	q_5	q_4	q_3	q_2	q_1	q_0	0

8.3.2   Shift Register Design

Each shift operation puts a shift register in a new state. Each present state has a next state. Apparently, the D flip-flop input equations are $d_j = q_{j-1}$ for left shifts and $d_j = q_{j+1}$ for right shifts. Also, the equations for the LSB and MSB are some variation of this. Does the ASM chart procedure yield these results? Let us verify our assumptions.

Suppose we want to design a four-bit ROTL shift register. How do we derive the ASM chart from the ROTL specification? The answer is that we pick any state and derive the next state from the specification. We repeat this step for each state. In this way a truth table is constructed from which the ASM chart is derived (Table 8.3). Now we can write to K maps (for example, the d_0 map in Figure 8.34). The design equations for ROTL and the other operators are as follows. (Reminder: $q_j^+ = d_j$.)

Shift operations define the next state.

ROTR	ROTL	SRL	SRA	SLL, SLA
$d_0 = q_1$	$d_0 = q_3$	$d_0 = q_1$	$d_0 = q_1$	$d_0 = 0$
$d_1 = q_2$	$d_1 = q_0$	$d_1 = q_2$	$d_1 = q_2$	$d_1 = q_0$
$d_2 = q_3$	$d_2 = q_1$	$d_2 = q_3$	$d_2 = q_3$	$d_2 = q_1$
$d_3 = q_0$	$d_3 = q_2$	$d_3 = 0$	$d_3 = q_3$	$d_3 = q_2$

TABLE 8.3   Truth Table for a Four-Bit ROTL Shift Register

PRESENT STATE (BEFORE ROTL)		NEXT STATE (AFTER ROTL)	
	q_{3210}	d_{3210}	Comment
0	0000	0000	A cycle of one state: 0
1	0001	0010	
2	0010	0100	
4	0100	1000	
8	1000	0001	A cycle of four states: 1 2 4 8
3	0011	0110	
6	0110	1100	
C	1100	1001	
9	1001	0011	A cycle of four states: 3 6 C 9
5	0101	1010	
A	1010	0101	A cycle of two states: 5 A
7	0111	1110	
E	1110	1101	
D	1101	1011	
B	1011	0111	A cycle of four states: 7 E D B
F	1111	1111	A cycle of one state: F

FIGURE 8.34
ROTL K map for d_0

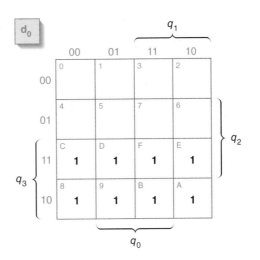

EXAMPLE 8.9   Johnson Counter Uses a Shift Register

The Johnson Counter sequence of states creates a set of squarewave output waveforms that are slipped one clock period, as shown here.

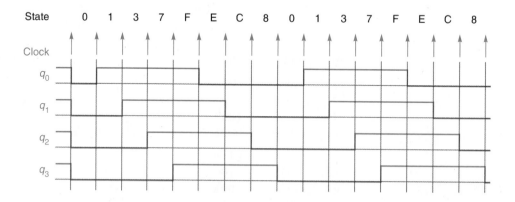

State decoding is simple. Any state, such as state 7, is detected by a two-input AND gate whose output waveform is as follows.

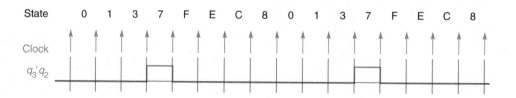

PS	q_{3210}	NS	$d_{3210} = q_{3210}^{+}$
0	0000	1	0001
1	0001	3	0011
3	0011	7	0111
7	0111	F	1111
F	1111	E	1110
E	1110	C	1100
C	1100	8	1000
8	1000	0	0000

By inspection:

$$d_0 = q_3{}'$$

$$d_1 = q_0$$

$$d_2 = q_1$$

$$d_3 = q_2$$

Note that the design is incomplete. The circuit needs a start control, and a clear line to be initialized to 0000. See the exercises.

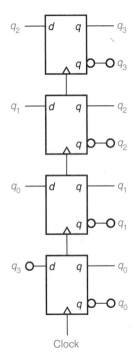

Clock

8.3.3   194A Universal Shift Register

All n-bit shift registers can be implemented by a chain of one or more 194A shift registers (Figure 8.35). The 194A data book schematic is at the gate level. We prefer the combinational block level, which uses multiplexers.

The 74LS194A schematic is more readable when modified to use equivalent combinational blocks and D flip-flops without changing

FIGURE 8.35
**Block diagram of
the 194A universal
shift register**

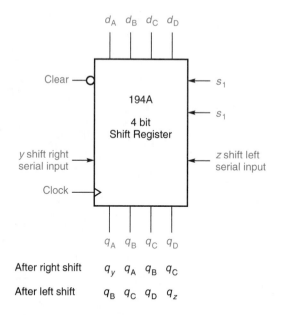

After right shift q_y q_A q_B q_C

After left shift q_B q_C q_D q_z

The 194 is a four bit
register that loads
input data or imple-
ments a programmed
shift operation on each
active clock edge.

functions (Figure 8.36). Note that the 194A clear operation is not synchronous. All other operations are synchronous. (This change of representation does not create a problem, because some data book schematics represent IC function, not necessarily the actual IC circuit.) The truth table for two mode select lines s_1 and s_0 follows. The z serial input sets the LSB value to q_z. The y serial input sets the MSB value to q_y.

s_1	s_0	Function	PS Before Operator	NS After Operator
0	0	hold value	$q_A q_B q_C q_D$	$q_A q_B q_C q_D$
0	1	right shift	$q_A q_B q_C q_D$	$q_y q_A q_B q_C$
1	0	left shift	$q_A q_B q_C q_D$	$q_B q_C q_D q_z$
1	1	load data	$q_A q_B q_C q_D$	$d_A d_B d_C d_D$

8.3.4   Special State Numbers

ASM with special state numbers reduce delays, because state number decoding and output decoding are not required.

FIGURE 8.36
**194A Universal
four-bit shift
register**

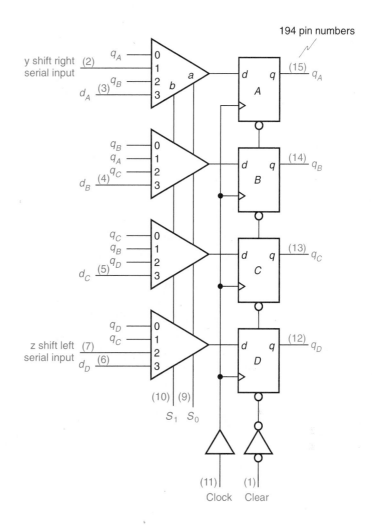

Normally, a fully encoded state number is decoded to provide
outputs during that state. We can avoid decoding the state number to
reduce output delays if we do not encode it in the first place. The basic
idea is for a special state to have only one bit in the state number equal
to 1 and all others equal to 0. Numbers equal to 2^n have this property.
This ASM form generates a state machine whose state register has the
property that each flip-flop corresponds to one state. This assignment
is often referred to as the "one-hot" assignment.

Deletion of state de-
coding results in very
fast state machines.

n	2^n	Binary Number	
		00000	Resting
0	1	00001	
1	2	00010	
2	4	00100	
3	8	01000	
4	16	10000	
and so forth			

Clearing all bits puts the ASM in state zero, which becomes a resting state.

A state machine with state numbers specified as 2^n is shown in Figure 8.37. This five-state ASM can be realized with three bits in a fully encoded physical realization. The unencoded state number version requires four bits. The price paid for avoiding state number decoding is more flip-flops in the state machine.

One informal state machine design process proceeds like this. Observe how the 1 "walks" through the state number. We interpret the walking 1 as being shifted through the state register. We add the 0000 start condition and immediately draw the circuit shown in Figure 8.38.

FIGURE 8.37

ASM with special state numbers

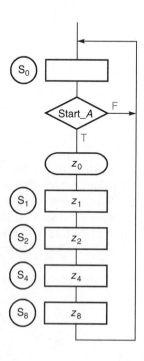

FIGURE 8.38
State machine with special state numbers

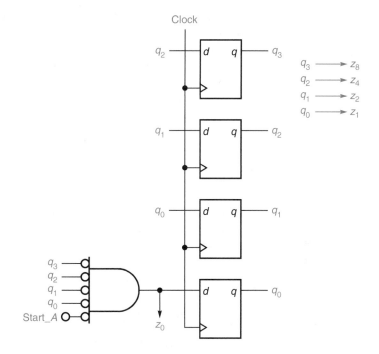

The intuitive process is readily implemented when the problem to be solved is not very complex. Nevertheless there are risks with this method. In order to be able to solve any problem we turn to a formal process.

One formal way to synthesize the state machine in the specified form is to proceed as follows.

STEP 1   This is the normal starting point: take information from the ASM chart and list present states PS, inputs, and next states NS (Table 8.4).

STEP 2   Four special states s_1, s_2, s_4, and s_8 in the ASM chart (Figure 8.37) imply the next-state number is a four-bit number $d_3d_2d_1d_0$ (Table 8.4).

STEP 3   The present outputs z_j are a function of the present states and present inputs (Table 8.4).

STEP 4   Use the special state number properties to simplify the state number equations. Given that s_1, s_2, s_4, s_8, and s_0 are the only allowed states s_j, the equations simplify as follows (because a bit is 1 only if all of the other bits are 0):

TABLE 8.4   **Truth Table Using Special State Numbers**

Step 1			Step 2	Step 3				
PS	Start_A	NS	$d_3d_2d_1d_0$	z_8	z_4	z_2	z_1	z_0
s_0	0	s_0	0 0 0 0					
	1	s_1	0 0 0 1					1
s_1	—	s_2	0 0 1 0				1	
s_2	—	s_4	0 1 0 0			1		
s_4	—	s_8	1 0 0 0		1			
s_8	—	s_0	0 0 0 0	1				

$$s_1 = q_3'q_2'q_1'q_0 = q_0q_0 = q_0 \text{ because } q_0 = q_3'q_2'q_1'$$

$$s_2 = q_3'q_2'q_1q_0' = q_1q_1 = q_1 \text{ because } q_1 = q_3'q_2'q_0'$$

$$s_4 = q_3'q_2q_1'q_0' = q_2q_2 = q_2 \text{ because } q_2 = q_3'q_1'q_0'$$

$$s_8 = q_3q_2'q_1'q_0' = q_3q_3 = q_3 \text{ because } q_3 = q_2'q_1'q_0'$$

STEP 5   Write the equations for the d_j using the truth table information in Table 8.4 and Step 4.

$$d_0 = s_0 \times \text{Start\_A} = q_3'q_2'q_1'q_0' \text{ Start\_A}$$

$$d_1 = s_1 = q_0$$

$$d_2 = s_2 = q_1$$

$$d_3 = s_4 = q_2$$

STEP 6   From the d_j equations, synthesize the state machine and outputs. [Do you recognize the shift register (Figure 8.38)?]

STEP 7   Recognize that the z_j outputs are directly related to the flip-flop outputs q_j (Figure 8.38).

$$z_8 = q_3, \quad z_4 = q_2, \quad z_2 = q_1, \quad z_1 = q_0,$$

$$z_0 = q_3'q_2'q_1'q_0' \text{ Start\_A}$$

Note: z_8, z_4, z_2, and z_1 are free of glitches; z_0 is not glitch-free; all z_j are synchronous.

8.4 ━━━━━	**Design Practice: Binary Multiplier**

Design is the art of converting a specification expressed in words, waveforms, sketches, or equations into circuits that work. Design specifications take many forms: "Please design a multiplier of two eight-digit BCD numbers . . . ," "Please design a special computer that . . . ," "Please design a two-level waveform generator that executes the Flim-Flam algorithm . . . ," and so forth.

Industrial designs are subjected to numerous constraints, such as minimizing cost or power or part count, only using certain types of parts, and so on. These days we usually are given very little time for product development, thorough test and evaluation, achieving zero defects, or use of competitive manufacturing processes, among many other considerations that also enter the designer's purview. The design we develop here does not address many of these concerns: the intention is merely to illustrate the design process.

This case study exemplifies the design of custom digital logic to perform a specific calculation. We begin with the problem statement. We use paper-and-pencil calculations to understand better the algorithm to be implemented. In the process, we recast the algorithm into a regular cyclic process, which will allow us to reuse one data path for the repetitive calculations it requires. With the calculation so defined, we select a set of hardware modules that can execute the calculations. The set constitutes the data path. Each hardware module has a set of control input and status output signals that will be controlled or monitored by the control state machine. A knowledge of the effects of the inputs and the meaning of the outputs, coupled with the insight into the algorithm gained in the initial paper-and-pencil calculation, allows us to construct an ASM chart specifying the needed behavior of the control state machine. The actual synthesis of the controller is straightforward, following the procedure outlined in Chapter 7. The completion of the design is left to the reader.

Binary multiplier design problem statement   Design a digital circuit that multiplies two four-bit binary numbers (x and y) using the shift-and-add algorithm. The result is the binary number c.

Nomenclature   Let

$$x = x_3x_2x_1x_0 = x_{3210}$$

$$y = y_3y_2y_1y_0 = y_{3210}$$

$$c = c_7c_6c_5c_4c_3c_2c_1c_0 = c_{76543210}$$

where x_3, y_3, and c_7 are the most significant bits.

8.4.1   Equations from Paper-and-Pencil Process

We calculate in binary the example $11 \times 6 = 66$ decimal.

$$
\begin{array}{r r l}
x_{3210} = & 0110 & \\
y_{3210} = & \underline{1011} & \textit{Line No.} \\
& 0110 & 11 \\
& 0110 & 12 \\
& 0000 & 13 \\
+ & 0110 & 14 \\
\text{Carry} & \underline{01111000} & \\
\text{Sum } c = & 01000010 &
\end{array}
$$

Line 11 is one times 0110. Line 12 is one times 0110, and the partial product is moved left one position relative to line 11 before we write it down. Line 13 is zero times 0110, with a two-position move left. Line 14 is one times 0110, with a three-position move left. The binary addition process creates the carries from digit to digit.

The first step forward is taken when 1011 is replaced by the digits of y. At the same time, 0110 is replaced with the variable x and each move left is recognized to be a multiplication by two. The resulting equations are shown below. Notice how the multiplier 1 is replaced by 2^0 in lieu of being dropped. Also notice that the power of two is the same as the y-digit subscript.

$$
\begin{array}{r l}
\text{line 11} & y_0 \times x \times 1 = y_0 \times x \times 2^0 \\
12 & y_1 \times x \times 2 = y_1 \times x \times 2^1 \\
13 & y_2 \times x \times 4 = y_2 \times x \times 2^2 \\
14 & y_3 \times x \times 8 = y_3 \times x \times 2^3
\end{array}
$$

and $c = \text{line } 11 + \text{line } 12 + \text{line } 13 + \text{line } 14$.

Attention is paid to the small details that make the equations regular. The result gives the designer confidence they are indeed correct.

8.4.2   Forming Identical Event Cycles

The paper-and-pencil process has one cycle of events per multiplier digit y_j. The idea is to make all cycles of events identical so that the same hardware can be used to execute each cycle. Seeking a standard cycle, we start with two types of cycles of events.

Type 1: Multiply x by digit y_j.
    Shift xy_j left j positions.
Type 2: Add the four lines.

There is one type 1 cycle per multiplier digit. The type 1 cycles are not identical, because j is different in each cycle. This is why they need modification in order to be converted into standard cycles. Even if these were identical, the totally different type 2 cycle remains standing alone. It needs to become part of the standard cycle. To achieve this goal, we apply the concept of partial sums: decompose the type 2 cycle into four partial sums.

Cycles are identical when hardware operations are not dependent on the subscript parameter j. The type 1 operation (shift xy_j left j positions) is j dependent and has to be replaced. So we do not shift xy_j. We add xy_j into the partial sum. Then we shift the partial sum left *one* position in each cycle. The result is that each xy_j is shifted left j times, which is what needs to happen. A subtle point is that the process starts with the most significant digit.

> **Cycle j:** Multiply x by digit y_j.
> Add the result to the partial sum.
> Shift the partial sum left one position.

We need to control how many of these cycles are executed. This involves counting the number of cycles, and stopping the process when the count equals the number of digits of the multiplier y. For this control purpose we use a counter. Counters offer us a choice: do we count up or do we count down?

Counting up implies that we initialize the counter to zero and stop the process when the count equals the number of digits in y. Counting down implies that we initialize the counter to the number of digits in y and stop the process when the count equals zero. The choice is ours. We prefer to count down, because loading a number into a 169 counter requires no external (glue) logic, for example, whereas counting up requires an external AND gate to detect the maximum allowable count. This circuit is not conveniently amenable to changes in the number of digits in y, whereas the number loaded into a 169 is. (This is an example of a proper use of bottom-up design.)

Counting down implies that the counter is decremented by one in each cycle. This raises the following question. When do we decrement by one? If we look back at line 11, we see that partial sum arising from the least significant digit or bit is not shifted. The LSB is the multiplier in the last cycle of the process. This is why the partial sum should not be shifted in the last cycle. An "if count > 0" phrase conditions the left-shift operation. This is effective if we recognize that the trick is to decrement the counter before invoking the left-shift operation.

Cycle j: Multiply x by digit y_j.
Add the result to the partial sum.
Decrement counter by 1.
Shift the partial sum left one position if the count is greater than zero.

A process is always initialized. In each cycle the xy_j partial product is added to the partial sum. In the first cycle the number added to must be zero. This is why the partial sum is initialized to zero. After the last digit is processed the partial sum becomes the sum c. So we use the variable c as the partial sum, and initialize c to zero.

8.4.3   Algorithm Walk-Through

Now we take a walk through the cycles of events. Observe that $b = j + 1$ in each cycle.

Start	Sum	Count b	Line No.
Initialize sum to 0, count to 4	00000000	0100	21
Cycle 1: Multiply x by digit y_3	0110		22
Add result to partial sum	00000110		23
Decrement counter by 1		0011	24
Shift left one position if $b > 0$	00001100		25
Cycle 2: Multiply x by digit y_2	0000		22
Add result to partial sum	00001100		23
Decrement counter by 1		0010	24
Shift left one position if $b > 0$	00011000		25
Cycle 3: Multiply x by digit y_1	0110		22
Add result to partial sum	00011110		23
Decrement counter by 1		0001	24
Shift left one position if $b > 0$	00111100		25
Cycle 4: Multiply x by digit y_0	0110		22
Add result to partial sum	01000010		23
Decrement counter by 1		0000	24
Omit shift left one position because $b = 0$			
Exit	01000010 = 66 decimal		

8.4.4   From Algorithm to Equations to Data Path

Here is event cycle j from the walk-through.

	Sum	Count b	Line No.
Cycle j:		$j + 1$	
Multiply x by digit y_j	0110		22
Add result to partial sum	00011110		23
Decrement counter by 1		j	24
Shift left one position if $b > 0$	00111100		25

The equations for, and implementation of, these operations are as follows.

Line 22: Multiply x by digit y_j

$$xy_j = x_3y_j,\ x_2y_j,\ x_1y_j,\ x_0y_j$$

The four terms are implemented with AND gates. Selection of y_j to feed to the AND gate inputs in cycle j is implemented with a multiplexer. The count variables represent j, so they have to be the multiplexer selection inputs. However, the algorithm walk-through shows that, entering the cycle, the count is $j + 1$. This implies feeding y_j to multiplexer input$_{j+1}$. Therefore a four-digit system requires an eight-to-one mux. That is one possible implementation. Another is to use a four-to-one mux so that y_j is fed to input$_j$. In this case we initialize the count to 3 (jmax $-$ 1 in this case) and stop on -1 (Figure 8.39).

FIGURE 8.39
Multiplier

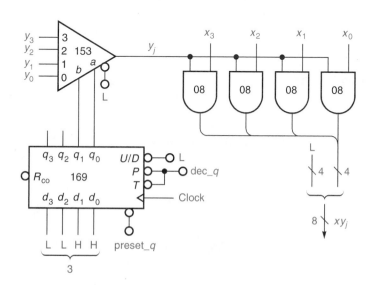

FIGURE 8.40

Reading and writing a register at the same time (see also Figure 8.41)

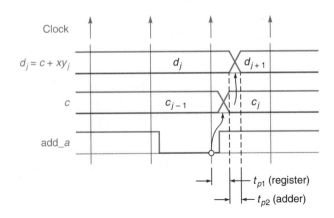

Line 23: Add result to partial sum

$$c = c \text{ plus } xy_j$$

First we need to know that a register can be read and loaded simultaneously. The register content c is the partial sum. Register c is read at its q outputs while $d = c$ plus xy_j at the d input is loaded into c. This is possible because output c and input $d = c$ plus xy_j are constant at the clock edge executing the addition (Figure 8.40). That is, they do not change until propagation times t_{p1} (from clock edge to register c output q) and t_{p2} (through the adder) elapse.

Standard 283 adders add c to xy_j. The sum is connected to the c register's d inputs. This implements the equation $c = c$ plus xy_j (Figure 8.41).

FIGURE 8.41

Partial sums adder

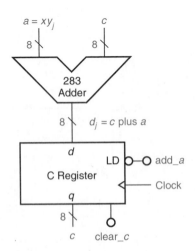

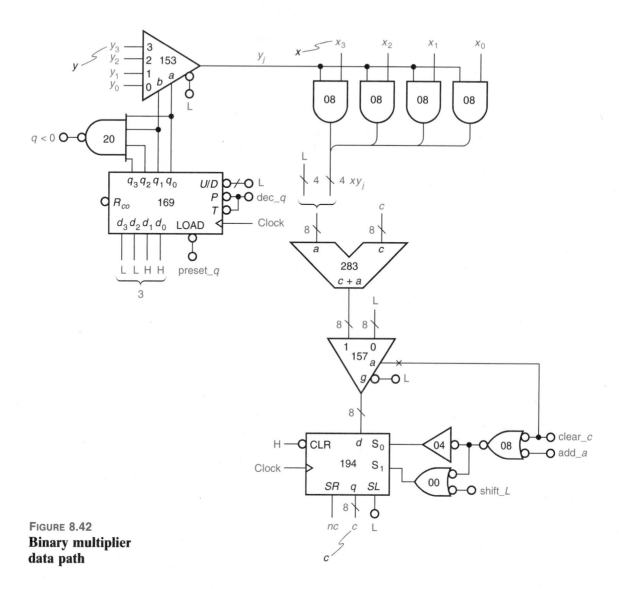

FIGURE 8.42
Binary multiplier data path

Line 24: Decrement count by 1

$$q = q - 1$$

Line 24 down-counts the counter that selects y_j (Figure 8.39).

Line 25: Shift left one position

$$c = c \times 2$$

This equation requires that the register storing the partial sum have a left-shift capability in addition to a load capability. The 194 shift register has this dual capability (Figure 8.42).

Initialize

$$c = 0 \qquad q = 0011$$

The 169 down counter is initialized to 3 by presetting it to 3 (Figure 8.42). The equation $c = 0$ is implemented by a multiplexer connected to the c register input. In this way zero or c plus xy_j can be loaded into c (Figure 8.42). The nonsynchronous 194 clear line is not used.

This completes the design of the data path shown in Figure 8.42.

8.4.5   From Algorithm and Data Path to ASM Chart

The data path implementing the multiplier operations has data inputs x and y, five control inputs, and one output (Figure 8.42). If we are using the multiplier, we need to be able to enter data x and y while the machine is at rest. Once the data is entered, we need to start the process, which executes once and stops. We have to design the controller.

First we make the system block diagram shown in Figure 8.43, using information from the data path design. At the system level, data path control inputs are controller outputs, and data path status out-

FIGURE 8.43
Multiplier block diagram

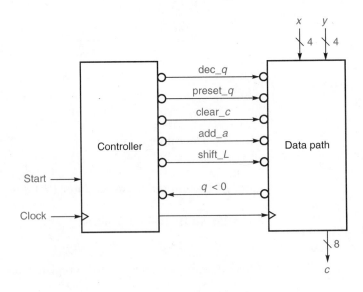

puts are controller inputs. Actual data are data path inputs, and results are data path outputs. Start is a controller input.

Here is the $c = x \times y$ process pseudocode as used up to this point.

Start

Initialize	$c = 0$ $q = 0011$
Cycle j:	Multiply x by digit y_j
	Add result to partial sum
	Decrement counter by 1
	Shift the partial sum left one position if $q > -1$
	Exit

Mechanization requires that a while-do-repeat construct implement the sequence of event cycles. An if-then construct omits the last shift left when count $q = -1$.

Start

Initialize $c = 0$ $q = 0011$

Cycle j:		Multiply x by digit y_j
	while	$q > -1$
	do	Add result to partial sum
		Decrement counter by 1
		If $q > -1$ then shift the partial sum left one position
	repeat	
		Exit

This sequence of pseudocode statements is translated into ASM blocks, a process that is not really obvious, because there is a great deal of hidden logic in the pseudocode. However, experience makes our next translation of words into a design more straightforward.

Resting state ASM block s_0 The ASM rests in state s_0 when input variable start is False (Figure 8.44). Start is asserted (True) synchronously. When this happens, the conditional outputs preset_q ($q = 3$) and clear_c ($c = 0$) are asserted for one clock period and executed at the next clock edge when the ASM steps from s_0 to s_1.

Event cycle ASM blocks s_1 and s_2 While input variable q is not less than 0 the DO path is taken (Figure 8.45), asserting the conditional

FIGURE 8.44
**Binary multiplier
ASM state s_0**

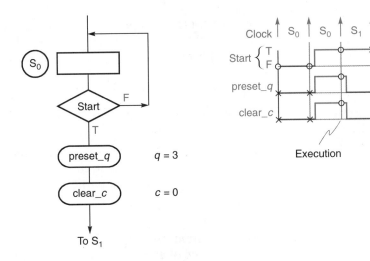

$q = 3$

$c = 0$

outputs dec_q and add_a. At the next clock edge the equations $q = q - 1$ and $c = c$ plus a are executed as the ASM steps to s_2. In state s_2, when the expression $q < 0$ is not asserted the conditional shift_L output is asserted to execute $c = c \times 2$ at the next clock edge.

Understanding the timing is very important. Outputs dec_q and add_a set up the equations for execution at the next clock edge. Dec_q and add_a do not execute while the ASM is in s_1. The equations are executed by the clock edge that moves the ASM from s_1 to s_2. This is why the expression $q < 0$ can be False (e.g., when $q = 0$) while in s_1 and true ($q = -1$) while in s_2. In s_2 the branch diamond implements the if-then construct that omits the last shift-left-one-position operation.

The timing diagram (Figure 8.46) is worth considerable study.

Exit ASM block s_3   When Start is a push button, a toggle switch, or a very long pulse, Start will be asserted for a time that is long when compared to the calculation process. The point is that if exit block s_3 is omitted, the ASM would return to s_0 while Start is still asserted, thereby starting a new calculation. This unwanted recycling is avoided when Start is used to hold the ASM in s_3 as long as Start is asserted (Figure 8.47).

The assembly of ASM blocks forms the complete binary multiplication ASM chart (Figure 8.48).

FIGURE 8.45
Binary multiplier ASM states s_1 and s_2

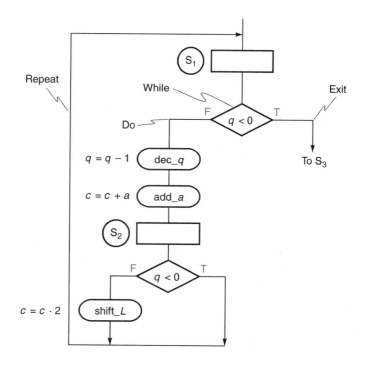

FIGURE 8.46
Binary multiplier ASM timing diagram

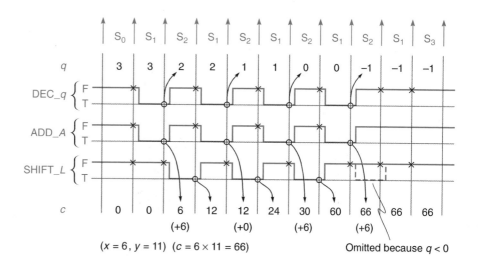

FIGURE 8.47
**Binary multiplier
ASM state s_3**

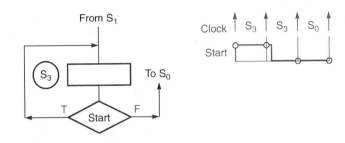

FIGURE 8.48
**Binary multiplier
ASM**

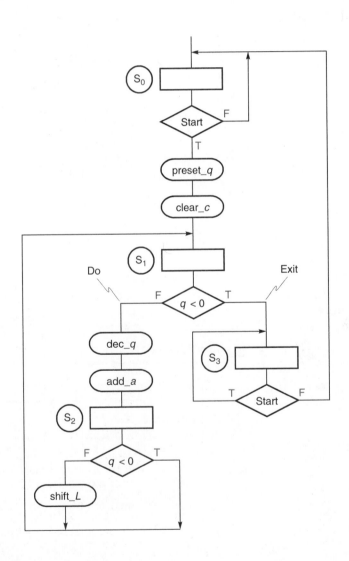

8.4.6   Data Path and State Machine Synthesis

The truth table is taken from the binary multiplier ASM chart by the ASM method of Chapter 7. The state machine and output circuits are designed from the truth table data. The truth table, state machine, and output circuits are left for the reader to design.

SUMMARY

Storage Register

The D flip-flop defining equation, $q^+ = d$, shows that at each active clock edge a D flip-flop stores the present value of d, which becomes the next value of q. Data is stored, however, with the intention of saving the data for future use. Clearly, a register cannot be a group of D flip-flops. We need something else.

The act of storing data is called loading. This implies the need for an input control line named load. The load function transforms a D flip-flop into a register flip-flop, also known as an enabled D flip-flop. A register is an assembly of enabled D flip-flops.

Counting Register

Synchronous look-ahead carry up and down counters are introduced. We start with two-bit counters with no controls, so that we can focus on basic principles outside the shadows of the multitude of details intrinsic to assemblies of larger numbers of bits. Then controls are added: count, load, and clear. Next a four-bit up counter with these control lines is synthesized. The result is the circuit of the 163 standard logic family member.

The fourth control line, up/down, is added and a four-bit up/down counter with four control lines is synthesized. Except for the clear line, which the 169 lacks, the result is the circuit of the 169 standard logic family member. The process shows one way to create equations from ASM chart fragments. Synchronous counters are used to synthesize precision waveforms for a variety of applications. One design process is presented.

Shift Register

A register is called a shift register when all bits stored in the register can move one or more positions to the left or right of their present position by asserting control lines. A shift register holds the bit values until a shift command is received. This is why the register flip-flops must be enabled D flip-flops. Commonly used one-bit shift operations are defined and implemented using the ASM method. Shift registers implement special state machines whose state numbers and outputs do not need decoding. Their significance is the increased performance resulting from no decoding delays.

REFERENCES

Blakeslee, T. R. 1979. *Digital Design with Standard MSI and LSI,* 2nd ed. New York: Wiley.

Breeding, K. J. 1989. *Digital Design Fundamentals*. Englewood Cliffs, N.J.: Prentice-Hall.

Roth, C. H. 1992. *Fundamentals of Logic Design,* 3rd ed. St. Paul, Minn.: West.

Wakerly, J. F. 1990. *Digital Design Principles and Practices*. Englewood Cliffs, N.J.: Prentice-Hall.

Name _____ SID# _____ Section _____

     (last)   *(initials)*

Approved by _____ Date _____

Grade on report _____

State Machine Design

1. Analyze the sequence of states in the ASM chart.

    Predicted timing diagram, 20 clock periods

    _____ CLOCK
    _____ GO

    _____ COUNT_163
    _____ CLEAR_163

    _____ CLKEN_166
    _____ LOAD_166

    _____ What does the circuit do?

2. Design the circuit.

    _____ GO circuit
    _____ State machine
    _____ Output circuits

3. Build the circuit.

    _____ circuit documents

4. Use the truth table signal generator to test the circuit.

    _____ input/output timing diagram

1. Analyze the sequence of states in the ASM chart in Figure PJ10.1.

    This is not a straightforward problem, because you will have to develop the waveforms interactively. By this we mean you plot the waveforms in parallel because they influence each other. Start by making a timing diagram showing 20 clock periods numbered 1 to 20. Let the GO signal go true in period 2 and then go False in period 15. Let the END_163 signal occur every eight clock periods when COUNT_163 is True. Let LOAD_166 load 10110100 in the 166 where the 0 LSB is 166 bit H.

    Show GO, END_163, and all output waveforms. Show the sequence of states, the count in the 163, and the binary number stored in the 166 for clock periods 1 to 20. Now tell what this circuit does.

FIGURE PJ10.1
ASM chart

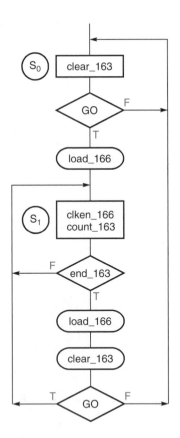

FIGURE PJ10.2
System data path

2. Design the circuit.

   Design a state machine and output circuits implementing the system data path in Figure PJ10.2 and the ASM chart in Figure PJ10.1. Use a toggle switch and debouncer circuit to generate the GO input.

3. Build the circuit. Use 74LS02, 74LS04, 74LS74, 74LS157, 74LS163, and 74LS166.

4. Test the circuit.

   Use the oscilloscope to recreate your timing diagram from Step 1 to prove the circuit works.

PROJECT 11 Demonstration Report

Name ______________________________ SID# ________________ Section _____
(last) (initials)

Approved by ________________________ Date ______________________________

Grade on report _____________________

Register designs generating sequences of numbers.

1. Design the circuit.

_____ State machine
_____ Start and end number circuits

2. ASM chart_____

3. Build the circuit.

_____ circuit documents

4. Use the truth table signal generator to test the circuit.

_____ input/output timing diagram

The 163 four-bit counter can generate the numbers 0 to F hex at the four bit outputs q_3, q_2, q_1, and q_0. A JK flip-flop is readily added as bit_4. The count in the five-bit counter ranges from 00 to 1F hex.

1. Design a circuit that generates any ascending sequence of numbers in the range 00 to 1F. For example: 5 6 7 8 9 5 6 7 8 9 . . . or C D E F 0 1 2 C D E F 0 1 2 . . . . Use H/L dip switch circuits (see Figure 12.2a) to select the start and end numbers.

2. Draw an ASM chart for this circuit.

3. Build the circuit. Use one 74LS163, one 74LS109, and other logic as required.

4. Connect signal generator outputs R_{co} to scope trigger, and q_0 to the clock input. Plot R_{co}, counter outputs q_0, q_1, q_2, q_3, q_4 (JK) and the 163 load input. Plot 33 or more clock periods.

Name _____   SID# _____ Section _____
　　　(last)　　　　　　*(initials)*

Approved by _____   Date _____

Grade on report _____

Synchronous three-bit up counter

1. Design the circuit.

   _____ state machine
   _____ start and end number circuits

2. Build the circuit.

   _____ circuit documents

3. Use the truth table signal generator to test the circuit.

   _____ input/output timing diagram
   _____ q_0
   _____ q_1 and delay from q_0
   _____ q_2 and delay from q_1

4. Maximum counting frequency

   _____ data
   _____ maximum counting frequency calculation

1. Design a synchronous three-bit D flip-flop up counter with look-ahead carry and a count control input p. The counter counts when p is True, and the counter holds the present count value when p is False. Do not gate the clock.

2. Build the circuit. Use 74LS109 and glue logic as required.

3. Connect signal generator outputs R_{co} to scope trigger input, clk to the clock input, and q_3 to the p input. Plot R_{co}, q_3, and counter outputs q_0, q_1, and q_2.

4. Estimate the maximum counting frequency of the counter. Design tests of what data to measure. Then measure the data that supports your conclusion. Show all calculations leading to your conclusion. *Hint:* What is the longest delay path from any flip-flop output to any flip-flop input?

Name _____ SID# _____ Section _____
    (last)   *(initials)*

Approved by _____ Date _____

Grade on report _____

Binary multiplication state machine design

1. ASM chart

    _____ main
    _____ linked ASM

2. System block diagram

    _____ state machine
    _____ data path

3. Circuit design

    _____ state machine
    _____ data path

4. Build the circuit.

    _____ circuit documents

5. Use the truth table signal generator to test the circuit.

    _____ input/output timing diagram
    _____ correct results

Using the repeated addition algorithm, calculate $c = a \times b$, where

$$a = a_3 a_2 a_1 a_0$$

$$b = b_3 b_2 b_1 b_0$$

$$c = c_7 c_6 c_5 c_4 c_3 c_2 c_1 c_0$$

Defining $+$ as plus instead of OR, we have

$$c = a + a + a + a + \cdots + a \; (b \text{ additions})$$

1. Create an ASM chart implementing the algorithm.

   The a and b inputs to the binary multiplier are set up while the state machine is at rest (stopped). Then a start button is pushed to execute the calculation. The state machine must stop when the calculation is executed. A problem arises when the start button is asserted for a longer time than the time to calculate a result. Therefore the calculation is repeated as the "still-started" ASM recycles through its states.

   If a stop input in the last state holds the state machine in that state, the ASM will not recycle. The key observation is that the complement of an asserted start input is a stop input. So, when the stop input is the complement of the start input, a fail-safe system results. Or the output of a single pulser is a reliable start input. Explain why.

2. Make a system block diagram from information on the ASM chart. *Hint:* use two black boxes—one box for the data path architecture, and the second box for the state machine.

3. From the ASM chart design a circuit to perform the calculation.

4. Build the circuit. Set up the a and b inputs with H/L dip-switch circuits. Use 74LS04, 74LS10, 74LS74, 74LS153, 74LS173, 74LS169, and 74LS283.

5. *Test.* Let $q_3 =$ start. Record the results.

Name _____ SID# _____ Section _____
      (last)   *(initials)*

Approved by _____ Date _____

Grade on report _____

Periodic pulse generator

1. Calculate the ranges of n_1 and n_2.

2. ASM chart

   _____ mainline ASM
   _____ linked ASM (?)

3. System block diagram

   _____ controller
   _____ data path

4. Circuit design

   _____ controller
   _____ data path

5. Build the pulse generator.

6. Test limiting values of n_1 and n_2 as well as intermediate values.

   _____ test data

Design a periodic pulse generator with the following output waveform:

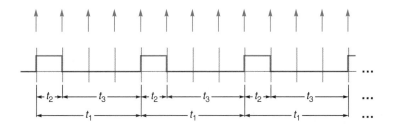

where t_1 ranges from 200 ns to 3.2 microsec, t_2 ranges from 100 ns to 3.1 microsec, and $t_3 = t_1 - t_2$ and $t_1 > t_2$ is guaranteed. If t_1 is not greater than t_2, then an LED is on and the pulse generator cannot start. Values for t_1 and t_2 are programmed by binary values. Use the signal generator 10-MHz clock.

Operation: to generate a new waveform, the user presses stop, selects n_1 and n_2, presses start.

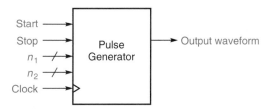

1. Convert t_1 and t_2 into counts n_1 and n_2 so that $t_1 = n_1 t_c$ and $t_2 = n_2 t_c$, where t_c = clock period. Calculate the ranges of n_1 and n_2.

2. Create an ASM chart for the pulse generator.

3. Create a system block diagram.

4. Design the data path and controller. Use 169 (bits 4, 3, 2, 1) and 74 (bit 0) in each 5-bit counter. (The purpose of the five-bit requirement is to enhance your skills. This is why you cannot use two 169 chips in any one counter.)

5. Build the pulse generator.

6. Test limiting values of n_1 and n_2 as well as intermediate values. Record the data.

Name _____   SID# _____   Section _____
       (last)   *(initials)*

Approved by _____   Date _____

Grade on report _____

Double-pulse generator

1. Calculate the ranges of n_1, n_2, and n_3.

2. ASM chart

    _____ mainline ASM
    _____ linked ASM (?)

3. System block diagram

    _____ controller
    _____ data path

4. Circuit design

    _____ controller
    _____ data path

5. Build the pulse generator.

6. Test limiting values of n_1, n_2, and n_3 as well as intermediate values.

    _____ test data

Design a double-pulse generator with the following output waveform:

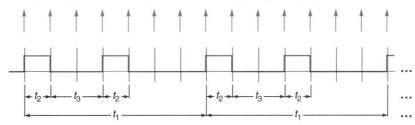

where t_1 ranges from 400 ns to 3.2 microsec, t_2 ranges from 100 ns to 900 ns, and t_3 ranges from 100 ns to 900 ns. If t_1 is not greater than $2t_2 + t_3$, then an LED is on and the pulse generator cannot start. Values for t_1, t_2, and t_3 are programmed by binary values. Use the signal generator 10-MHz clock.

   Operation: to generate a new waveform, the user presses stop, selects n_1, n_2, and n_3, presses start.

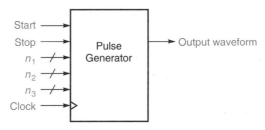

1. Convert t_1, t_2, and t_3 into counts n_1, n_2, and n_3 so that

$$t_1 = n_1 t_c \qquad t_2 = n_2 t_c \qquad t_3 = n_3 t_c$$

   where t_c = clock period. Calculate the ranges of n_1, n_2, and n_3.

2. Create an ASM chart for the pulse generator.

3. Create a system block diagram.

4. Design the data path and controller. Use 169 (bits 4, 3, 2, 1) and 74 (bit 0) in any 5-bit counter. Use a 169 in any 4-bit counter. (The purpose of the five-bit requirement is to enhance your skills. This is why you cannot use two 169 chips in any one counter.)

5. Build the pulse generator.

6. Test limiting values of n_1, n_2, and n_3 as well as intermediate values. Record the data.

PROBLEMS

8.1 Draw the K maps for d_1 and d_0 flip-flop inputs in a two-bit synchronous up counter. Derive the d_1, d_0 equations from the maps:

(a) without a count control p
(b) with a count control p

8.2 Refer to Figure 8.12. Replace the s_j with q_1q_0 minterms and derive the d_j equations in XOR format.

8.3 Refer to Example 8.3 on page 422. Find the q_1^+, q_0^+ equations by reading the K maps with p as a map-entered variable.

8.4 Draw the K maps for d_1 and d_0 flip-flop inputs in a two-bit synchronous down counter. Derive the d_1, d_0 equations from the maps:

(a) without a count control p
(b) with a count control p

8.5 Refer to Figure 8.17. Replace the s_j with q_1q_0 minterms and derive the d_j equations in XOR format.

8.6 Draw the K maps for d_1 and d_0 flip-flop inputs in a two-bit synchronous up/down counter. Derive the d_1, d_0 equations from the maps:

(a) without a count control p
(b) with a count control p

8.7 Design a three-bit up counter using D flip-flops and with count control p. Use K maps.

8.8 Design a three-bit up counter using JK flip-flops in T format and with count control p. Use K maps.

8.9 Refer to Figure 8.23.

(a) Draw a timing diagram showing the four ripple carries.
(b) Draw a timing diagram showing the four look-ahead carries.

8.10 Refer to Figure 8.24. Draw a timing diagram that shows how the 163 R_{CO} waveform relates to q_3 and q_0.

8.11 Refer to Figure 8.25. Write to a K map the $q_j^+ = d_j$ column. Read the K map and derive the equation for d_j.

(a) d_3
(b) d_2
(c) d_1
(d) d_0.

8.12 Refer to Example 8.6 on page 440.

(a) Do Step 1.2.
(b) Do Step 1.3.
(c) Do Step 1.4.
(d) Do Step 1.5.

8.13 Refer to Example 8.6 on page 440 and Problem 8.12a.

(a) Convert the expression for d_1 to $(q_3 q_0)'(q_1 \text{ xor } q_0)$.
(b) Convert the expression for d_3 to $(q_3 q_0)'(q_3 \text{ xor } q_2 q_1 q_0)$.

8.14 Refer to Example 8.7 on page 442. Design a modulo-24 up counter with state 0-to-23 sequence using 163 chips.

8.15 Refer to Example 8.7 on page 442. Design a modulo-8 down counter with state F-to-8 sequence using a 169 chip.

8.16 Refer to Figure 8.25. Make a truth table for a four-bit down counter. Write to a K map the $q_j^+ = d_j$ column. Read the K map and derive the equation for d_j.

(a) d_3
(b) d_2
(c) d_1
(d) d_0.

8.17 Refer to Example 8.7 on page 442. Design a modulo-24 down counter with state 23-to-0 sequence using 169 chips.

8.18 Ring counter sequence of states $(q_3 q_2 q_1 q_0)$ is 0001, 0010, 0100, 1000. The sequence of states implies use of a shift register. The counter is self-correcting if the d_0 input is the term $q_2' q_1' q_0'$. Make a truth table. Draw the 16-state ASM chart, and design a self-correcting ring counter. Explain the self-correcting feature. Draw the timing diagram.

8.19 Ring counter sequence of states $(q_3 q_2 q_1 q_0)$ is 1110, 1101, 1011, 0111. The sequence of states implies use of a shift register. The counter is self-correcting if the d_0 input is the term $q_2' + q_1' + q_0'$. Make a truth table. Draw the 16-state ASM chart, and design a self-correcting ring counter. Explain the self-correcting feature. Draw a timing diagram.

8.20 Johnson counter sequence of states $(q_3 q_2 q_1 q_0)$ in hex is 0, 1, 3, 7, F, E, C, 8, 0, 1, 3, etc. The sequence implies use of a truth table. The counter is self-correcting if the load input is the term $q_3' q_0'$. Make a truth table. Draw the 16-state ASM chart, and design a self-correcting Johnson counter. Explain the self-correcting feature. Draw the timing diagram.

8.21 Analyze the waveform generator in Figure 8.30.

(a) Find the equations for n_F, n_8, and w.
(b) Make timing diagrams, and show that the circuit works correctly.

8.22 Analyze the waveform generator in Figure 8.33.

(a) Find the equations for n_{4F}, n_{FF}, and w.
(b) Make timing diagrams, and show that the circuit works correctly.

8.23 Refer to Figure 8.28. Design the waveform generator.

8.24 Refer to Figure 8.31. Design the waveform generator.

8.25 Refer to Table 8.1 on page 451. Use mathematics to calculate the error and the number's magnitude when the remainder of a positive number is dropped. Repeat for negative numbers.

8.26 Draw the ASM chart for a four-bit shift register inplementing ROTL.

8.27 Refer to Example 8.9 on page 454. Use the present-state table to explain why $q_3' q_2$ decodes state 7. Explain why the waveform is glitch-free.

(a) Repeat for state F.
(b) Repeat for state E.
(c) Repeat for state C.
(d) Repeat for state 8.
(e) Repeat for state 0.
(f) Repeat for state 1.
(g) Repeat for state 3.

8.28 Refer to Example 8.9 on page 454. The n-bit Johnson counter has $2^n - 2n$ unspecified states (e.g., in the truth table of Example 7.9, $n = 4$, $2^n - 2n = 16 - 8 = 8$). List the eight states for $n = 4$, and let 0001_2 be the next state for the eight newly listed states. Design a circuit.

8.29 Refer to Section 8.3.2 (page 452) and Table 8.4 (page 460). Make the PS, NS table and the ASM chart for four-bit registers implementing shift operators:

(a) SRL
(b) SRA
(c) SLL

8.30 Refer to Figure 8.34. Write K maps for shift operators.

(a) SRL
(b) SRA
(c) SLL

8.31 Build a four-bit shift register with two modes: (a) hold, (b) shift left. Show control lines and serial input/output. The four-bit number in the register is $q_3q_2q_1q_0$.

8.32 Build a four-bit shift register with four modes: (a) hold, (b) rotate left, (c) rotate right, (d) shift right. Show control lines and serial input/output. The four-bit number in the register is $q_3q_2q_1q_0$.

8.33 Design a programmable circuit using the 194, combinational blocks, and gates to implement the following:

Code	0	1	2	3	4	5
Operator	HOLD	ROTL	SLL/SLA	SRA	SRL	LOAD

8.34 A four-bit LFSR (linear feedback shift register) has the following d input equations: $d_3 = q_2$, $d_2 = q_1$, $d_1 = q_0$, $d_0 = q_3$ xor q_2. Draw the circuit schematic. What is the sequence of states?

8.35 A four-bit LFSR (linear feedback shift register) has the following d input equations: $d_3 = q_3$ xor q_2, $d_2 = q_1$, $d_1 = q_0$, $d_0 = q_3$. Draw the circuit schematic. What is the sequence of states?

9 MEMORY

OVERVIEW

In this chapter we add memory devices to our repetoire of complex building blocks. These widely used devices are found in controllers as well as in data paths. A useful way to think of these devices is as implementing an array of registers, an array in which only one register at a time can be accessed.

We start with memory device categories and the important properties of memory devices: access time, volatility, static vs. dynamic, and programmability. The basic device structure is shown to be a group of registers, each with a unique address. How data is read from, and written into, any register is then explained. The explanation is facilitated by use of static and dynamic device read/write timing diagrams. This is followed by some examples of memory devices.

When memory size exceeds memory device size, an address decoder organizes access to an array of memory devices. The ideas basic to address decoder design are presented in the context of a hypothetical memory chip containing eight four-bit words. A specific decoder design illustrates a basic method. Then we study more complex decoders of larger address ranges and the consequences of word sizes with more bits.

Next, a typical memory block diagram is discussed. We learn that the basic memory module components are address register, address decoder, RAM array, data register, the controlling state machine, and error detection and correction circuits. (The latter subject is not studied.) At this point the stage is set for memory system design.

INTRODUCTION

This chapter is about random-access memory devices and the design of various types of memories. After discussing memory device categories, structure, and timing, we give specific examples of devices. Then a method for address decoder design is presented. This is followed by an example of memory design that illustrates the essential design issues.

Combinational- and sequential-logic circuit applications of these and other programmable logic devices are discussed in Chapter 10.

9.1 Memory Devices

Only one memory register at a time can be accessed.

Memory devices provide a cost-effective means for implementing a large number of registers. However, only one register at a time can be accessed in a memory chip array. This is a serious limitation that may not be obvious.

The following discussion of memory device categories, structure, and timing emphasizes the essentials; the details are left to examples.

9.1.1 Memory Device Categories

Memory devices are classified as read-only memory (ROM) or as read-write memory, also known as random-access memory (RAM). In turn, RAMs divide into two categories: static and dynamic. ROMs also divide into two categories: reprogrammable and one-time-programmable. An alternative categorization for ROMs is field-programmable vs. factory-programmable (Figure 9.1). Both read–write memory devices and read-only memory devices are random-access devices. Next we discuss important concepts applicable to the categories of memory devices presented in Figure 9.1.

Memory chips do not include timing circuits

Access time All memory devices allow access to any internal register by giving each register a unique address. *Access time* is the time required to locate and read a word from a memory device. A memory device is a random-access device when the time to access data in any location is less than or equal to some specified maximum access time, and the access time is independent of the sequence of locations being read. In this sense, each register is equally accessible, because access time is independent of the address. On the other hand, magnetic tapes and disks are not random-access devices, for the time required to access data on a tape or disk depends on where the data is located on the tape or disk.

Access time is the most important memory chip specification.

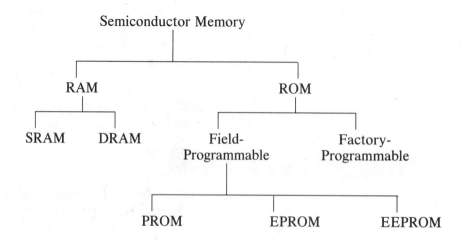

FIGURE 9.1
Memory device categories

Data is permanently
stored in ROMs.

Volatility ROM devices are *not volatile*. That is, we can turn the power off and on without destroying the data stored there. ROMs use fuses or transistors permanently altered to be in the on or off state to store patterns of 1's and 0's as words. In contrast, all RAM devices are *volatile*, meaning data is not preserved in a RAM when the power is turned off and on.

Data is volatile in
RAMs.

Static vs. dynamic RAMs divide into static RAMs (SRAMs) and dynamic RAMs (DRAMs). An SRAM one-bit cell is a latch (Figure 9.2a). A DRAM one-bit cell is a transistor and a capacitor (Figure 9.2b). In essence, static memory devices store data in arrays of latches, whereas dynamic memory devices store data in arrays of capacitors

FIGURE 9.2
Typical memory cells and device structure

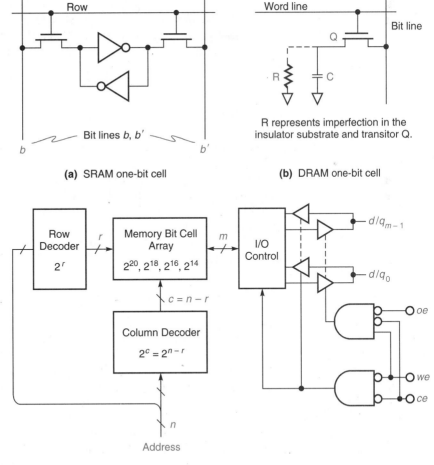

(a) SRAM one-bit cell

(b) DRAM one-bit cell

R represents imperfection in the insulator substrate and transitor Q.

(c) SRAM structure

(shunted by parasitic resistors). Thus, static devices are arrays of some stable circuit that can hold data indefinitely, and dynamic devices are arrays of some unstable circuit that needs constant refreshing because it cannot hold data indefinitely.

Factory-programmability ROM programming that takes place in a factory is a part of the manufacture of integrated circuits. Programming information is in the form of masks employed in the manufacturing process. The significant factory setup charges for masks are justified when the ROM lot size is large and the program to be stored is proven. A program etched in silicon cannot be changed.

Field-programmability Programmable ROMs (PROMs) are field programmed by vaporizing fusible links. These programmed-once devices are cost effective in low-volume applications. The erasable programmable read-only memory (EPROM) is reprogrammable because its contents can be erased with ultraviolet light. EPROMs can be reprogrammed an indefinite number of times. Strictly speaking, EPROMs are volatile; but as a practical matter, EPROMs are not volatile because they retain the stored charge differentiating 1's from 0's for over 10 years. EPROMs must be removed from any printed circuit board to be programmed in a separate programming instrument.

The electrically eraseable programmable read-only memory (EEPROM) can be reprogrammed only a limited number of times compared to the EPROM. As a practical matter, however, this limitation is not serious because the EEPROM can be programmed over 10,000 times. We will not discuss how ROMs are physically programmed because programming is not a digital design issue here. Designers use various software and hardware tools to program the multiplicity of ROM types.

9.1.2 Memory Device Structure

A memory device consists of a number of registers addressed by n binary-encoded address lines (Figure 9.2c). Consequently, the number of registers is 2^n. A number in the range from 0 to $2^n - 1$ is assigned to each register. For $n = 16_{10}$, the range is 0 to $65,535_{10}$. A number is the register's address. (The register address is like a house address.) Each register is the same size and stores some number of data bits m.

The size of a memory chip is usually stated as 2^n by m ($2^n \times m$), where m is the number of data bits per address. The total number of bits on the chip is thus $2^n m$. For example, a memory device with 12 address lines and 4 data bits per register, or address, contains 4096×4

n address lines address 2^n registers.

bits. For convenience, 1024 (2^{10}) is denoted as 1K (see the following table). The 4096 (2^{12}) × 4 (2^2) device contains 16,348 (2^{14}), or 16K, bits.

n	2^n	K (= thousands)
10	1,024	1K
11	2,048	2K
12	4,096	4K
13	8,192	8K
14	16,384	16K
15	32,768	32K
16	65,536	64K
17	131,072	128K
18	262,144	256K
19	524,288	512K
20	1,048,576	1024K = 1M

The $2^n \times m$ device has n address lines and m data lines plus control lines. Typical static read-write RAM (SRAM) device control lines are write-enable (WE), chip-enable (CE), and output-enable (OE). A device is in the read mode when WE is False. ROM devices omit the WE line. In programmable ROMs (PROMS, EPROMS, and EEPROMS), however, the write-enable line is replaced by a program-the-bits line. In dynamic RAM devices, the chip-enable line is replaced by row address strobe (RAS) and column address strobe (CAS) lines. (The various line functions are explained later.)

The manufacture of a packaged circuit with many pins per package is a significant cost issue. In practice, when $m = 1$ there are two data lines: one input line and one output line. When $m > 1$, m pins are saved by assigning data input and output to the same pins, because additional input/output control circuits on the chip cost less than m additional pins.

RAS stores $n/2$ address bits. CAS stores $n/2$ address bits. n address bits address 2^n registers.

Also, in DRAMs where n is large the address pin count is halved by sharing $n/2$ address pins between $n/2$ high-order and $n/2$ low-order address bits. This involves additional expense, because the designer needs to add circuits external to the DRAMs that multiplex the address lines. The external multiplexer presents the row address to the DRAM chip. Then the row address strobe stores the row address in the memory chip. Next, the external multiplexer switches and presents the column address to the memory chip. Finally, the column address strobe stores the column address in the memory chip.

The memory device addressed by n bits has 2^n registers. An n-bit decoder can be implemented with 2^n gates, one per register. When $n = 20$, the decoder requires 2^{20}, or about one million, gates.

There is a better way to implement the decoder. We partition the n bits into two sets of row bits r and column bits $c = n - r$. Then we decode each set. Now the number of decoder gates required is 2^r plus $2^c = 2^{n-r}$. When $n = 20$ and $r = 9$, the number of row gates is $2^9 = 512$, and the number of column gates is $2^{11} = 2048$, for a total of 2560 (not one million) complex multi-input gates. We are not home free yet, for there is still another requirement.

The row and column sets of decoder gates are arranged as inputs to a row/column matrix of memory device registers. Any register is selected by the logical AND of one row line and one column line. This requires adding one two-input AND gate to each of one million registers. Nevertheless, this solution still uses significantly less hardware. A typical SRAM memory structure is shown in Figure 9.2c.

EXERCISE 9.1   A static RAM has 14 address lines and eight data lines. Calculate the number of bits in the RAM.

Answer: 128K   ■

EXERCISE 9.2   How many address lines does a one-megabit RAM have when there are four data lines?

Answer: 18   ■

EXERCISE 9.3   How many address lines does one 512K × 32 bank of RAMS consisting of four 512K × 8 devices with addresses wired in parallel have?

Answer: 19   ■

9.1.3   Memory Device Timing

The basic simplicity of memory timing is obscured by the multitude of timing parameters found in data books. Once the basics are understood, the details are readily assimilated. The timing parameters arise from propagation delays through the logic within the devices.

Memory devices do not include any timers. One principal reason is the lower chip cost. Another reason is that the impact on system cost is minimal because the timing is readily implemented by the system. (This section closes by showing how memory devices are controlled by timing that is provided by the memory system.)

RAM—present the address to read (fetch) the data from the addressed register.

FIGURE 9.3
Reading a static memory device

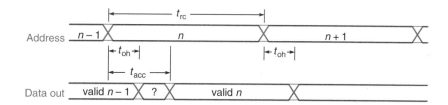

t_{rc} = read cycle time (minimum elapsed time between read requests)

t_{acc} = read access time (maximum elapsed time from address change to valid data out)

t_{oh} = output data hold time (maximum elapsed time data is held valid from address change)

A combinational device has the following property: present the inputs, and the outputs appear after propagation times elapse. Static memory read operations are similar: present the address, and valid data appears at the outputs after access time t_{acc} expires (Figure 9.3). This is why the combinational property can be attributed to memory device read operations despite the presence of latches in the memory device. (The latches are controlled by enable signals, not by clocks, which is why memory chips do not have clock inputs.) Observe how the valid data is held for t_{oh} after a change to a new address. Also note that the data remains valid if the address does not change. The data is held constant for $t_a + t_{oh}$, where t_a is the time the address is held constant.

RAM—present the address and data to write (store) the data in the addressed register.

Static memory write operations shown in Figure 9.4 can be described as follows. Present the address, wait t_{as}, start the write pulse WE_L, present the data, and then terminate the write pulse. Data is strobed into the memory register latches when WE returns to the inactive high level (the write L-to-H "clock edge").

In general, the speed of any memory device is limited by access time and/or read/write cycle time. The *read access time* is the time delay from presentation of address until valid data is available at the output. The *write access time* is the delay from presentation of address until valid data is strobed into the latches by the write pulse trailing edge. Cycle times, such as t_{rc} in Figure 9.3, are the elapsed times between address changes. Cycle times are determined by the particular system using the memory.

Parameter Definitions

Read cycle time (t_{rc}) is the minimum elapsed time required between read requests.

Read access time (t_{acc}) is the maximum elapsed time from a read cycle start to data delivery.

Write cycle time (t_{wc}) is the minimum elapsed time required between write requests.

Write access time ($t_{as} + t_{wp}$) is the maximum elapsed time from write cycle start to data strobed in.

EXERCISE 9.4

In Figure 9.3, calculate the length of time, in nanoseconds, that data is valid when $t_{rc} = 95$ ns, $t_{oh} = 10$ ns, and $t_{acc} = 60$ ns.

Answer: 45 ns   ∎

EXERCISE 9.5

In Figure 9.4, calculate how long it takes, in nanoseconds, from the start of the write cycle to the start of valid data in when $t_{wc} = 105$ ns, $t_{ah} = 15$ ns, $t_{as} = 10$ ns, and $t_{ds} = 10$ ns.

Answer: 80 ns   ∎

FIGURE 9.4
Writing a static memory device

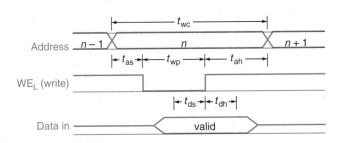

t_{wc} = write cycle time (minimum time between write requests)

t_{as} = address setup time (minimum time from address change to write pulse assertion)

t_{ah} = address hold time (minimum time from write pulse negation to address change)

t_{wp} = write pulse time (minimum time write pulse is asserted)

t_{ds} = data setup time (minimum time from data valid to write pulse negation)

t_{dh} = data hold time (minimum time from write pulse negation to data invalid)

Read/write cycle time, t_{rc} or t_{wc}, is greater than or equal to access time. For static devices, cycle time t_{cycle} can be reduced until it equals access time t_{access}, thereby minimizing the cycle time. The situation is different for dynamic devices. There, cycle time cannot be reduced to access time because reading or writing any DRAM register requires that the register contents be refreshed. This is why the RAS waveform has a "precharge" time t_{rp} (Figure 9.5). This register refresh process (which actually refreshes a DRAM row) is different from the process of refreshing each DRAM row on a regular basis (see Section 9.1.5).

$$\text{SRAM:} \quad t_{access} \leq t_{cycle}$$

$$\text{DRAM:} \quad t_{access} < t_{cycle}$$

FIGURE 9.5

Reading a dynamic memory device

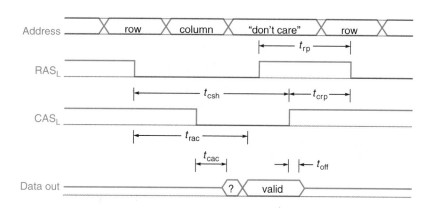

t_{rp} = RAS precharge time (minimum time from RAS negation to RAS assertion)

t_{csh} = CAS hold time (minimum time from RAS assertion to CAS negation)

t_{crp} = CAS to RAS precharge time (minimum time from CAS negation to RAS assertion)

t_{rac} = access time from RAS (minimum time from RAS assertion to valid data out)

t_{cac} = access time from CAS (minimum time from CAS assertion to valid data out)

t_{off} = output buffer turn-off (data hold) time (minimum time from CAS negation to data invalid)

FIGURE 9.6
**Writing a dynamic
memory device**

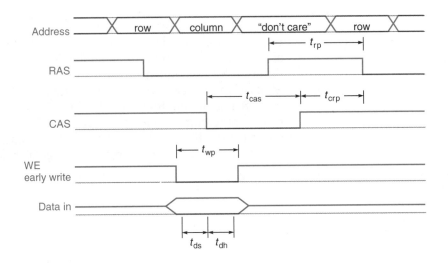

t_{rp} = RAS precharge time (minimum time from RAS negation to RAS assertion)

t_{cas} = CAS pulse time (minimum time CAS is asserted)

t_{crp} = CAS to RAS precharge time (minimum time from CAS negation to RAS assertion)

t_{wp} = write pulse time (minimum time WE is asserted)

t_{ds} = data setup time (minimum time from data valid to CAS assertion)

t_{dh} = data hold time (minimum time from CAS assertion to data invalid)

RAS and CAS enable the DRAM in addition to storing the address bits.

Dynamic memory read operational details are different. Two strobes store the multiplexed address in the DRAM chip. The row address strobe, RAS, stores the row address. The column address strobe, CAS, stores the column address.

Valid data appears at the outputs after access time t_{rac} expires (Figure 9.5). Observe that if CAS is presented "late," the cycle access time is lengthened, because t_{cac} also needs to expire before data is valid.

Dynamic memory write operations shown in Figure 9.6 are similar to read operations. We present the multiplexed address with the row and column strobes, RAS and CAS. In parallel, we start the write pulse WE so that it straddles the CAS falling edge. Then we present the data. Data is strobed into the addressed register latches when the CAS falls to the True level (the early write "clock edge"). The RAS/CAS/WE sequence shown in Figure 9.6 is only one of several possible write sequences.

Note: In Figures 9.5 and 9.6, setup and hold time restrictions on row and column addresses are not shown. Row address must meet setup and hold times for RAS H-to-L fall. Similar requirements hold for the column address. See any data book.

FIGURE 9.7

Synchronizing SRAM and DRAM write cycles to clock

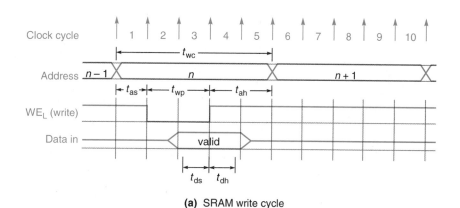

(a) SRAM write cycle

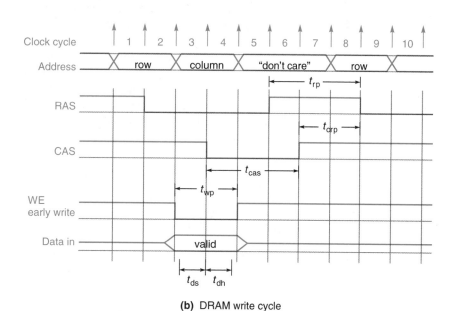

(b) DRAM write cycle

EXERCISE 9.6

Refer to Figure 9.5. Calculate the read cycle time, in nanoseconds, when data is valid for 25 ns, $t_{rac} = 70$ ns, $t_{rp} = 40$ ns, $t_{off} = 5$ ns, $t_{crp} = 20$ ns, and $t_{rise} = t_{fall} = 5$ ns.

Answer: 115 ns   ∎

Clock   The read and write cycle timing diagrams in Figures 9.3 to 9.6 do not include clocks. We have also pointed out that memory devices do not include any timers. However, memory devices must satisfy many timing requirements. The no-timers-in-a-chip cost-saving reality forces a designer to provide external timing and the associated clock. One positive consequence is that the resulting system is synchronous. Another consequence is that minimum values for all parameters are not achievable.

Let us assume that the specified minimum read/write cycle time t_{cmin} is 140 ns. If the system clock period τ is 30 ns, the minimum possible t_c is 5τ, or 150 ns. Furthermore, all other waveform transitions follow clock edges, as shown in Figure 9.7.

EXAMPLE 9.1   Read Cycle—80386 Timing

Any microprocessor, e.g., the 80386, is a state machine that specifies memory system timing. An 80386 read or write cycle consists of one T_1 cycle followed by one or more T_2 cycles, as shown below. The 80386 emits an address, after a delay, in the T_1 cycle. The address remains constant for the associated T_2 cycles. T_j cycles have two phases, ϕ_1 and ϕ_2.

In a read cycle, the 80386 is prepared, at the end of a T_2 cycle, to strobe into its input buffer the data read from memory. The 80386 READY input line specifies in which T_2 cycle the data-in strobe is asserted. The READY signal is a memory system output that reports the end of a read or a write cycle. The longer the memory cycle time, the longer the 80386 has to wait for the READY signal. This is why extra T_2 cycles are called *wait states*. Wait states are T_2 cycles added to the bus read or write cycle.

In this very flexible way the 80386 specifies the memory chip access time t_{acc}. The number of wait states is set automatically by the time the READY signal occurs. The following waveforms include one wait state.

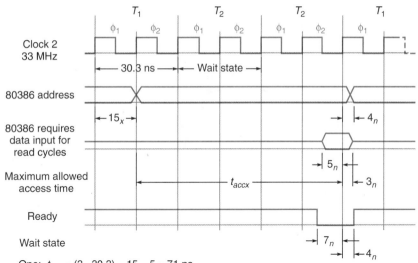

One: $t_{acc} = (3 \cdot 30.3) - 15 - 5 = 71$ ns

None: $t_{acc} = (2 \cdot 30.3) - 15 - 5 = 40.6$ ns

EXAMPLE 9.2   Write Cycle—80386 Timing

The discussion in Example 9.1 continues here. In a write cycle the 80386 emits a write "pulse" on the write/read line, as well as the data word to be written into memory and the address. The data word straddles the rising edge of the write pulse, as required by memory chips. The 80386 write "pulse" is active high. Memory chips require an active low write pulse. The extra 10-ns delay shown in the write waveform accounts for the necessary inverter.

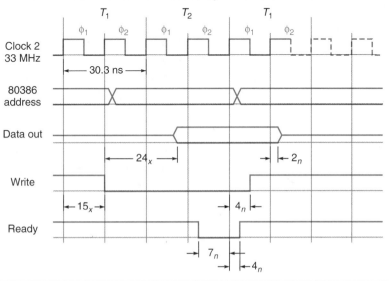

TABLE 9.1 **Potential Commercial Megabit RAM Chip Configurations**

* M = megabit				
2^n	64 Megabit (2^{26})	16M* (2^{24})	4M (2^{22})	1M (2^{20})
2^n	64M $\times$ 1	16M $\times$ 1	4M $\times$ 1	1M $\times$ 1
2^{n-2}	16M $\times$ 4	4M $\times$ 4	1M $\times$ 4	256K $\times$ 4
2^{n-3}	8M $\times$ 8	2M $\times$ 8	512K $\times$ 8	128K $\times$ 8

9.1.4 Memory Device Examples

Commercial memory chips have 1, 4, or 8 data lines.

Designers of integrated circuits have learned to lay out memory chip designs so they are readily reconfigured into the following formats: $2^n \times 1$, $2^{n-2} \times 4$, and $2^{n-3} \times 8$. These formats are commercially available from many manufacturers. For example, the formats for various megabit RAMs are illustrated in Table 9.1. Static and dynamic RAM generic block diagrams are shown in Figures 9.8 and 9.9, respectively.

FIGURE 9.8
Generic static RAM block diagrams

FIGURE 9.9
Generic dynamic RAM block diagrams

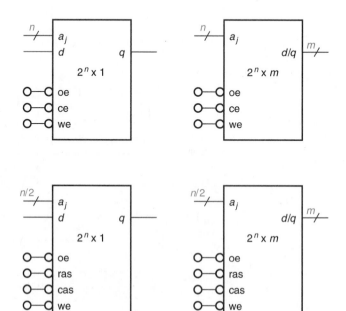

9.1.5 DRAM Refresh

The simple DRAM storage cell requires constant refresh.

DRAMs must be refreshed within a fixed time T in order to maintain data integrity. The refresh process divides time into an unending sequence of time intervals of duration T.

Refresh uses refresh cycles, which are a third type of memory cycle (read and write cycles being the other two types). One refresh cycle refreshes all the cells in one row of the DRAM row/column cell structure. All rows must be refreshed once in every fixed time interval T. There are no constraints regarding *when* they are refreshed in any T interval as long as no row waits for refresh longer than interval T.

There are three types of DRAM memory cycles: read, write, and refresh.

The refresh process proceeds regardless of the read and write cycles that are executed. Refresh cycles interweave with read and write cycles. This is known as *interleaved refresh*. The refresh cycle has priority over read and write cycles. But as a practical matter, there is no need to interrupt ongoing read or write cycles. The process waits for a current read/write cycle to complete before starting a refresh cycle. A refresh cycle in process is reported to the user system as yet another "memory busy" message. Therefore the system does not know what type of memory chips are used, nor does it need to know. More to the point, the system ought to be able to use any memory subsystem with an insignificant time-out penalty for refresh if DRAMS are used.

A 1% time-out penalty for refresh cycles is acceptable.

Industrial practice sets the refresh time-out penalty to about 1% for all DRAM sizes. Let us assume memory cycle time t_c is 150 ns. A 1% penalty implies a $100t_c$ interval t_{ir} between refresh cycles. Thus the interval t_{ir} = 15,000 ns [or 15 microseconds (μs)] when t_c = 150 ns.

Since each refresh cycle refreshes only one row, the refresh interval $T = kt_{ir}$, where k equals the number of rows. Modern 1meg $\times$ 1 semiconductor technology allows T to be 8 milliseconds (ms) (8,000 μs) or less. Therefore we set $k = 512$, which is the greatest power of 2 less than 8,000/15 = 533. Prior technology limited T to 4 ms, resulting in $k = 256$. The limit on k influences DRAM chip layout.

A 1meg $\times$ 1 DRAM has 10 column address bits and 10 row address bits. Ten address bits imply 1024 (2^{10}) rows and 1024 columns. The 1024 number contradicts the $k = 512$ limit. Therefore the 1meg $\times$ 1 DRAM cell array is laid out to have 2048 columns and 512 rows. This is transparent to the user.

Another form of refresh is *burst refresh,* which simply refreshes all rows without pausing for read or write cycles. Burst refresh makes the memory system unavailable for kt_c seconds. When $k = 512$ and $t_c = $ 150 ns, the time to burst refresh is 76,800 ns, 76.8 μs, or 0.076 ms. This memory time-out is not acceptable in many systems. DRAM manufacturers have developed various types of refresh cycles.

RAS-only refresh cycle   The original refresh cycle is the RAS-only refresh cycle. This type of cycle presents a row address to the address pins, an RAS waveform on the ras pin, and no CAS waveform on the cas pin. The DRAM's internal state machine recognizes this set of inputs as a refresh request. Consequently, the addressed row is refreshed. An external counter and extra multiplexers are required to generate and place the address-of-the-row-to-be-refreshed on the address pins.

CAS-before-RAS refresh cycles are easily implemented.

CAS-before-RAS refresh cycle   DRAMs with this special capability have an internal row address counter. The DRAM's internal state machine recognizes this type of cycle by the combination of waveforms where CAS precedes RAS (Figure 9.10).

Hidden refresh   The CAS waveform in a read/write access cycle is extended while RAS goes low again. In effect, this combines a read/write cycle with a CAS-before-RAS cycle.

Refresh with scrubbing cycle   An error detection and correction operation (EDC) is performed during a refresh cycle. The EDC circuits must be part of the memory system. (EDC circuits are not covered in this text.)

The actual refresh cycle is typically under the control of the dynamic RAM controller. Section 9.3 describes one such controller; others are available in single chip form from a number of integrated circuit manufacturers.

EXERCISE 9.7   Calculate how many milliseconds are required to refresh a DRAM with 256 rows with a 2% penalty and 120-ns read, write cycles. How many clock cycles are needed between refreshes?

Answer: 1.536 ms for $50t_c$ between refreshes   ■

FIGURE 9.10
CAS-before-RAS dynamic RAM refresh

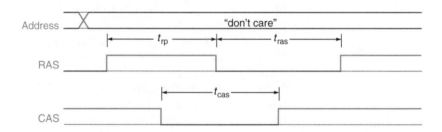

EXAMPLE 9.3 DRAM Refresh Request Generator

Design a refresh request generator with one request signal output. The request signal is used by the DRAM controller to initiate the appropriate refresh cycle defined by the controller design. Assume:

The system clock is 33 MHz.

The memory cycle time t_{cycle} is 150 ns.

Each DRAM row is refreshed every 8 ms or sooner.

The memory chip has 512 rows.

Refresh cycles are interleaved with read and write cycles.

Interleaving requires one row to be refreshed every 8,000,000 ns/512, or 15,625 ns. This means a refresh cycle occurs every 15,625/150, or $104t_c$. Therefore the memory is unavailable for about 1% of the time.

The 33-MHz clock period is 30 ns. A counter is required to count clock periods in order to generate a refresh request every 15,625 ns. The number of clock periods to count is 15,625/30 = 520 after rounding down to the nearest integer. The number 520 would usually be changed to 512 in practice in order to simplify the circuit required. However, let us design a divide-by-520 circuit.

The first power of 2 greater than 520 is 2^{10} = 1024. The exponent 10 specifies a 10-bit counter. Let us use three 163 chips, which form a 12-bit counter. Now the maximum count is 4096 (2^{12}), not 512. In order to eliminate the need for a 520-number detector (see Section 8.2.7 on page 446), we preset the counter to 4096 − 520 = 3576, which is 10111111000_2, or $DF8_{16}$. Then the R_{co} output can be used as the number detector. The inverted R_{co} signal is the refresh request. R_{co} also loads $DF8_{16}$ into the counter, allowing for continuous recycling operation.

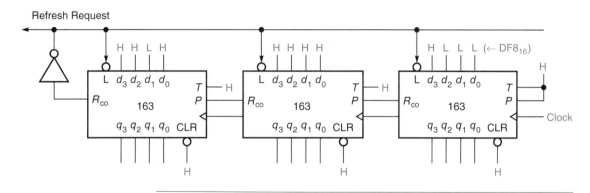

9.2 Address Decode

The address decoders internal to memory chips select the addressed register. No other decoding is necessary when system memory size is less than or equal to the memory device size. Address decoding external to the memory chips is necessary, however, when system memory size exceeds the memory device size. As a practical matter, memory size is some power of 2 times the size of the memory device (a chip). When memory *word* size is greater than chip word size, address decoding is *not* involved. More chips are involved, nothing else.

A decoder is necessary when memory range exceeds memory chip size.

Let us suppose chip size is $2^n \times m$ and memory size is $2^{n+\alpha} \times \beta_m$. For example, if chip size is 128K $\times$ 8 and memory size is 1meg $\times$ 32, then $\alpha = 3$ and $\beta = 4$.

$$2^{n+\alpha} \times \beta m = 2^\alpha (2^n \times \beta m)$$

$$1\text{meg} \times 32 = 2^{17+3} \times 4 \cdot 8$$

$$= 2^3[(2^{17} \times 8) + (2^{17} \times 8) + (2^{17} \times 8) + (2^{17} \times 8)]$$

$$= 8 \quad \text{Equivalent bank of } 2^{17} \times 32$$

$$= 32 \quad \text{chips of size } 2^{17} \times 8$$

Address lines partition into three groups: constant, bank, and device.

The 17 address lines a_0 to a_{16} are what we call *device lines*. They go to all RAM chips. The 3 ($= \alpha$) address lines a_{17}, a_{18}, and a_{19} are *bank-select lines*, which are decoded into 8 (2^α) *bank-enable lines*.

In addition, there are *constant address lines,* whose values are constant over the range of addresses spanned by the memory. This is best explained by an example. Suppose there are 32 address lines. In our example there are 17 device lines and three bank lines. Thus the 12 most significant address lines, a^{20} to a^{31}, are constant when the memory is addressed over the range 0 to $2^{20} - 1$.

| 12 constant lines | 3 bank lines | 17 device lines |
| a_{31} to a_{20} | a_{19}, a_{18}, a_{17} | a_{16} to a_0 |

Our address decoding design method is based on these three categories of lines.

EXERCISE 9.8   How many 64K $\times$ 8 RAM chips does a 256K $\times$ 24 memory require?

Answer: 12   ∎

9.2.1 A Specific Decoder Design

Our introductory vehicle for describing address decoding is a hypothetical static RAM memory chip containing eight four-bit words. The RAM is an 8 $\times$ 4 memory chip.

A memory chip has three classes of signal lines: address, data and control.

address lines	select the desired register
data lines	route words in and out of the memory
control lines	manage the read and write processes

Each word stored in our hypothetical chip must be individually addressed. The eight words require eight addresses, one for each word. Furthermore, the addresses are encoded in binary to minimize the number of address lines. Since $8 = 2^3$, three encoded address lines are used in lieu of eight unencoded lines. The address lines are decoded by an internal decoder on the chip.

Reading or writing four-bit data words requires four memory device input/output lines. An asserted device write line stores words in the device. An asserted output-enable line selects data movement to be out of the device. An asserted device chip-enable line activates the device. The black box representation of our hypothetical memory device is shown in Figure 9.11.

If the system controlling the memory has six address lines, addresses presented to the memory range from binary 000000 to 111111. In hex this range is 00 to 3F. How many addresses are there in this address range? The answer is $2^6 = 64$. This is calculated by adding 1 to the high address and subtracting the low address from the sum:

$$
\begin{array}{lr}
\text{High address} & \text{3F} \\
\text{Plus 1} & \underline{+\ \ 1} \\
\text{Sum} & \text{40} \\
\text{Low address} & \underline{-00} \\
& \text{40} \quad \text{hex (which is 64 decimal)}
\end{array}
$$

The device address line group selects memory chip registers.

How do we design a memory using our hypothetical static RAM that occupies the address range 20 to 2F, which is only a part of the

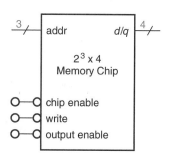

FIGURE 9.11

Hypothetical 8 × 4 memory chip

The bank address line group partitions memory range into chip size.

full range 00 to 3F? Notice that the address range 20 to $2F_{16}$ includes 16 addresses. This implies a memory of size 16. The hypothetical chip contains eight registers; therefore two chips are needed in the memory system. Additional chips are required if the memory system word has more than four bits.

Study the address binary bit patterns. Table 9.2 shows the bit pattern in two formats: in binary numbers, and with bits partitioned into three groups. The least significant group, $a_2a_1a_0$, repeats the cycle from 000 to 111 twice over the range 20 to 2F hex. This is consistent with the need for two memory chips. Lines $a_2a_1a_0$ are designated as device lines. When bit a_3 is 0, one cycle is addressed; when a_3 is 1, the other cycle is addressed. Bit a_3 must be a bank line, because it discriminates between the two cycles of 000 to 111 (20 to 27 and 28 to 2F). Bits a_5a_4 equal 10 over the range. This is a typical partitioning of address lines over a range of addresses.

The constant address line group defines the start of the memory range.

constant lines: a_5a_4

bank lines: a_3

device lines: $a_2a_1a_0$

TABLE 9.2   **Partitioning Address Lines**

a_j lines	a_{543210}	$a_{54}a_3a_{210}$
20 hex	100000	10 0 000
	100001	10 0 001
	100010	10 0 010
	100011	10 0 011
	100100	10 0 100
	100101	10 0 101
	100110	10 0 110
27	100111	10 0 111
28	101000	10 1 000
	101001	10 1 001
	101010	10 1 010
	101011	10 1 011
	101100	10 1 100
	101101	10 1 101
	101110	10 1 110
2F	101111	10 1 111

We connect the three address lines $a_2a_1a_0$ to the 8×4 RAM chips' address pins. Observe that for any address, such as $a_2a_1a_0 = 101$, one register is addressed in *each* chip. A set of address lines connected to many chips addresses the register with the same address in all chips (Figure 9.12a). When $a_2a_1a_0 = 101$, for example, register number 5, in

FIGURE 9.12

Memory spanning 20-to-2F address range

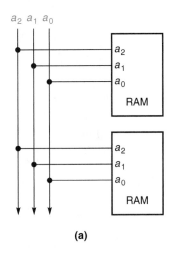

(a)

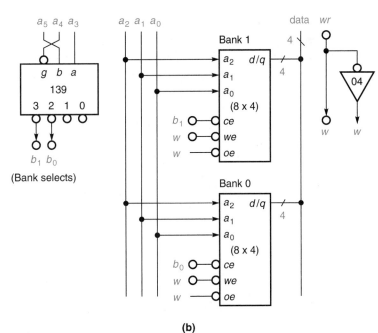

(b)

every chip, is accessed in parallel. Parallel access is a potential problem, requiring control signals to prevent more than one chip from responding to the same address.

A typical memory has only one data bus. Consequently, the four d/q pins of every memory chip are connected to the four-wire data bus. When data is written to the memory, the d/q pins are in the d (input) mode. In d mode, there is no problem connecting the data bus to the d/q pins of all chips.

We know any address selects a register with the same address in each chip in the RAM array. So the words in all of these registers are available for placement on the data bus. In our two-bank memory this means we must pick only one of the two banks at a time. This is where a_3 enters the scene. The memory design address range is 20 to 2F. Consistent with this we must use a_3 in addition to $a_2a_1a_0$ to assign four-bit addresses 0 to 7 to one bank and addresses 8 to F to the second bank. Next, we use a_5a_4 to "add" 20 to each bank of eight addresses. But our chip has only three address lines. This why we now use the chip-enable (CE) pin (Figure 9.12b).

A memory device is activated when its CE is asserted. This is why a_3 indirectly activates CE as follows. Since a_5 and a_4 are constant over the range, they are used to activate the entire memory over the range. The memory has two banks of addresses: 20 to 27 ($a_3 = 0$) and 28 to 2F ($a_3 = 1$). Therefore, to enable one bank at a time, the bank-select equations are as follows:

$$\text{bank zero} = b_0 = a_5a_4{}'a_3{}' \quad \text{True when } a_5a_4a_3 = 100$$

$$\text{bank one} = b_1 = a_5a_4{}'a_3 \quad \text{True when } a_5a_4a_3 = 101$$

Within a bank, the memory chip input address lines $a_2a_1a_0$ are decoded inside the memory chips to select the addressed register.

The memory circuit is shown in Figure 9.12b. Output-enable pins are asserted when write is not asserted. The memory chip output is not enabled until the OE *and* CE pins are asserted *and* the WE pin is not asserted. A 139 is used in lieu of several AND gates to create the bank-enable lines b_0 and b_1.

Let us leave our hypothetical memory device and move on to larger address ranges and words with more bits.

EXAMPLE 9.4   Address Decoder

A memory is essentially a group of registers, each with a unique address. Suppose we wish to address simultaneously any one of eight subgroups of 32 registers out of the single group of 256 registers. Here is how decoders implement this function.

Let the group be addressed by address $a = a_7a_6a_5a_4a_3a_2a_1a_0$. Each 32-register subgroup is spanned by the address range 00000 to 11111. In turn, the address of each subgroup is 000----- to 111-----, where - means "don't care."

Subgroup	Range in Binary	Range in "Don't Care" Format
000	00000000 to 00011111	000-----
001	00100000 to 00111111	001-----
.	.	.
.	.	.
.	.	.
110	11000000 to 11011111	110-----
111	11100000 to 11111111	111-----

Notice that address bits $a_7a_6a_5$ are constant in each range while device bits $a_4a_3a_2a_1a_0$ vary from 00000 to 11111. A control line for each subgroup that is True in the corresponding range is needed. Therefore bank bits $a_7a_6a_5$ need to be decoded.

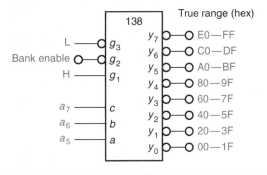

9.2.2 Memory Decoder Design

Now that we have used an elementary example to explore decoder design concepts, we consider the effects on design of large address ranges, where memory size is greater than chip size and memory register (word) size is greater than chip register (word) size.

As mentioned earlier, the bits on an n-line address bus are interpreted as a binary number with n digits. An n-bit address spans the range of addresses from 0 to $2^n - 1$. For example:

n Bits	Address Range, in Hex	CPU Example
4	0 to F	none
8	00 to FF	none
12	000 to FFF	pdp8 (the first mini)
16	0000 to FFFF	pdp11, 8085, 6800
20	00000 to FFFFF	68008
24	000000 to FFFFFF	80286 68000
32	00000000 to FFFFFFFF	mainframes, 68030, 80386

Suppose we are asked to design a memory with 32-bit word registers and with a range spanning the addresses 80000000 to 8003FFFF. First we calculate the number of addresses in the range:

$$\begin{array}{r} 8003FFFF \\ +1 \\ \hline 80040000 \\ -80000000 \\ \hline 40000_{16} \text{ registers} = 2^{18} \end{array}$$

$$2^{18} = 2^8 \times 2^{10} = 2^8 \times 1K = 256K \text{ nominal value}$$

Eighteen address lines is the minimum number required to span the range 00000 to 3FFFF, which consists of 40000_{16} addresses. This becomes clear when we convert hex to binary:

$$8003FFFF = 1000\ 0000\ 0000\ 0011\ 1111\ 1111\ 1111\ 1111$$

$$80000000 = 1000\ 0000\ 0000\ 0000\ 0000\ 0000\ 0000\ 0000$$

The first 14 bits are constant in this range, leaving 18 bits to partition into bank and RAM bits. The partition is made after a RAM chip is specified.

If a static RAM with 2^{18} registers is used, we would design a memory with only one bank and be done with it. Then the address lines partition as follows:

constant lines: a_{31} to a_{18} to select the memory range

bank lines: none   to select the banks

device lines: a_{17} to a_0 to select the registers

EXERCISE 9.9   How many addresses (in decimal) are there in the range $A800_{16}$ to $BFFF_{16}$?

Answer: $1800_{16} = 6144_{10} = 6K$ ■

FIGURE 9.13

32K × 8 memory chip

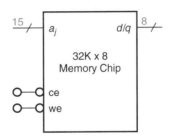

If we assume there is no 2^{18} static RAM available, memory size exceeds memory device size and a memory design is required.

Let us assume the largest chip size is 32K × 8 (Figure 9.13). Also, let's assume the chip does not have an external output-enable line because the OE line equation (oe = we'ce) is implemented internally.

Specifying the 32K RAM chip implies partitioning the address lines into bank and RAM lines as follows:

$$32K = 2^{15}$$

$$40000_{16} = 2^{18} = 2^3 \times 2^{15} = 8 \times 32K = 256K \text{ memory size}$$

Memory size is the number of banks multiplied by chip size.

So, eight 32K banks are needed to implement 2^{18}, or 40000 hex, registers. And the 32-bit address partitions into constant, bank, and RAM lines as follows (the decoder is shown in Figure 9.14):

8003FFFF = 1000 0000 0000 00	11 1	111 1111 1111 1111
80000000 = 1000 0000 0000 00	00 0	000 0000 0000 0000
constant	bank	RAM

Word size has nothing to do with address decoding.

Memory chips usually have one, four, or eight bits per word. If the system word size is 32 bits, then memory chips are wired in parallel to build 32-bit-memory words. For example, four 32K × 8 chips wired in parallel are equivalent to a 32K × 32 chip (Figure 9.15). A 256K × 32 memory is shown in Figure 9.16. The RAM array may require additional drivers for various inputs to satisfy the RAM array's address, data, and control line input current demands. If the RAM chips have three-state *d/q* pins, only the bank-selected chips are connected to the data bus, thereby reducing the output current demand at those *d/q* pins.

The 32K × 32 assembly of chips is usually called a *memory bank of words*. Specifically, this is a 32K bank of 32-bit words.

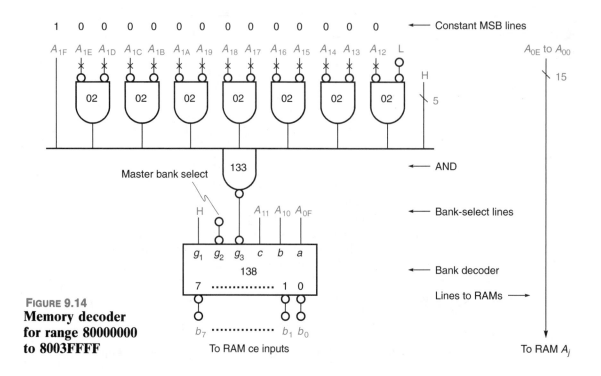

 (labels within figure)

Constant MSB lines

A_{0E} to A_{00}

Master bank select

133 — AND

Bank-select lines

Bank decoder

Lines to RAMs →

To RAM ce inputs

To RAM A_j

FIGURE 9.14
Memory decoder
for range 80000000
to 8003FFFF

9.3 Memory Design

A memory system
consists of ram array,
decoder, mar, mbr,
mbw, and ASM con-
troller.

Every digital system partitions into the controller and the controlled.
The controller is defined by an ASM chart, which is implemented by a
state machine and output circuits. The controlled are defined by the
task and implemented by the data path. A minimum memory data path

FIGURE 9.15
Assembled bank of
32-bit words

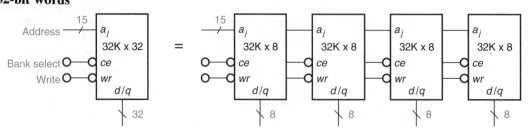

FIGURE 9.16

256K × 32 memory array

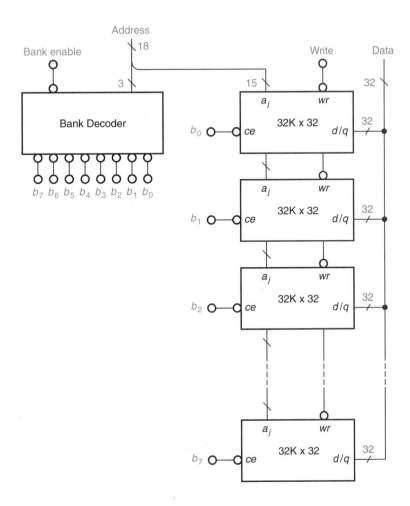

includes registers to store address and data, the address decoder, the memory chip array, and a timer defining access time. [Some controllers have ROM devices to hold addresses used by next-address calculation logic as well as firmware (also known as *microcode*) for data path and state machine control.]

We are now in a position to design a complete system. In fact, we have already designed the major components.

9.3.1 Memory Block Diagram

We have learned that when system memory size exceeds memory chip size, an address decoder is needed to select and activate the bank storing the addressed word. Thus the core of a memory data path includes the memory chip arrays and the address decoder. An array is a group of memory chips assembled with only one type of chip. There is one array for each type of memory chip in the memory system. Reading and writing the array require addresses and data. This is why a memory system is always a slave of some main system that provides addresses and data. A more basic reason why a memory is a slave is that the main system *uses* the data, whereas the memory only *stores* the data. Memory does not process data.

Modern microprocessor address, data, and control signals are designed to minimize memory system requirements. Consequently, memory system design for microprocessors is relatively straightforward. Microprocessor memory systems do not begin to hint at the complex control requirements that arise in memory system designs when microprocessors are *not* used.

Usually, a main system outputs an address or data for one clock period. However, the synchronized memory write cycle in Figure 9.7 shows that the address must be available for five clock cycles, and that the data must be available for at least two clock cycles in order to straddle a clock edge. This means the address and data must be stored in registers. The generic names for these registers are *memory address register* (MAR) and *memory buffer write* (MBW) register (Figures 9.17 and 9.18). (A detailed discussion of timing and the assertion of control lines is found in the next section.)

In microprocessor systems, the MAR and MBR are inside the microprocessor. In such systems the MAR we use is typically omitted. In some systems a register akin to the MAR is required to latch addresses when strobed by a CPU address latch strobe. Here we have simplified this design by not including buffers that may in practice be required.

In general, the main system is unable to use immediately the data read from the memory array. Therefore a memory buffer read register is used. This MBR is loaded by the memory module ASM load output because the ASM "knows" when data is available. (The memory module ASM is discussed in Section 7.5 on page 367 and illustrated in Figure 7.15 on page 365.) In many systems one register is used for both purposes. This is made possible by adding three-state gates. At this point, two registers seem to be a more straightforward solution.

Error detection and correction (EDC) logic is a necessary part of any reliable memory system. (We mention this even though EDC design is outside the scope of this discussion.) The essential elements of a practical memory system without EDC are shown in Figure 9.17.

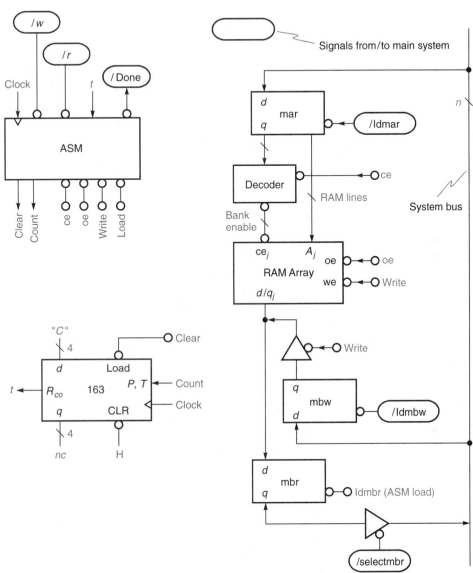

FIGURE 9.17
**Memory module
block diagram**

FIGURE 9.18
**Main system block
diagram**

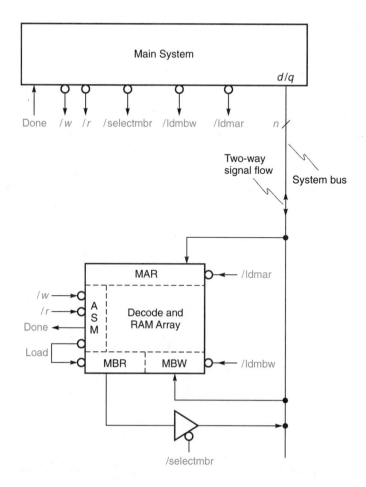

9.3.2 Memory Circuit Design

Except for the omitted EDC block, most of the blocks in Figure 9.17 have been designed earlier in this book. A typical address decoder and RAM array are found in Figures 9.13 to 9.16. The ASM chart was presented in Figure 7.15 on page 365. The state machine and output circuits implementing the ASM chart of Figure 7.15 are shown in Figures 7.20 (page 375) and 7.21 (page 377). (The ASM is also shown in upcoming Figure 9.21.) But the timer generating the *t* signal has not yet been designed.

In general, a memory system is the slave of another system—the main system. The main system sends addresses and data to this memory system via the system bus shown in the main system diagram in Figure 9.18. The main system also sends commands such as /w,

/ldmar, and /ldmbw via separate control lines. The memory address and data storage registers are shown here as part of the memory system. However, they could be placed anywhere in the main system. Data read from the memory is returned to the main system via the system bus.

The /w signal is an input to the ASM controller, which has been waiting in state s_0 for a read or write command (see upcoming Figure 9.21). The write command /w advances the ASM to the next state, s_2, which is designed to implement a memory write operation. The ASM dwells in s_2 until it receives a t input signaling the end of the memory's specified access time. Then the ASM returns to the resting state, s_0. The ASM input and output timing waveforms for a write cycle are shown in Figure 9.19.

FIGURE 9.19

Synchronous memory-write timing diagram

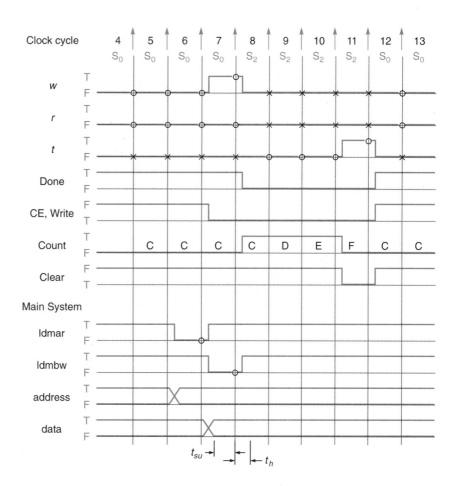

Writing the memory requires the main system state machine to present the address with output /ldmar in, say, clock cycle 6 of Figure 9.19. This is followed up with data, /ldmbw, and write command /w in cycle 7. The done report goes False when the write cycle starts. The done report is a memory module ASM output. The main system knows the memory is available again when done returns to the True state, thereby reporting the end of the write cycle. The done report provides what may be called a "handshake" capability; main and memory systems agree to interact according to the state of the done report.

A timing circuit is necessary to account for access time.

Our discussion of memory device timing mentioned that memory devices do not include timers. Nevertheless, valid data is not available in read cycles, for example, until the *specified* access time t_{acc} expires. This is why the memory system must include a timer to measure, in effect, t_{acc} and other parameters. The timing diagram in Figure 9.19 shows that the state machine dwells in state s_2 during a write cycle. The state machine dwells in s_2 for four clock periods (cycles). The time occupied by four clock periods (cycles 8 to 11) is the actual write cycle time. In a system, the clock period is usually set by other considerations, and the clock period multiplier, which is 4 in this case, is determined by dividing the specified write cycle time t_{wc} by the clock period and rounding up to the next integer.

Timers are implemented by counters, which count clock edges when count control p is True (Section 8.2.4 on page 430). Then the stored count represents a number of clock periods. A 163 counter is readily slaved to the state machine by connecting the state machine COUNT output to the 163 p input, as shown in Figure 9.20. Furthermore, the counter is effectively measuring t_{wc} during the write cycle because the COUNT output is True as the state machine dwells in s_2. And, when the counter reports, by asserting output t, that a clock period is the fourth period (cycle 11 in Figure 9.19), the clock edge ending cycle 11 steps the state machine to s_0.

The counter output signal is an end-of-write-cycle report, labeled t in Figure 9.20. Report t performs two functions. First, t is used as a state machine input, so that when the state machine is in s_2, an as-

FIGURE 9.20

Memory module *t* report generator

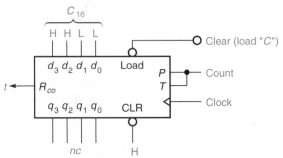

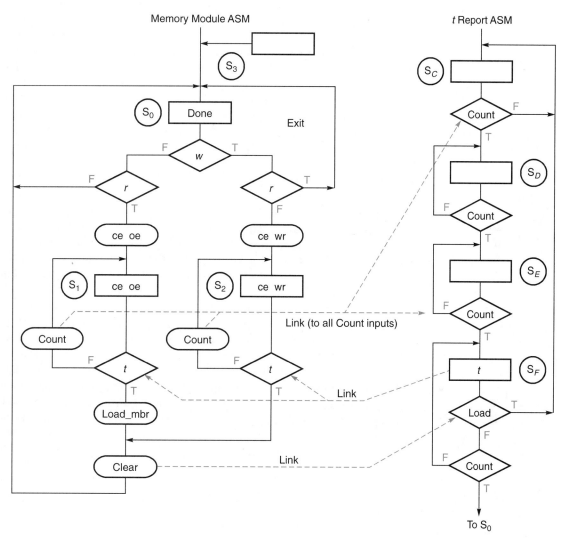

FIGURE 9.21

Memory module *t* report generator as a linked ASM

serted *t* asserts the state machine CLEAR output. The CLEAR output is connected to the 163 load input in the circuit (Figure 9.20). Second, asserted *t* allows the state machine to step to state s_0 after the clock edge ending cycle 11. Notice that *t* is active high and CLEAR is active low. The 163 specification dictates that CLEAR has to be active low, whereas *t* originates at the 163 R_{CO} output, which is active high.

Perhaps a restatement of t's functions is useful. The timing diagram in Figure 9.19 shows the t report ending the write cycle according to the ASM chart in Figure 9.21 and the timing diagram in Figure 9.19. In effect, the t report allows the system to satisfy memory chip time specifications. For example, the timing diagram is in state s_2 for three extra clock cycles (extra because the state machine would remain in s_2 for only one clock cycle if t was always True).

Design is simplified when a memory timing circuit is a separate linked state machine.

A 163 counter can count 1, 2, 3 and generate the t report (Section 8.2.5, page 432) when it advances to 4. The circuit without extra gates in Figure 9.20 results if CLEAR loads state C into the 163 in lieu of clearing the 163 to state zero. Then the 163 counts hex C, D, E and generates the t report when it advances to state F_{16} (Section 8.2.7, page 446).

In fact, the 163 is a linked state machine (Figure 9.21). Upon receipt of a read or write input, the memory module ASM steps to state s_1 or s_2 and dwells there, because t is False. In a write cycle the ASM advances to state s_2, asserting CE, WR, and COUNT. Initially, $t = F$ because the linked ASM is in state s_C. The COUNT output is a linked input to the t report generator ASM that allows the ASM to step from state s_C to s_D to s_E to s_F. While in state s_F, the t output is asserted. The t output is a linked input to the memory module ASM.

Upon receipt of asserted t, the memory module conditional output CLEAR is asserted *in the same clock period*. This is cycle 11 in Figure 9.19. CLEAR is a linked input to the t report generator ASM that asserts the 163 load input. An asserted load input disables the asserted count input (see Figure 8.21 on page 431 or a data book). At the next clock edge the ASM steps to state s_C. (The linked state machine concept is presented in Section 7.8.2, page 388.)

SUMMARY

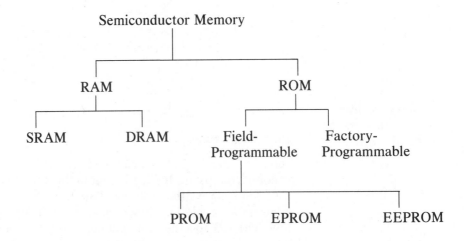

Semiconductor Memory — RAM (SRAM, DRAM), ROM (Field-Programmable, Factory-Programmable), Field-Programmable (PROM, EPROM, EEPROM)

A memory device is a random-access device when the time to access data in any location is less than or equal to some specified maximum access time, and the access time is independent of the sequence of locations being read.

A memory device consists of a number of registers addressed by n binary-encoded address lines. Consequently, the number of registers is 2^n. Each register is the same size and stores some number of data bits m. The size of a memory chip is usually stated to be 2^n by m ($2^n \times m$).

A memory chip has three classes of signal lines:

address lines	select the desired register
data lines	route words in and out of the memory
control lines	manage the read and write processes

The basic simplicity of memory timing is obscured by the multitude of timing parameters found in data books. Once the basics are understood, the details are readily assimilated. Memory devices do not include any timers.

DRAMs must be refreshed within a fixed time T in order to maintain data integrity. The refresh process divides time into an unending sequence of time intervals of duration T.

Address decoding is necessary when memory size exceeds the memory device size. As a practical matter, memory size is some power of 2 times the size of the memory device (a chip). When memory word size is greater than chip word size, address decoding is not involved. More chips are involved, nothing else. A typical partitioning of address lines over a range of addresses is:

constant lines   bank lines   device lines

Memory design includes design of the following components: ASM chart, state machine and output circuits, data and address registers, address decoder(s), and memory device array(s).

REFERENCES

BiCMOS/CMOS Data Book. 1990. San Jose, Calif.: Cypress Semiconductor.

Blakeslee, T. R. 1979. *Digital Design with Standard MSI and LSI,* 2nd ed. New York: Wiley.

Clare, C. R. 1971. *Designing Logic Systems Using State Machines.* New York: McGraw-Hill.

Dynamic Memory Design, #11580A. 1990. San Jose, Calif.: Advanced Micro Devices.

Seidensticker, R. B. 1986. *The Well-Tempered Digital Design*. Reading, Mass.: Addison-Wesley.

Wakerly, J. F. 1990. *Digital Design Principles and Practices*. Englewood Cliffs, N.J.: Prentice-Hall.

Name _____   SID# _____   Section _____
     (last)   *(initials)*

Approved by _____   *Date _____

Grade on report _____

LIFO (Last-in-First-out) Static RAM Module

1. ASM chart and timing

   _____ push timing
   _____ pop timing
   _____ push-pop timing
   _____ empty/full reports
   _____ mainline ASM chart
   _____ linked (?) ASM charts

2. System block diagram

   _____ controller
   _____ data path

3. Circuit design

   _____ controller
   _____ data path

4. Build the LIFO.

5. Test the circuit by pushing data equal to FF. Then pop all registers to verify contents.

   _____ test circuit
   _____ data in/out

A last-in-first-out, or LIFO, memory module is one implementation of a stack.

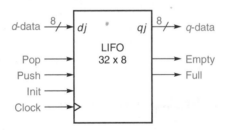

A LIFO stack has two basic operators: push and pop. We explain by using an analogy. We *push* when we put a plate on top of a stack of plates. We *pop* when we remove a plate from the top of the stack of plates. Note that the last plate onto the stack is the first plate removed. A third operator initializes the stack.

A hardware LIFO includes a controller and a data path. The data path consists of a group of registers, a stack pointer (sp) that addresses the *last* register into which data was pushed, and empty and full detectors. In a down-memory stack, the push and pop operators are defined as follows.

A push decrements the address by one (sp ← sp − 1) and writes the new data into register with the decremented address.

A pop reads the data from the register pointed to by the sp, and then increments the sp by 1 (sp ← sp + 1).

A third operator initializing the stack simply resets the sp to zero (one possibility). This empties the stack. Note that it is unnecessary to erase data memory registers.

The system using a LIFO stack in its data path needs to know when the stack is full or when it is empty. When the stack is full, pushes are inhibited; when the stack is empty, pops are inhibited.

SRAM memory chips provide a low-cost means of implementing the group of registers in a stack.

In a synchronous digital system, the system controller is simplified when operator commands (such as push, pop, init) are synchronous and have a duration of one clock period. Assume this is the case for this project.

1. Create an ASM chart. Use information based on the block diagram, an understanding of what the LIFO operators do, and a knowledge of SRAM read and write cycle timing.

   Suggestions:

   Draw a timing diagram for the shortest possible static RAM write (push) cycle. Show two cycles. Show address, write, and data waveforms. Add the push command waveform.

   Repeat the above for two pop cycles.

   Create a timing diagram showing the following sequence of operation cycles: push, push, pop, push, pop, pop, push. Add empty and full reports.

2. Create a system block diagram.

3. Design the data path and controller.

4. Build the circuit.

5. Test the circuit by pushing only data equal to FF. Then pop all registers to verify contents. Use eight LEDs to monitor the data lines.

Name _____   SID# _____   Section _____
     (last)   *(initials)*

Approved by _____   Date _____

Grade on report _____

FIFO (first-in-first-out) Static RAM Module

1. ASM chart and timing

   _____ write timing
   _____ read timing
   _____ write-read timing
   _____ empty/full reports
   _____ mainline ASM chart
   _____ linked (?) ASM charts

2. System block diagram

   _____ controller
   _____ data path

3. Circuit design

   _____ controller
   _____ data path

4. Build the FIFO.

5. Test the circuit by writing 32 data words ranging from 00 to 1F. Then read all registers to verify contents.

   _____ test circuit
   _____ data in/out

A first-in-first-out 32×8 FIFO memory module has two ports: one for writing data in, another for reading data out.

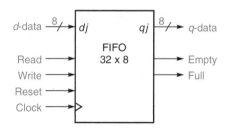

A FIFO dual port memory has two basic operators: read and write. A third operator resets the read and write pointers to zero. Read and write commands may be interleaved. They cannot be simultaneous.

A hardware FIFO includes a controller and a data path. The data path consists of a group of registers (SRAM), a write pointer (wp) that addresses the last+1 register data was written to, a read pointer (rp) that addresses the last+1 register data was read from, as well as empty and full detectors. In a FIFO operators are defined as follows.

A write writes the new data into the register and increments the address by 1 (wp = wp + 1).

A read reads the data from the register and increments the address by one (rp = rp + 1).

A reset resets the wp and rp to zero. This empties the FIFO. Note that erasing data memory registers is not necessary.

The system using a FIFO stack in its data path needs to know when the FIFO is full or empty. When the FIFO is full writes are inhibited, and when the FIFO is empty reads are inhibited.

SRAM memory chips provide a low cost means of implementing the group of registers in a FIFO.

In a synchronous digital system the system controller is simplified when operator commands (such as read, write, reset) are synchronous and have a duration of one clock period. Assume this is the case for this project.

1. Create an ASM chart.

   Use information based on the block diagram, an understanding of what the FIFO operators do, and a knowledge of SRAM read and write cycle timing.

   Suggestions:

   Draw a timing diagram for the shortest possible static ram write cycle. Show two cycles. Show address, write and data waveforms. Add the write command waveform.

   Repeat the above for two read cycles.

   Create a timing diagram of showing a sequence of operation cycles: write, write, read, write, read, read, write. Add empty and full reports.

2. Create a system block diagram.

3. Design the data path and controller.

4. Build the circuit.

5. Test the circuit by writing 32 data words ranging from 00 to 1F. Then read all registers to verify contents. Use eight LEDs to monitor the data lines.

Name _____   SID# _____ Section _____
       (last)   *(initials)*

Approved by _____   Date _____

Grade on report _____

Dynamic RAM Module

1. Memory block diagram

    _____ data path
    _____ controller

2. Design the data path.

    _____ data path

3. Create an ASM chart for the controller.

    _____ mainline ASM
    _____ linked ASM (?)

4. Design the controller circuit.

    _____ state machine
    _____ output circuits

5. Build the memory module.

6. Design tests for the memory module.

    _____ test plan

7. Test the module.

    _____ test data

This project has three parts: data path, controller, and test.

Specification and block diagram

Mem map: 3AD70000 to 3AD7FFFF

Banks: four each 64K × 8

Chips: TMS4464 64K × 4 or equal

Refresh: CAS before RAS

Write: Delayed (after CAS)

Registers: mar (memory address)
mbr (data out)
mbw (data in)

Timing:

All commands are synchronous and one clock period wide.

aj and *dj* are available for duration of access command and one clock period after command time period.

On write cycles, write command is available at time of access command.

Refresh is generated inside the module.

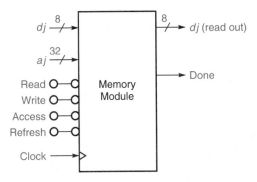

Access is a fail-safe feature.

The access command *and* the write command run one write cycle that implements the write cycle algorithm.

The access command *and* the read command run one read cycle that implements the read cycle algorithm.

The refresh command runs one refresh cycle that implements the refresh cycle algorithm after any ongoing read or write cycle completes.

An asserted done report indicates that the module will accept a command.

In a synchronous digital system, the system controller is simplified when operator commands are synchronous and have a duration of one clock period. Assume this is the case for this project.

Refer to Figure PJ18.1 and DRAM data book specification.

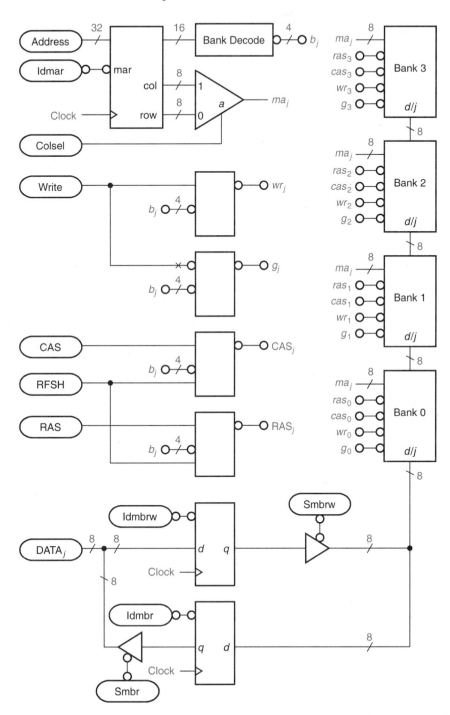

Project Tasks

1. Make a block diagram showing data path and controller with suitable input and output signals.

2. Design the data path. Use the DRAM pinout as a guide and checklist.

8 pins	a_j
4 pins	d/q_j
1 pin	ras
1 pin	cas
1 pin	write
1 pin	enable
2 pins	5 V, 0 V
18 total	

3. Create an ASM chart for the controller.

4. Design the controller circuit: state machine and outputs.

5. Build the memory module.

6. Design tests for the memory module.

7. Test the module.

PROBLEMS

9.1 *Reading a static RAM.* Plot the data-out waveform. Mark invalid and valid data. Show intervals t_0 to t_5.

t_0 120-ns read cycle time

t_1 70-ns access time from address

t_2 10-ns output data hold time from address

t_3 40-ns from chip enable H to L to data

t_4 10-ns from chip enable H to L to output low Z

t_5 20-ns from chip enable H to L to output high Z

Address

CE   30 ns

9.2 *Writing a static RAM.* Assume the RAM chip has separate data d and q pins.

(a) Plot the data-out waveform as being in low Z or high Z state.
(b) Plot the valid-data-in waveform required for write.
(c) Mark invalid and valid data.
(d) Show intervals t_0 to t_6.

t_0 90-ns write cycle time

t_1 10-ns from write H to L to output high Z

t_2 20-ns from write L to H to output low Z

t_3 10-ns from chip enable H to L to output low Z

t_4 60-ns from chip enable H to L to output high Z

t_5 30-ns data-in setup time

t_6 20-ns data-in hold time

Address

CE   20 ns

Write   20 ns

9.3 Reading a dynamic RAM. Plot the data-out waveform. Mark invalid and valid data. Show intervals t_0 to t_5.

t_0 120-ns read cycle time

t_1 50-ns RAS precharge time

t_2 20-ns CAS-to-RAS precharge time

t_3 60-ns access time from RAS

t_4 25-ns access time from CAS

t_5 10-ns output buffer turn-off (data hold) time.

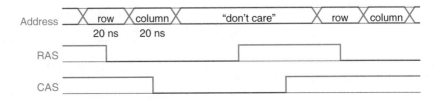

9.4 *Writing a dynamic RAM.* Assume the ram chip has separate data d and q pins.

(a) Plot the data-out waveform as being in low Z or high Z state.
(b) Plot the valid-data-in waveform required for write.
(c) Mark invalid and valid data.
(d) Show intervals t_0 to t_8.

t_0 100-ns write cycle time

t_1 40-ns RAS precharge time.

t_2 60-ns CAS pulse time.

t_3 20-ns CAS-to-RAS precharge time.

t_4 30-ns write pulse time.

t_5 10-ns data setup time.

t_6 10-ns data hold time.

t_7 10-ns output buffer turn off time

t_8 20-ns access from CAS

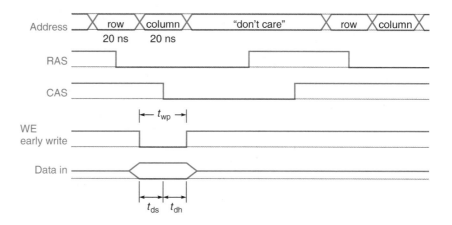

9.5 Refer to Figures 9.7 and 9.3. Synchronize a static RAM read cycle to a 20-MHz clock. Use the time values given in Problem 9.1.

9.6 Refer to Figure 9.5. Synchronize a dynamic RAM read cycle to a 50-MHz clock. Use the time values given in Problem 9.3. The minimum RAS pulse width t_{ras} is 70 ns.

9.7 Refer to Figure 9.6. Synchronize a dynamic RAM write cycle to a 50-MHz clock. Use the time values given in Problem 9.4.

9.8 Refer to Figure 9.10. Synchronize a dynamic RAM CAS-before-RAS refresh cycle to a 50-MHz clock. Use the time values given in Problem 9.4. CAS before RAS setup time is $t_{csr} = 10$ ns min.

9.9 Refer to Example 9.3. Design a DRAM refresh request generator. Assume:

The system clock is 20 MHz.

The memory cycle time t_c is 180 ns.

Each DRAM row is refreshed every 4 ms or sooner.

The memory chip has 256 rows.

Refresh cycles are interleaved with read and write cycles.

9.10 Design a four-bank memory map decoder. Show the address range for each bank-select line.

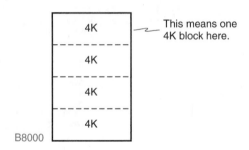

9.11 Design a decoder given the following memory map. You must decide how many address lines you need. Each map segment is called a block. Show address range of each decoder output line.

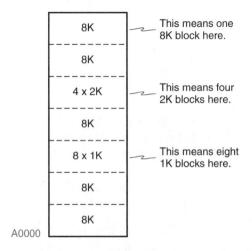

9.12 Given six address lines $a_5a_4a_3a_2a_1a_0$, derive the address decode circuit whose 16 enable outputs are true for the following (hex) addresses:

10 11 12 13 18 19 1A 1B 30 31 32 33 38 39 3A 3B

9.13 A printer control interface needs four ports. Three ports are write only and one port is read or write. Design a decoder to select these ports. The addresses are 9AC0, 9AC1, 9AC2, and 9AC3. Decode all address lines. Port 9AC1 is the r/w port.

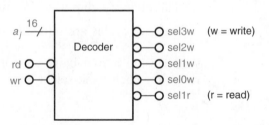

9.14 Design a decoder so that four output bank-enable lines are true over the address ranges as listed (*caution:* read addresses carefully).

Bank Enable	Address Range, Hex
b_0	9EC00000 to 9EC3FFFF
b_1	9ED00000 to 9ED3FFFF
b_2	9EE80000 to 9EEBFFFF
b_3	9EF80000 to 9EFBFFFF

Decoder black box:

inputs: 32 address lines a_j

outputs: n address lines directly to RAMS
four low True bank-enable lines b_j

Specify which lines are included in the n lines. All 32 address lines a_{1F} to a_{00} must be decoded. Hex numbers must be used for address lines. Calculate the number of words in each bank.

Specifications common to the following decoder design problems.

Decoder black box:

inputs: 32 MAR address lines a_j

outputs: n address lines directly to RAMS
m low True bank-enable lines

Specify which address lines are included in the n lines. Must decode all 32 MAR address lines: a_{1F} to a_{00}.

9.15 *Decoder analysis and synthesis.* Constant, bank, and RAM address lines are numbered from 00_{16} to $1F_{16}$. A convenient way to show the numbers on drawings is as follows. The two-digit numbers now occupy one position.

1	1	1	1	0	0	0	0
FEDC	BA98	7654	3210	FEDC	BA98	7654	3210

Address line bit values when bank-enables are True are as follows (x means a_j can be 0 or 1, r means a_j to RAM banks)

1x00 11x0 010x 1111 rrrr rrrr rrrr rrrr

(a) What is the RAM bank size? Why?
(b) What address ranges does this a_j specification imply?
(c) Design a decoder so that the output bank-enable lines are true over the address ranges.

9.16 *Decoder synthesis*. Design a decoder so that the output bank-enable lines are true over the following address ranges:

4D0A0000 to 4D0BFFFF

4D4A0000 to 4D4BFFFF

What is the size of the RAM bank?

10 PROGRAMMABLE LOGIC

Programmable logic devices (PLDs) are another step up the ladder of complexity. The PLD allows designers to implement logic functions with fewer chips. In effect, a custom circuit is "wired" on one chip.

The new PLD notation that makes the higher level of integration tractable is presented. The notation clearly shows that no new theory or logic design processes are required for circuits implemented with the PLD. We imply that an understanding of switching theory is necessary to appreciate the generality of the PLD.

Next the PLD logic structure is shown to be AND/OR, but with a difference. The difference is in the numbers; there may be 64 or more AND gates, each with 32 inputs. Variations of the structure include addition of D flip-flops for state machines and registers, and replacement of OR with XOR to simplify counting state machine implementation.

A review of commercially available PLDs informs us of what is available for use in designs. We learn that practical implementations require physical PLD programming instruments, as well as some programming language to convert designs into files suitable for controlling those programming instruments. However, the variety of languages and instruments puts these topics outside the scope of the discussion.

Finally, we cover several specific combinational and sequential designs so that as a designer you can start using PLDs.

INTRODUCTION

A designer has the option to replace the usual multitude of logic family chips in a circuit with one or more programmable logic devices (PLDs). Designers invariably invoke this cost-effective option, which provides a flexible, custom solution. By *flexible* we mean that design changes are readily implemented by programming a new part or a reprogrammable part. The process truly is flexible, because design changes do *not* require changes to the supporting printed circuit board.

The basic reason programmable logic is commercially feasible is that practical logic functions of n variables are the sum of just a few product terms, not 2^n terms or any number remotely approaching the maximum of 2^n. In fact, sales of large numbers of programmable logic devices with multiple seven-term or eight-term structures in effect define "just a few" as eight or fewer. Nevertheless, solutions are available for any function, because a ROM with n address lines provides 2^n terms.

Programmable logic has a place on the ladder of complexity, whose bottom rung is occupied by discrete digital gates. A very important factor in the choice of design procedure is the amount of time needed to achieve a result. In most cases programmable logic offers the shortest time to a result. Once the design is proven, lower-unit-cost methods that are higher up the ladder of complexity are used as production volume increases.

Discrete gate logic   Implementation of relatively simple logic functions requires many parts. Each part may have a low unit cost; however, the relatively large and complex printed circuit board and the structure housing the functions are relatively expensive.

Medium-scale integrated (MSI) combinational and sequential logic MSI chips dramatically reduce the parts count and the board size because they replace most of the discrete gate chips.

Programmable logic   A significant cost reduction results when the usual multitude of logic family chips in a circuit is replaced by one or more programmable logic devices. Furthermore, the indirect costs of design are reduced, because design errors are usually corrected by simply reprogramming the PLD. This is why, in practice, the PLD provides a cost-effective method of proving designs. However, we must have the correct list of inputs and outputs for the black box that the PLD represents. The relatively low setup charges involved in the use of PLDs in production make PLDs an attractive option in initial production lots. As production volume increases, a move up the ladder is usually made.

Gate-array-on-a-chip logic   Individual gates in a gate array on a chip need to be "wired" into a circuit in the factory. The design and manufacturing setup charges are substantial. Gate arrays are used after the design is proven and a competitive cost is made possible by adequate production volume. Adequate volume allows for amortization of the initial costs over many units.

Standard-cell-array-on-a-chip logic   This format is like a gate array except that cells of many gates are "wired" into a function. Cells are proven designs of various basic functions, such as a JK flip-flop or a 4-to-1 multiplexer. Initial costs are higher than those for a gate array.

Custom logic   The entire chip is designed from scratch. Of all approaches, this one usually yields the smallest chip size. Clearly, even higher initial costs require yet higher volumes.

10.1

Programmable Logic Notation

A circuit is "wired" by blowing fuses to program the minterms in an OR of ANDs.

Programmable logic devices are integrated circuits whose function(s) are field programmable. A circuit is "wired" on a PLD chip by blowing fuses or by storing 1, 0 bit patterns. A PLD has a much higher level of integration than either a small-scale integrated (SSI) circuit or a medium-scale integrated (MSI) circuit. Not only are there more gates per chip, but there are more inputs per gate. For example, the 16L8 PLD has 64 AND gates, each with 32 inputs.

In a logic diagram the numbers 64 and 32 imply a large number of wires. A new notation is required to simplify PLD logic diagrams and schematics. Whereas traditional schematics flow horizontally from inputs on the left to outputs on the right (Figure 10.1a), the new PLD notation retains inputs and outputs on the left and right, respectively. One new feature is that input flow into a gate is vertical (Figure 10.1b). Another new feature is the "x," which represents an intact programmable fuse at each intersection with the line flowing into the AND gate. (Absence of a "x" represents a blown fuse.) The one line into the AND gate is understood to be a bus of $2n$ lines, where n is the number of input variables. There are n lines for variables and n lines for the variables' complements. [In electrically programmed (EPLD) devices the fuse is replaced by an EPROM cell and a gate.]

The new notation permits the entire logic structure of the device to be presented concisely in a matrixlike format, such as the unprogrammed matrix in Figure 10.2, with a full field of x's representing intact fuses. Prior to being programmed, a PLD has all fuses intact. All the "wires" are in place. Programming deletes wires. This wiring-on-a-chip process is the inverse of wiring logic family gates together.

The programmed matrix implementing the three-input gate in Figure 10.1b is shown in Figure 10.3. This matrix of seven programmable AND gates driving one OR gate builds functions with as many as seven product terms. The three-input AND gate corresponds to one term, and so the other six terms must be programmed for logical 0 output. This is achieved by leaving all x's (fuses) in place. In this case the drawing of the six AND gates is simplified by replacing all x's on

FIGURE 10.1

Traditional and PLD logic schematics for a gate

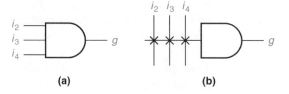

(a)   (b)

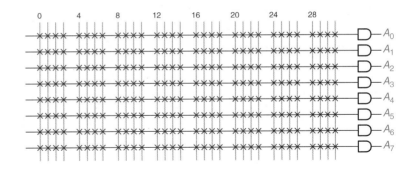

FIGURE 10.2

Typical unprogrammed PLD logic schematic

An AND term is deleted from a function when all fuses are intact.

The AND output is asserted when all fuses at the AND input are blown.

All inputs are buffered to simplify system design.

any input line by an X inside the corresponding AND symbol. An X inside the AND shape means this gate is deleted. The three-literal term $i_2 i_3 i_4$ is programmed by removing all connections to the input line of the seventh AND gate, except the connections for wires corresponding to $i_2 i_3 i_4$.

There are no x's in the line feeding the top AND gate. This means there are no connections to the input line. The output of a PLD AND gate with no connections is always asserted. The top AND gate's asserted output enables the output buffer.

The PLD schematic fragment in Figure 10.3 shows that every input i is buffered and inverted in parallel, producing i and i' for internal use. The same is true for every output. This allows for internal feedback when output buffers are enabled. When the output buffer is disabled, so-called "output" pins may be used as input pins. The output OR gate has seven inputs accommodating a maximum of seven product terms. This fragment is a basic PLD building block.

The 16R4 PLD schematic in Figure 10.4 illustrates the intrinsically regular PLD structure. The x's are omitted here, and in data books, to simplify the schematic.

FIGURE 10.3

Programmed gate PLD logic schematic

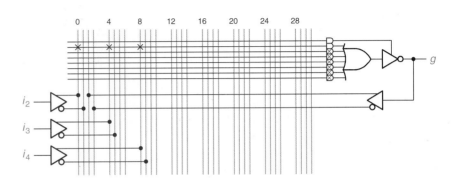

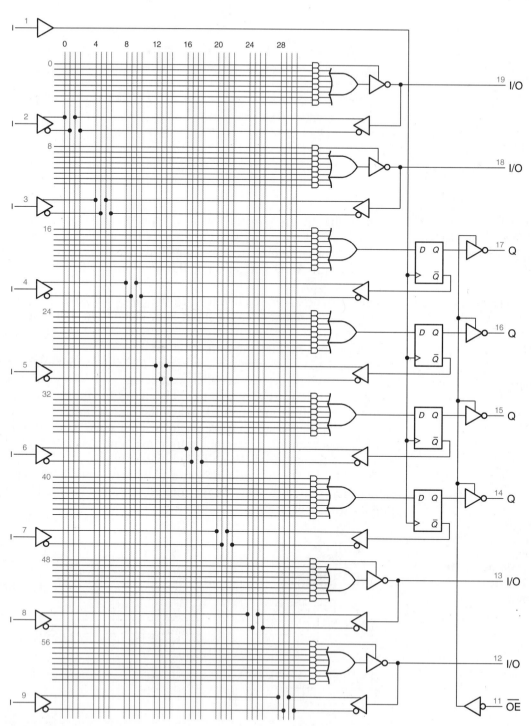

FIGURE 10.4

PAL16R4 logic schematic

Programmable Logic Structures

There are variations on the theme of the PLD building block. The major ones exploit the fact that any function can be expressed as a sum of product terms. Sum and product give rise to two opportunities: selection of inputs to the AND gates, creating product terms, and selection of inputs to the OR gates, creating sums of product terms. The two opportunities result in four basic commercially available types of devices (Table 10.1)

10.2.1 PLD Structure Types

ROM generates all minterms of one function.

The ROM has a fixed AND matrix and a factory-programmable OR matrix, which is also considered as fixed. *Fixed* means "not field programmable." The ROM is very different from other PLDs, because the inputs are fully decoded into 2^n minterms. The ROM's n address lines represent n input variables. The internal decoder may be perceived as 2^n n-input AND gates. In PLD terminology, the ROM's m data-output lines are the outputs of m 2^n-input OR gates. An x connection (Figure 10.5) is represented as a stored 1 (the minterm is included). An omitted x represents a stored 0 (the minterm is excluded). In this way a ROM implements m switching functions, each with 0 to 2^n minterms. The stored 1's and 0's are programmed in various ways that we will not delve into.

PLD generates several function each with 7 or 8 minterms.

The PROM is a ROM with a field-programmable OR matrix. In all other respects, the PROM is the same as a ROM. The structure for PROMs and ROMs is illustrated in Figure 10.5.

The *programmable logic array (PLA)* has field-programmable AND and OR matrices (Figure 10.6). PLA commercial activity is essentially nil. However, the *programmable logic sequencer (PLS)* usually includes a PLA structure.

TABLE 10.1 **Types of Programmable Logic Devices**

AND Matrix	OR Matrix	PLD Device Type
fixed*	fixed*	ROM
fixed*	programmable	PROM
programmable	fixed*	PAL
programmable	programmable	PLA, PLS

*Not field-programmable.

FIGURE 10.5
ROM or PROM structure

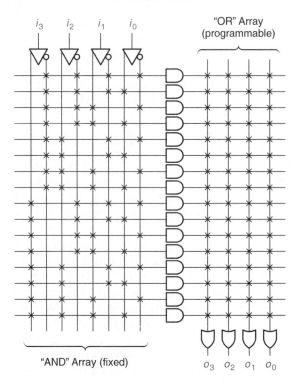

PROM
16 words x 4 bits

"OR" Array
(programmable)

"AND" Array (fixed)

o_3 o_2 o_1 o_0

The *programmable array logic (PAL[1])* is where the action is (Figure 10.7). The remainder of this chapter discusses the various types of these PALs, which are extremely popular PLDs.

10.2.2 PLD Structure Variations

PLD structure is a set of OR of ANDs.

First we establish a reference by using the 16R4 member of the 20-pin PAL16R8 family. The PAL20R8 family is a 24-pin version of the PAL16R8 family. The 16R4 PLD logic schematic in Figure 10.4 illustrates the circuits found in most PLD devices. Each input is isolated from the AND matrix by input inverter/buffers. The buffer is mandatory because each buffer has to drive 64 inputs when all x's are intact. Each input i_j and the complement i_j' drive the AND matrix. The AND outputs drive an OR gate, which feeds a three-state gate. The three-

---

[1] PAL is a registered trademark of Advanced Micro Devices, Inc.

FIGURE 10.6
PLA structure

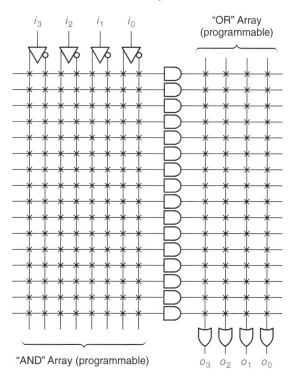

FPLA
4 in-4 out-16 products

"AND" Array (programmable)

"OR" Array (programmable)

i_3 i_2 i_1 i_0

o_3 o_2 o_1 o_0

state gate is an output buffer that may or may not invert the input signal. A separate AND output drives the output buffer enable input.

The output pins can also serve as input pins when the output buffer is disabled, because an input inverter/buffer drives the AND matrix from the output. On the other hand, when the output buffer is enabled, the output may be fed back into the AND matrix.

An output may be *registered;* that is, the OR output drives the *d* input of a D flip-flop. Then the flip-flop active high output drives an output buffer. The output buffer is enabled from a dedicated enable pin instead of being programmed. The flip-flop active low output drives an input inverter/buffer, which drives the AND matrix. A dedicated buffered clock pin drives the D flip-flop clock inputs.

XOR replaces OR of ANDs for counter applications.

One major variation is (Figure 10.8) to replace the output OR gate with an XOR gate to facilitate counter implementations (see Section 8.2 on page 417 and upcoming Section 10.6.2). XOR gates are found in the 20X10, 20X8, and 20X4 family.

FIGURE 10.7
PAL structure

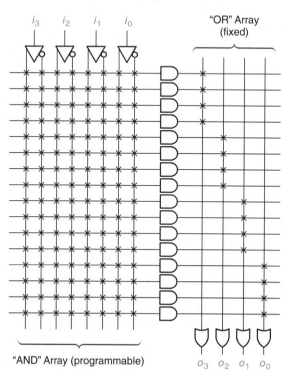

PAL
4 in-4 out-16 products

FIGURE 10.8
**XOR registered
PLA20X8 circuit
fragment**

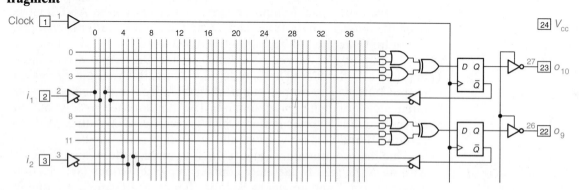

More sophisticated PLD circuits have an output circuit that is called a *macrocell,* or I/O architecture. A typical example is the PAL22V10 device, whose macrocell is shown in Figure 10.9. A 4-to-1 multiplexer drives the inverting output buffer. The four mux inputs are the AND/OR gate output p, the complement p', the registered output q, and the complement q'. Two fuses drive mux input-select lines. Feedback into the AND matrix is from the output pin or directly from the D flip-flop active low output. The latter is designated as *high-speed feedback,* because the mux and output buffer propagation delays are bypassed.

A PLD becomes a PLS (a sequencer) when a *buried register* is included. Buried register outputs are not available at any device pin (Figure 10.10), nor do they need to be. They are buried inside the PLD, because they are used as a state machine state register whose outputs act internally to feed back the present state. ASM conditional and unconditional outputs are generated by using AND terms whose outputs drive registers whose outputs are connected to output pins via buffers.

Buried registers implement state machines.

FIGURE 10.9
PAL22V10 macrocell

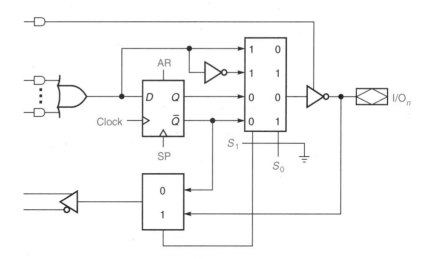

S_1	S_0	Output Configuration
0	0	Registered/Active Low
0	1	Registered/Active High
1	0	Combination/Active Low
1	1	Combination/Active High

0 = Unprogrammed fuse or programmed EE bit.
1 = Programmed fuse or erased (charged) EE bit.

FIGURE 10.10
Programmable logic sequencer block diagram

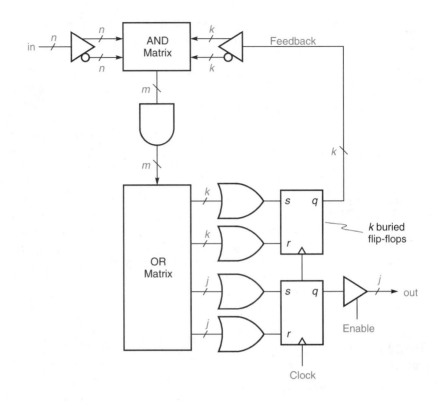

10.3 Programmable Logic Devices

The choice of a specific device is influenced by the functions' equations to be implemented. Widely used PLDs are listed in Table 10.2. Device schematics are found in most PLD data books.

10.4 Programmable Logic Languages

You need to learn a specialized language to program PLDs.

Effective use of PLDs requires us to learn one or more of the commercially available PLD programming languages. This is not a formidable task because a PLD language's purpose is limited when compared, for example, to the C language. In fact, some PLD languages have many C-like features and syntax. What follows is a small sample. The program that "wires" the three-input AND gate inside the PAL shown in Figure 10.3 consists of three statements such as the following from Texas Instrument's proLogic[2] language.

---

[2] proLogic is a trademark of proLogic Systems Inc.

TABLE 10.2 **Widely Used Combinational/Registered PLD Devices**

Device	Pins*	Dedicated Inputs	Combo Out	Output Registers	Feedback	Output Enable	Product Terms/Output
16L8	20	10	6	0	yes	prog	7 to OR
			2	0	no	prog	7 to OR
R4	20	8	4		yes	prog	7 to OR
				4	yes	pin	8 to OR
R6	20	8	2		yes	prog	7 to OR
				6	yes	pin	8 to OR
R8	20	8	0	8	yes	pin	8 to OR
20L8	24	14	6	0	yes	prog	7 to OR
			2	0	no	prog	7 to OR
R4	24	12	4		yes	prog	7 to OR
				4	yes	pin	8 to OR
R6	24	12	2		yes	prog	7 to OR
				6	yes	pin	8 to OR
R8	24	12	0	8	yes	pin	8 to OR
20L10	24	12	8		yes	prog	3 to OR
			2		no	prog	3 to OR
20X4	24	10	6		yes	prog	3 to OR
				4	yes	pin	4 to XOR
X8	24	10	2		yes	prog	3 to OR
				8	yes	pin	4 to XOR
X10	24	10	0	10	yes	pin	4 to XOR
22V10	24	11 and clock		2	prog	prog	8 to OR
				2	prog	prog	10 to OR
				2	prog	prog	12 to OR
				2	prog	prog	14 to OR
				2	prog	prog	16 to OR

* Two pins dedicated to V_{cc}, V_{dd} (+5 V and ground).

```
include p16r4;
!pin19 = pin2 & pin3 & pin4;   (! = NOT, & = AND)
pin19.oe = 1;
```

The first line includes a header file defining the PAL16R4. The next two lines are assignment statements. (An *assignment* is an expression using signal names and logical operators.) A simplifying feature of one compiler is that signal names can be pin numbers. In that language, predefined signal names are shown on PLD logic diagrams. The language's compiler recognizes these names. Other compilers require that a pin declaration list precede the actual program.

The compiler is a software development design tool, and its output is an industrial-standard file describing the fuse map that can be downloaded to a device programmer.

Programs also require a certain amount of "boiler plate," such as title blocks. Designs may be entered into a program using Boolean equations, state (truth) tables, and schematic diagrams.

Next we briefly discuss several widely available software packages (no recommendation is implied). In effect, the design process for all software products proceeds from left to right, as follows, where one or more items per column are executed.

INPUT $\longrightarrow$ OUTPUT

Boolean	Syntax checking	Fuseplot assembly	Fusemap
Truth Table	Logic reduction	Simulation	Jedec file
Schematic	Error detection		Test Vectors

PALASM 90[3] evolved from the original PALASM[3] software that supported the first PAL devices. Equations and truth tables may be entered.

ABEL[4] (Advanced Boolean Expression Logic), like all other PLD programming languages, is based on a compiler that translates the user's input equations and truth tables into a fuse map.

FutureDesigner[4] includes an industry-standard schematic capture feature. The user can describe a design via any combination of equations, truth tables, or state diagrams.

10.5   Combinational Circuit Designs

Design using PLDs has a high art-form content. This is especially true when trying to "fit" equations into a PLD. The representation of switching functions found in PLDs is a sum of product terms that do

---

[3] PALASM 90 is a registered trademark of Advanced Micro Devices.
[4] ABEL and FutureDesigner are registered trademarks of Data I/O Corp.

not have to be minterms. Except for the ROM, a function does not have to be in canonical form.

The twelve-to-one multiplexer in Figure 5.7 on page 234 is a typical candidate for a PLD implementation.

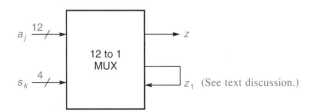

The design process starts with an assessment of pin requirements. Pins are required for 12 inputs a_j, four input-select lines s_k, and one output z. The 20L8 (Figure 10.11) meets these requirements.

Pins available

Device	Number of Pins	Dedicated Inputs	Combo Out	Output Registers	Feedback	Output Enable	Product Terms/Output
20L8	24	14	6	0	yes	prog	7 to OR
			2	0	no	prog	7 to OR

Pins required	My Design Needs:	PLD Provides:
Data inputs (a_j)	12	12
Select inputs (s_k)	4	4
Function output (z)	1	1 combo, no feedback
z_1 output	1	1 combo with feedback

Functions with a greater number of terms than the PLD functions provide for are implemented by cascading PLD functions.

Practical circuit implementation   There is a problem: the multiplexer function z is the sum of 12 product terms, but the 20L8 only supports functions with seven or fewer terms. One solution is first to form z_1

from the last six terms and to pass output z_1 back into the 20L8 input matrix. Then z is formed from the first six terms and z_1. The disadvantage is that z now has twice the propagation time (e.g., 2×7 ns) for six of the 12 terms. However, the resulting propagation time is still shorter than that offered by TTL or ALS parts.

$$z = z_1 + a_0 s_3' s_2' s_1' s_0'$$
$$+ a_1 s_3' s_2' s_1' s_0$$
$$+ a_2 s_3' s_2' s_1 s_0'$$
$$+ a_3 s_3' s_2' s_1 s_0$$
$$+ a_4 s_3' s_2 s_1' s_0'$$
$$+ a_5 s_3' s_2 s_1' s_0$$
$$z_1 = + a_6 s_3' s_2 s_1 s_0'$$
$$+ a_7 s_3' s_2 s_1 s_0$$
$$+ a_8 s_3 s_2' s_1' s_0'$$
$$+ a_9 s_3 s_2' s_1' s_0$$
$$+ a_A s_3 s_2' s_1 s_0'$$
$$+ a_B s_3 s_2' s_1 s_0$$

The AND matrix now has 17 inputs: 12 a_j, 4 s_k, and z_1. Two combo outputs are enabled: one for z and one for z_1. The z_1 output must have feedback capability. The programmed PAL20L8 is illustrated in Figure 10.11.

10.5.2 Address Decoder

The address decoding method discussed in Section 9.2 (page 510) partitioned address lines into three classes: constant, bank, and RAM. The constant lines are inputs to an AND circuit producing a master bank-enable report. Bank lines drive a decoder to produce individual bank-enable reports, and the RAM lines are connected to RAM address pins. The propagation delay through logic family parts can seriously degrade memory performance by increasing access time on read and write operations. The positive statement is that PLDs can significantly improve memory performance by minimizing address decoder delays. To this end, specialized PLDs are available.

A 64-input AND can decode a 32-bit address bus.

The 18N8 PLD, with a 7-ns propagation time, is designed for address decoding applications. Each output includes only one product term, eliminating the need for an OR output circuit, which reduces

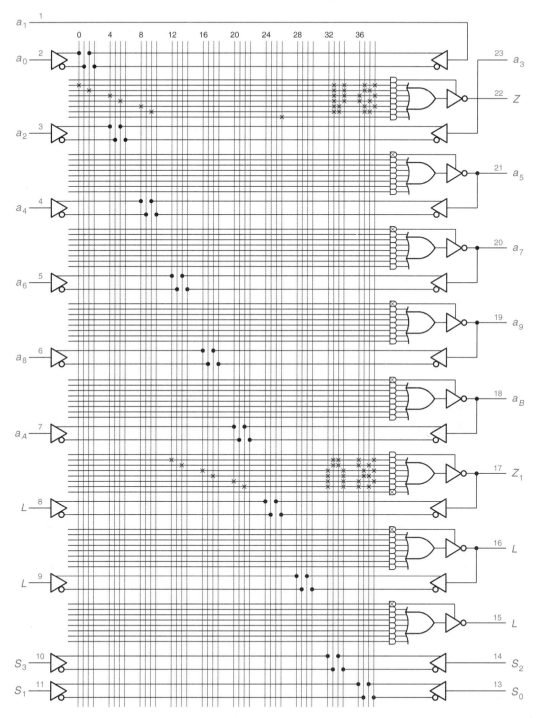

FIGURE 10.11
Twelve-to-one multiplexer using a PAL20L8

propagation time. A PAL, such as the 18N8, replaces the traditional AND of the constant lines and the bank lines' decoder to produce the bank-enable outputs directly from equations, including constant and bank address lines as inputs.

EXAMPLE 10.1   Eight-Bank Address Decoder Using a PAL18N8

In a 32-bit address space, the partition for a 32 meg (8×2^{22}) memory space (5C000000 to 5DFFFFFF) using 4-megabyte (2^{22}) RAM is as follows.

RAM	22 lines a_{21} to a_0
bank	3 lines a_{24} to a_{22}
constant	7 lines a_{31} to a_{25}

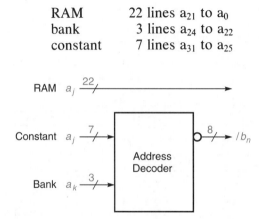

Pins Available

Device	Number of Pins	Dedicated Inputs	Combo Out	Output Registers	Feedback	Output Enable	Product Terms/Output
18N8	20	10	8	0	yes	prog	1 (no OR)

Pins required	My Design Needs:	PLD Provides:
Constant (a_j)	7	7
Bank (a_j)	3	3
Bank-enable outputs (b_n)	8	8 combo

Practical Circuit Implementation
18N8

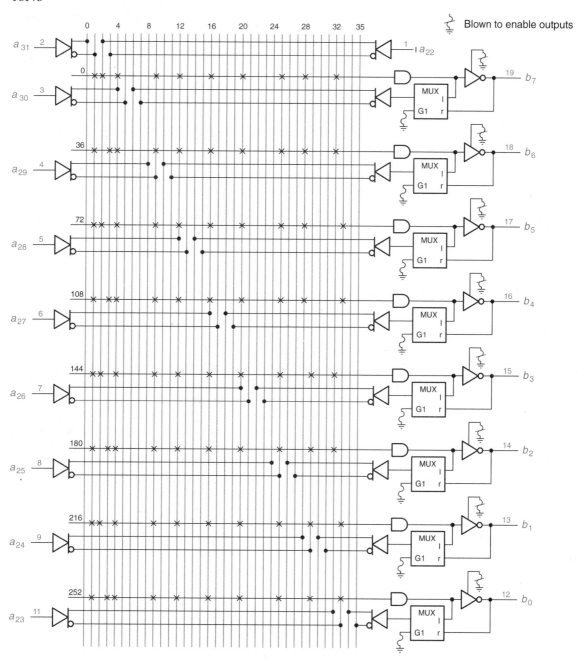

Design

The constant lines in the example are defined by the high seven bits 0101110_2 of $5C000000_{16}$. The bank lines are defined by the bank numbers 0 to 7.

	constant		bank	

$$b_7 = a_{31}'a_{30}a_{29}'a_{28}a_{27}a_{26}a_{25}' \qquad a_{24}\ a_{23}\ a_{22}$$

$$b_6 = a_{31}'a_{30}a_{29}'a_{28}a_{27}a_{26}a_{25}' \qquad a_{24}\ a_{23}\ a_{22}'$$

$$b_5 = a_{31}'a_{30}a_{29}'a_{28}a_{27}a_{26}a_{25}' \qquad a_{24}\ a_{23}'a_{22}$$

$$b_4 = a_{31}'a_{30}a_{29}'a_{28}a_{27}a_{26}a_{25}' \qquad a_{24}\ a_{23}'a_{22}'$$

$$b_3 = a_{31}'a_{30}a_{29}'a_{28}a_{27}a_{26}a_{25}' \qquad a_{24}'a_{23}\ a_{22}$$

$$b_2 = a_{31}'a_{30}a_{29}'a_{28}a_{27}a_{26}a_{25}' \qquad a_{24}'a_{23}\ a_{22}'$$

$$b_1 = a_{31}'a_{30}a_{29}'a_{28}a_{27}a_{26}a_{25}' \qquad a_{24}'a_{23}'a_{22}$$

$$b_0 = a_{31}'a_{30}a_{29}'a_{28}a_{27}a_{26}a_{25}' \qquad a_{24}'a_{23}'a_{22}'$$

10.5.3   Designs Using a ROM

Designs using a ROM require equations to be in canonical form. In practice, truth tables serve this purpose, because every 1 in a truth table output column represents a minterm.

In a memory, the ROM data outputs are related to each other; they are a data word. In PLD designs, it is useful to view the ROM data outputs as independent outputs. Each ROM output line represents a truth table output column, and the ROM address input lines now represent truth table input variables. Hence a ROM "stores" a truth table. This point is illustrated in Example 10.2.

A ROM requires equations in sum-of-minterms format.

Note: ROMS are not supported by PLD software packages. We can manually enter values into a device programmer or write our own software that creates a file of values to be downloaded to a device programmer.

EXAMPLE 10.2   ROM or PROM as a Function Generator

First the given equations are converted into canonical form as a sum of minterms. Then the minterms of three variables are entered into the output column representing the equation.

Full-adder sum $\quad d_3 = a_2 \text{ xor } a_1 \text{ xor } a_0 \qquad = m_1 + m_2 + m_4 + m_7$

Full-adder carry $\quad d_2 = a_1 a_0 + a_2 (a_1 + a_0) \qquad = m_3 + m_5 + m_6 + m_7$

Whatever $\qquad\qquad d_1 = a_1 a_0 + a_2 a_1 a_0 \qquad\quad = m_3 + m_7$

XOR $\qquad\qquad\quad d_0 = a_2 \text{ xor } a_0 \qquad\qquad = m_1 + m_3 + m_4 + m_6$

INPUTS			OUTPUTS			
a_2	a_1	a_0	d_3	d_2	d_1	d_0
0	0	0	0	0	0	0
0	0	1	1	0	0	1
0	1	0	1	0	0	0
0	1	1	0	1	1	1
1	0	0	1	0	0	1
1	0	1	0	1	0	0
1	1	0	0	1	0	1
1	1	1	1	1	1	0

TI TBP18S030

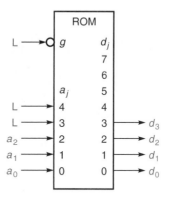

The size of a ROM doubles each time a variable is added to an equation: three variables have eight minterms, four variables have 16 minterms, and so forth. Equations with many variables and only a few terms have large truth tables with many zeros (see Table 10.3). Let us assume we want to implement the equations for q_1^+ and q_0^+ in a ROM.

$$q_0^+ = q_1'q_0'w'r + q_1'q_0 t' = s_0 w'r + s_1 t'$$

$$q_1^+ = q_1'q_0'wr' + q_1 q_0't' = s_0 wr' + s_2 t'$$

The equations include five variables, which are represented by the 32-row truth table in Table 10.3. The s_0 term in q_0^+ expands into two minterms because $w'r = w'r(t + t')$. The s_1 term expands into canonical form as the sum of four minterms because $t' = t'(wr + wr' + w'r + w'r')$. The terms are listed under the "TERM = 1" columns. The other terms create 1 entries in a similar manner.

TABLE 10.3 **Memory Module State Machine ROM Contents**

INPUTS q_1q_0wrt	OUTPUTS q_1^+	q_0^+	TERM = 1 q_1^+	q_0^+
00000	0	0		
00001	0	0		
00010	0	1		$s_0w'rt'$
00011	0	1		$s_0w'rt$
00100	1	0	$s_0wr't'$	
00101	1	0	$s_0wr't$	
00110	0	0		
00111	0	0		
01000	0	1		$s_1w'r't'$
01001	0	0		
01010	0	1		$s_1w'rt'$
01011	0	0		
01100	0	1		$s_1wr't'$
01101	0	0		
01110	0	1		s_1wrt'
01111	0	0		
10000	1	0	$s_2w'r't'$	
10001	0	0		
10010	1	0	$s_2w'rt'$	
10011	0	0		
10100	1	0	$s_2wr't'$	
10101	0	0		
10110	1	0	s_2wrt'	
10111	0	0		
11000	0	0		
11001	0	0		
11010	0	0		
11011	0	0		
11100	0	0		
11101	0	0		
11110	0	0		
11111	0	0		

TABLE 10.3   (continued)

TI TBP18S030

```
                        ROM
                    ┌──────────────┐
    L ───────○──── │ g          dⱼ │
                    │             7 │
                    │             6 │
                    │ aⱼ          5 │
(feedback) q₁ ───── │ 4          4 │
          q₀ ────── │ 3          3 │
           w ────── │ 2          2 │
           r ────── │ 1          1 │──── q₁
           t ────── │ 0          0 │──── q₀
                    └──────────────┘
```

Note: the q_j outputs are present-state values. The state machine implements the q_j^+ equations.

The large number of zeros in the output columns imply that input variables are "don't cares" under certain conditions (see Table 7.1 on page 371). In a pencil-and-paper truth table, the "don't care" entry is a dash. In a ROM, the "don't care" does not exist, for in physical devices an entry is either a 1 or a 0.

10.5.4   Latch Structures

The latch implemented in a PLD is the fastest circuit known to be able to convert pulses to levels. For example, one pulse sets an output to the H level, and later another pulse resets the output to the L level. Furthermore, the latch in the PLD readily supports set and reset inputs from multiple sources with no performance degradation. First we build the latch in a PLD structure.

When the pulses are outputs of an ASM that uses special state numbers (see Section 8.3.4, page 456), they have a duration of one clock period. Furthermore, the elapsed time from pulse to pulse is some number of clock periods. In this way waveforms with levels having precise time durations of some number of clock periods and minimum possible delay may be generated. The PLD in combination with an ASM using special state numbers is a fierce competitor. For example, a latch can replace the JK flip-flop in the output circuit of a waveform generator (see Figure 8.30 on page 448 and Figure 8.33 on

Feedback from output to input creates latch structures in PLDs.

page 449). One idea is to associate one latch s or r input pulse with each rising and falling edge of the level-waveforms generated. We implement this idea with the RS latch.

In Section 6.2 (page 297), latches are assembled from 00 and 02 gates plus feedback from the output to an input, and the latch defining equations were derived.

$$\text{RS latch:} \quad q^+ = s + r'q$$

A latch using two 00 gates is symmetrical: either output is delayed one t_p from the corresponding input, and either output switches to H in response to an input. The return to L at the output is delayed two t_p from the resetting input. So without loss of generality, we say the s input will "start" the output and the r input will "end" the output. In most applications the crucial delay is at the start of the output waveform. If the desired waveform is at the H level, and if the crucial delay is at the end of the waveform, the z_L output of a latch with two 02 gates provides such an output at the gate driven by the r input (Figure 10.12).

A latch using two 02 gates has similar properties, except that the output switches to L at the start time and returns to H at the end time. The active high inputs require input pulses that are positive-going instead of negative-going. ASM state register flip-flops can supply both types of pulses.

Next we add multiple r and s inputs to the latch in a PLD. The RS latch is readily adapted to multiple sources of R and S inputs (Figure

FIGURE 10.12
RS latch generates waveforms

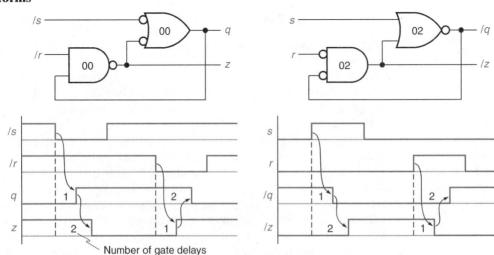

Number of gate delays

10.13) without introducing additional delays. This scheme avoids the need to use multiplexers to switch input sources. In this way, (significant) multiplexer delays are avoided. The trick is to replace s and r in the defining equations with expressions.

$$q^+ = s + qr'$$

We let

$$s = s_1 + s_2 + s_3 \quad \text{and} \quad r = r_1 + r_2 + r_3$$

Then

$$q^+ = (s_1 + s_2 + s_3) + q(r_1 + r_2 + r_3)'$$

$$q^+ = (s_1 + s_2 + s_3) + q(r_1'r_2'r_3')$$

The latch circuit r_j and s_j inputs are at the H level when not asserted (Figure 10.13a). When q_2 is high, the OR gate output q_1 is low. The latch is reset. When any s_j input falls to active low, q_1 is driven active high one t_p later. This sets the output waveform. Later, when any r_j falls to active low, q_1 is driven inactive low $2t_p$ later. This resets the waveform.

10.6 ### Sequential Circuit Designs

Registered PLDs include D flip-flops. The flip-flops are driven by a common clock supplied by the system incorporating the PLD. Two types of circuits feed a flip-flop's D input: AND/OR and AND/XOR. The AND/OR and AND/XOR configurations are suitable, respectively, for general state machines and specific state machines such as counters. Examples illustrate issues peculiar to PLDs. In what follows the state machine design shows how a registered PLD with AND/OR circuits implements all the necessary circuits. The waveform generator shows how counter equations are modified to suit the XOR circuit.

FIGURE 10.13a
Latch with multiple inputs using a PAL 20L8

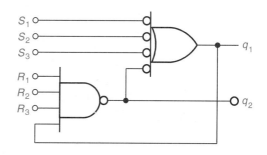

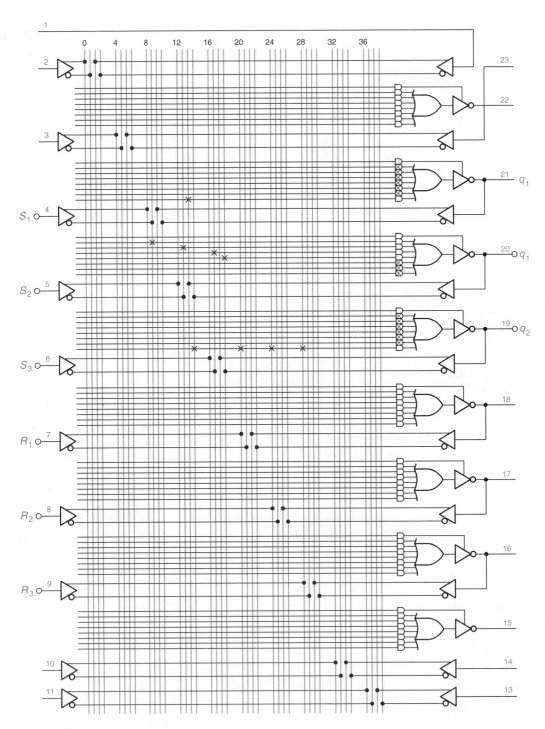

FIGURE 10.13b

The binary-to-BCD converter deals with some of the PLD pin limitation problems and other design issues.

10.6.1 State Machine

Fast, high performance state machines can be implemented in PLDs.

This design implements the standard four-state state machine circuit shown in Figure 7.17b on page 368. The truth table with table-entered variables defines the circuit (Figure 10.14a). Pins are required for four inputs a_j, four inputs b_j, one clock input, and two outputs q_k. The 16R4 satisfies these requirements (Figure 10.14b).

Pins available

Device	Number of Pins	Dedicated Inputs	Combo Out	Output Registers	Feedback	Output Enable	Product Terms/Output
16R4	20	8	4		yes	prog	7 to OR
				4	yes	pin	8 to OR

The "standard" next-state calculator in a PLD is in OR of ANDs format.

$$q_1^+ = b_3 q_1 q_0 + b_2 q_1 q_0' + b_1 q_1' q_0 + b_0 q_1' q_0'$$

$$q_0^+ = a_3 q_1 q_0 + a_2 q_1 q_0' + a_1 q_1' q_0 + a_0 q_1' q_0'$$

FIGURE 10.14a

Four-state state machine using a PAL16R4

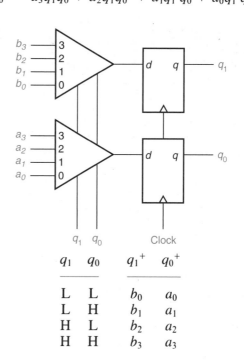

q_1	q_0	q_1^+	q_0^+
L	L	b_0	a_0
L	H	b_1	a_1
H	L	b_2	a_2
H	H	b_3	a_3

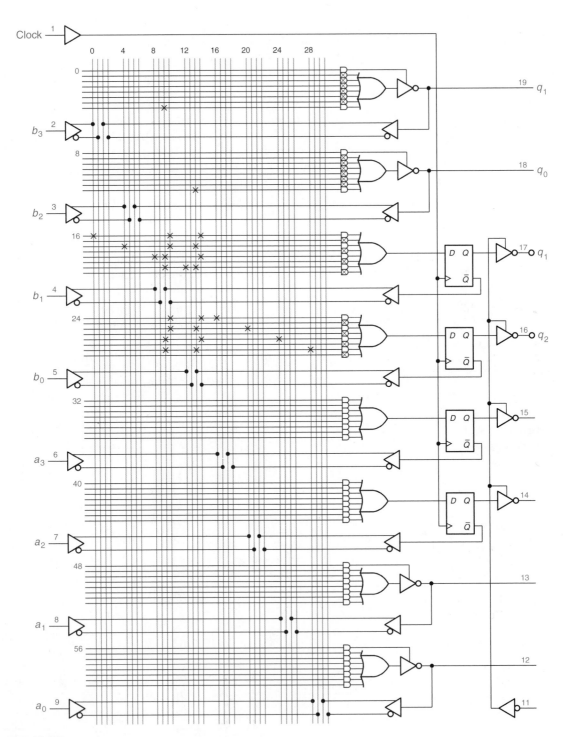

FIGURE 10.14b

10.6.2 Waveform Generator

The key idea discussed in Section 8.2.7 (page 446), and illustrated below in Figure 10.15, is decoding counter state numbers in any clock period prior to a desired output event's occurrence. A waveform transition from H to L or L to H is an event that occurs one propagation time after a clock edge in a synchronous system. Number decoders produce "pulses" one clock period in duration. The "pulses" are converted to levels by flip-flops or latches, which produce waveforms without hazards. The "pulses" are also used to load numbers into the counter. For example, in what follows the n_{4F} output asserts the counter load input so that EF_{16} is loaded into the counter at the clock edge ending clock period $4F_{16}$. [A review of Section 8.2.7 (page 446) may be in order at this point.]

Suppose the waveform in Figure 10.15 is specified for a cathode ray tube screen raster scan display application. (The waveform is not drawn to scale.) Two hex digits represent eight bits, labeled h_j ($j = 0$ to 7). h means horizontal. The set of h_j address 80_{10} columns, from 00_{16} to $4F_{16}$, to display their contents as the electron beam sweeps from left to right across the screen. The sweep is blanked and retracing from right to left as the h_j scan from EF_{16} to FF_{16}. The numbers 00 to 4F are used to fetch the graphics bits to be displayed. When the h_j are scanning from EF to FF, nothing is displayed: the screen is blank. This means we do not care what numbers are scanned by the counter during the blank interval, which is why we can use the numbers EF_{16} to FF_{16}. We want to use these numbers because they result in a "minimum" circuit, which is not obvious.

The waveform generator requires an eight-bit counter, two number detectors (4F and FF), and a latch. The design process starts from a

FIGURE 10.15
Horizontal periodic waveform

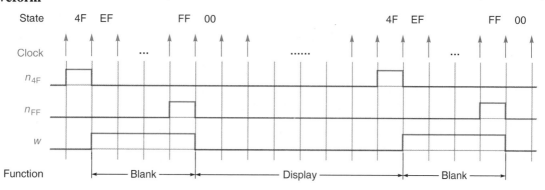

TABLE 10.4   **Counter Function Table**

Function	oe	clk	c	g	p
high-Z	H	–	–	–	–
clear	L	↑	L	–	–
load	L	↑	H	L	–
hold	L	↑	H	H	L
count	L	↑	H	H	H

black box description; PLD pins are assigned; equations are generated, and then implemented in a PLD. The PLD circuit is presented in Figure 10.16.

Counter design   The 20X– – parts implement the following equation:

$$q^+ = (X + T) \text{ xor } (W + Y)$$

Mathematics shows us why PLD XOR format implements a counter equation.

The following up-counter equation was derived in Section 8.2.5 (page 437).

$$q_j^+ = c'z = c'[gd_j + g'q_j \text{ xor } g'pa_j]$$

where

$$a_j = q_{j-1} \cdot \ldots \cdot q_1 q_0 \quad \text{and} \quad a_0 = 1$$

Start with z:

$$z = gd_j + g'q_j \text{ xor } g'pa_j$$

Substitute variables:

$$z = w + x \text{ xor } y$$

where

$$w = gd_j, \qquad x = g'q_j, \qquad\qquad y = g'pa_j,$$
$$wx = 0, \qquad wy = 0 \ (gg' = 0)$$

Expand the XOR function:

$$z = w + xy' + x'y$$

Expand with $w + w' = 1, x + x' = 1$:

$$z = w(x + x') + xy'(w + w') + x'y$$
$$z = wx + wx' + xy'w + xy'w' + x'y$$

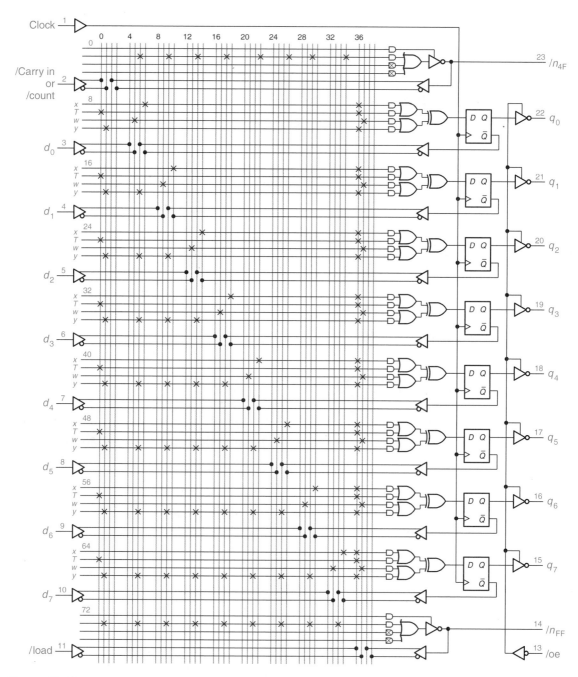

FIGURE 10.16
Waveform generator using a PLA20X8

Invoke $wx = 0$,

$$z = 0 + wx' + 0 + xy'w' + x'y$$
$$= xw'y' + x'(w + y)$$

Invoke De Morgan's Theorem in the first term:

$$z = x(w + y)' + x'(w + y)$$
$$= x \text{ xor } (w + y)$$
$$= (x + 0) \text{ xor } (w + y)$$
$$= (\text{hold} + 0) \text{ xor } (\text{load} + \text{count})$$

Compare to q^+:

$$q^+ = (X + T) \text{ xor } (W + Y)$$

where

$$X = c'x = c'g'q_j$$
$$T = 0$$
$$W = c'w = c'gd_j$$
$$Y = c'y = c'g'pa_j$$

Substitute these X, T, W, and Y equivalents into q_j^+:

$$q_j^+ = \underset{\text{hold}}{(c'g'q_j} + \underset{\text{clear}}{0)} \text{ xor } (\underset{\text{load}}{c'gd_j} + \underset{\text{count}}{c'g'pa_j})$$

Replace j by 0, 1, . . . , 7 and a_j by q_{j-1} . . . q_0, a_0 by 1:

$$q_0^+ = (c'g'q_0 + 0) \text{ xor } (c'gd_0 + c'g'p)$$
$$q_1^+ = (c'g'q_1 + 0) \text{ xor } (c'gd_1 + c'g'pq_0)$$
$$q_2^+ = (c'g'q_2 + 0) \text{ xor } (c'gd_2 + c'g'pq_1q_0)$$
$$\vdots$$
$$q_7^+ = (c'g'q_7 + 0) \text{ xor } (c'gd_7 + c'g'pq_6q_5q_4q_3q_2q_1q_0)$$

Pins available

Device	Number of Pins	Dedicated Inputs	Combo Out	Output Registers	Feedback	Output Enable	Product Terms/Output
20X8	24	10	2		yes	prog	3 to OR
				8	yes	pin	2 & 2 to XOR

Pins required	My Design Needs:	PLD Provides:
Counter q_j outputs	8	8 reg
Carry out (n_{FF})	1	1 combo
Detector (n_{4F})	1	1 combo
output enable in	1	1
clock in	1	1
carry in (count)	1	1
clear in	1	0 (This is a problem.)
load in	1	1
data d_j in	8	8
Waveform w out		Use q_7, because no other outputs are available or needed.

The PLD cannot provide a pin for the clear input. This is not catastrophic, because the counter can be cleared by loading zero via the d_j data-input lines. So we let $c' = 1$ ($c = 0$) in the counter equations.

Number detector design

$$n_{4F} = q_7'q_6q_5'q_4'q_3q_2q_1q_0$$

$$n_{FF} = q_7q_6q_5q_4q_3q_2q_1q_0 \times c_{in} \quad \text{To generalize the design}$$

$$w = q_7 \quad \text{Another advantage of using counts EF to FF}$$

Practical circuit implementation The 20X8 register outputs are inverted and then routed to the output pins (Figure 10.16). The 20X8 registered q_j active low outputs are fed back into the AND matrix, and they are equal to the outputs at the pins. Instead of manipulating the counting equation to account for the inversion at the outputs, let us intuitively decide the active levels to use in the AND matrix. The counting equation is as follows when $c' = 1$:

$$q_j{}^+ = \underset{\text{hold/tog}}{(g'q_j + 0)} \text{ xor } (\underset{\text{load}}{gd_j} + \underset{\text{count}}{g'pa_j})$$

The inversion at the output means active high q_j outputs require active low registered q_j. To hold active low q_j, the complements of the pin outputs are fed back into the AND matrix. Inversion also means active low d_j must be loaded in the load mode of operation. However, any carry-forward pa_j term to the next bit must be active high to

FIGURE 10.17
PLD binary-to-BCD serial converter

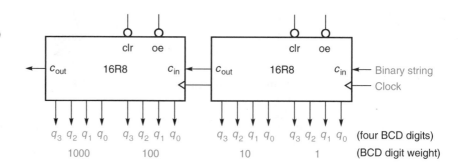

generate an asserted AND gate output, which puts the bit in the toggle mode.

10.6.3 Binary-to-BCD Serial Converter

A bit string representing a binary number is shifted left into the converter (Figure 10.17) that converts it to BCD digits. A left shift implies the most significant bit is converted first. The process created by Couleur (see References) is rather an elegant one because after each clock edge the BCD report is correct for bits shifted into the converter so far. The number of bits in the string can be as small as one. The maximum number depends on the number of BCD digits implemented in the converter. Two one-BCD-digit converters are implemented in one 16R8 PLD. Each one-BCD-digit converter provides for input and output carries. This means $2n$-digit converters for BCD digits are assembled from n 16R8 PLD chips. (See Section 1.5.2 (page 36), "BCD numbers.")

Algorithm Do a left shift of all the BCD digit bits while shifting in a new binary string bit from the right. Correct any BCD digits that are greater than nine by subtracting 10_{10} and forwarding a carry $= 1$ up to the next digit.

In a binary shifter, a left shift is multiplication by two, and an input carry (c_{in}) adds zero or one. The bits are grouped by fours when they represent BCD digits. Each group of four bits forms what may be called a BCD-left-shift state machine with carry input from the prior digit and carry output to the next digit. When a BCD digit shifts left, the value doubles and the carry input is added to the result. In Section 1.5.2 (page 36) a result greater than 9 was adjusted by adding 6 and passing the carry out to the next digit. The adjustment may also be perceived as subtracting 10_{10} and passing the carry out to the next digit. This results in the following present-state (PS) and next-state (NS) list, where + means plus.

PS	NS	C_{out}
0	$0 + c_{in}$	0
1	$2 + c_{in}$	0
2	$4 + c_{in}$	0
3	$6 + c_{in}$	0
4	$8 + c_{in}$	0
5	$0 + c_{in}$	1
6	$2 + c_{in}$	1
7	$4 + c_{in}$	1
8	$6 + c_{in}$	1
9	$8 + c_{in}$	1

For PS from 0 to 4: $NS = 2PS + c_{in}$ and $C_{out} = 0$

For PS from 5 to 9: $NS = 2PS - 10_{10} + c_{in}$ and $C_{out} = 1$

EXAMPLE 10.3 Binary-to-BCD Conversion

n_{10}	Four BCD Digits				Binary Input String
0	0000	0000	0000	0000	11010111011 (1723_{10})
1	0000	0000	0000	0001	1010111011
3	0000	0000	0000	0011	010111011
6	0000	0000	0000	0110	10111011
13	0000	0000	0000	1101	0111011
13	0000	0000	0001	0011	0111011 (adjusted)
26	0000	0000	0010	0110	111011
53	0000	0000	0100	1101	11011
53	0000	0000	0101	0011	11011 (adjusted)
107	0000	0000	1010	0111	1011
107	0000	0001	0000	0111	1011 (adjusted)
215	0000	0010	0000	1111	011
215	0000	0010	0001	0101	011 (adjusted)
430	0000	0100	0010	1010	11
430	0000	0100	0011	0000	11 (adjusted)
861	0000	1000	0110	0001	1
1723	0000	10000	1100	0011	end (carry-out shown)
1723	0001	0111	0010	0011	end (adjusted)

Shift in the first four bits and adjust if greater than 9.

After each subsequent bit is shifted in, adjust if the lowest four bits are a number greater than 9.

Pins available   (Figure 10.18)

Device	Number of Pins	Dedicated Inputs	Combo Out	Output Registers	Feedback	Output Enable	Product Terms/Output
16R8	20	8	0	8	yes	pin	8 to OR

Pins required	My Design Needs:	PLD Provides:
output enable in	1	1
clock in	1	1
clear in c	1	1
Carry in c_{in}	1	1
First BCD digit q_j outputs	4	4 reg
Carry out to carry in (from first to second digit)	0	internal "wire"
Second BCD digit q_j outputs	4	4 reg
Carry out c_{out}	1	0 (This is a problem.)

The circuit requires one combinational output pin for c_{out}. But the 16R8 has no combinational output pins (neither does the 20R8). We can eliminate the carry-out pin if we make several observations. (The following is an example of PLD design as an art form.)

The 16R8 has eight input pins, and so far only four are allocated. As will be shown, the c_{out} equation is a function of the clear c variable and the four q_j variables. Pins are allocated for these variables. In lieu of passing c_{out} to the next digit, let us pass the four q_j. The revised pin allocations follow.

Pins required (revision 1)	My Design Needs:	PLD Provides:
output enable in (oe)	1	1
clock in (clk)	1	1
clear in (c)	1	1
new BCD digits in (b_j)	4	4
Carry in c_{in}	1	1
First BCD digit q_j outputs	4	4 reg
Carry out to carry in (from first to second digit)	0	internal "wire"
Second BCD digit q_j outputs	4	4 reg

State machine design   The PS and NS list is converted from decimal to binary in Table 10.5 so that we can design a BCD-left-shift state machine using flip-flops. Sixteen states are listed because four bits have 16 possible states. Here is how the next-state decision is made for states 10 through 15.

When power is turned on, the present state may be any state. Therefore the converter must be cleared to zero output. This is why we included the clear input variable c. The state machine may enter states 10 to 15 when power is turned on or when an error occurs (noise may generate the error). We do not care what their next state is, so we choose state zero as their next state. These considerations lead to the next-state table in Table 10.5.

Next the truth table is written to five K maps for q_3^+, q_2^+, q_1^+, q_0^+, and c_{out}. The simplified next-state equations are read from the K maps. (This is more art form.) Intuition tells us the clear variable c' is a common factor because the q_j must be zero when clear c is asserted. We leave the K maps to an exercise.

TABLE 10.5   Binary-to-BCD Serial Converter Truth Table

PS $q_3 q_2 q_1 q_0$	INPUT c	NS $q_3^+ q_2^+ q_1^+ q_0^+$	c_{out}
0 0 0 0	0	0 0 0 c_{in}	0
0 0 0 1	0	0 0 1 c_{in}	0
0 0 1 0	0	0 1 0 c_{in}	0
0 0 1 1	0	0 1 1 c_{in}	0
0 1 0 0	0	1 0 0 c_{in}	0
0 1 0 1	0	0 0 0 c_{in}	1
0 1 1 0	0	0 0 1 c_{in}	1
0 1 1 1	0	0 1 0 c_{in}	1
1 0 0 0	0	0 1 1 c_{in}	1
1 0 0 1	0	1 0 0 c_{in}	1
1 0 1 0	0	0 0 0 0	0
1 0 1 1	0	0 0 0 0	0
1 1 0 0	0	0 0 0 0	0
1 1 0 1	0	0 0 0 0	0
1 1 1 0	0	0 0 0 0	0
1 1 1 1	0	0 0 0 0	0
– – – –	1	0 0 0 0	0

$$q_3^+ = c'q_3'q_2q_1'q_0' + c'q_3q_2'q_1'q_0$$

$$q_2^+ = c'q_3'q_2'q_1 + c'q_3'q_1q_0 + c'q_3q_2'q_1'q_0'$$

$$q_1^+ = c'q_3'q_2'q_0 + c'q_3'q_2q_1q_0' + c'q_3q_2'q_1'q_0'$$

$$q_0^+ = c'c_{in}(q_3' + q_2'q_0')$$

$$c_{out} = c'q_3'q_2(q_0 + q_1) + c'q_3q_2'q_1'$$

Practical circuit implementation Digit$_0$ needs c_{in}. For $j > 0$, digit$_j$ needs $c_{inj} = c_{outj-1}$. A 16R8 package contains two BCD-digit circuits. Unless we take special action, two different 16R8 packages are needed for more than two digits: one for digits 0/1, and another for digits 2/3. Each additional pair of BCD digits requires another 16R8. A desirable feature of any design is to make all BCD-digit circuits identical at the package level.

The sole difference concerns carry input, which appears only in the q_0^+ equation. For any digit, $q_{0j}^+ = c_{in} + c_{outj-1}$. The next state of q_{0j} is c_{in} *or* c_{outj-1}. The c_{outj-1} equation is implemented in the digit$_j$ circuit. This equation needs the q_{3210} from the *prior* digit. Therefore, the $q_3q_2q_1q_0$ from the prior digit are connected to the BCD digits in (b_j) pins. Then the c_{outj-1} equation can be rewritten as follows:

$$c_{outj-1} = c'b_3'b_2(b_0 + b_1) + c'b_3b_2'b_0'$$

This is satisfactory if users are told to ground the b_j pins when the chip is used digit$_0$ and digit$_1$, and to ground the c_{in} pin for all other digit pairs. The revised q_0^+ equation and the other three q_j^+ follow, where q_{0j} and q_{1j} represent the two BCD digits.

FIGURE 10.18a
Binary-to-BCD converter using PAL 16R8

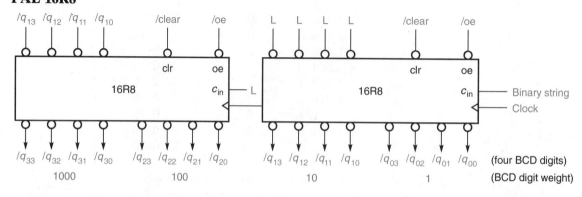

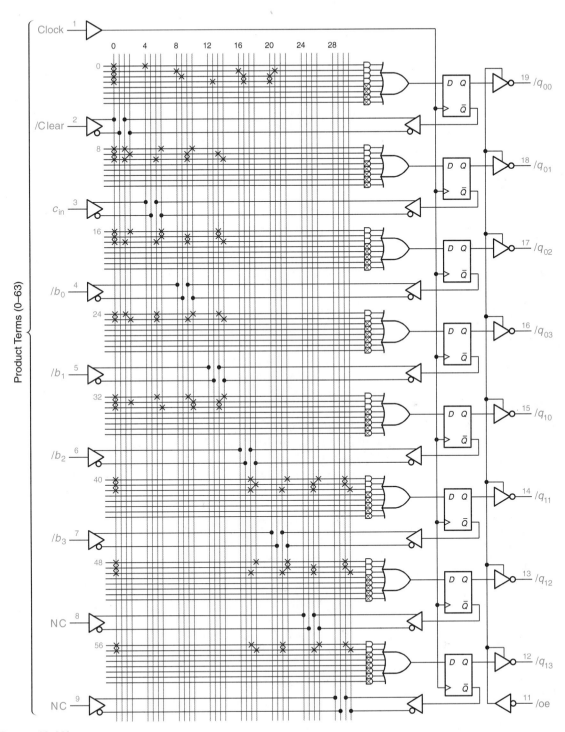

FIGURE 10.18b

$$q_{00}^+ = c'c_{in} + c_{out\ j-1}$$
$$= c'c_{in} + c'b_3'b_2b_0 + c'b_3'b_2b_1 + c'b_3b_2'b_0'$$
$$q_{01}^+ = c'q_{03}'q_{02}'q_{00} + c'q_{03}'q_{02}q_{01}q_{00}' + c'q_{03}q_{02}'q_{01}'q_{00}'$$
$$q_{02}^+ = c'q_{03}'q_{02}'q_{01} + c'q_{03}'q_{01}q_{00} + c'q_{03}q_{02}'q_{01}'q_{00}'$$
$$q_{03}^+ = c'q_{03}'q_{02}q_{01}'q_{00}' + c'q_{03}q_{02}'q_{01}'q_{00}$$

The carry into the second BCD digit is simply the carry out of the prior digit.

$$q_{10}^+ = c_{out\ j-1} = c'q_{03}'q_{02}q_{00} + c'q_{03}'q_{02}q_{01} + c'q_{03}q_{02}'q_{01}'$$
$$q_{11}^+ = c'q_{13}'q_{12}'q_{10} + c'q_{13}'q_{12}q_{11}q_{10}' + c'q_{13}q_{12}'q_{11}'q_{10}'$$
$$q_{12}^+ = c'q_{13}'q_{12}'q_{11} + c'q_{13}'q_{11}q_{10} + c'q_{13}q_{12}'q_{11}'q_{10}'$$
$$q_{13}^+ = c'q_{13}'q_{12}q_{11}'q_{10}' + c'q_{13}q_{12}'q_{11}'q_{10}$$

SUMMARY

Programmable logic devices (PLDs) fit in between standard logic and large-scale integration in the hierarchy of integrated circuits. PLDs allow a designer to replace a multitude of standard logic parts in a design, and allow for circuit changes without changes to the circuit's printed circuit board; the designer replaces the existing PLD with a reprogrammed PLD.

Programmable Logic

Programmable logic devices are integrated circuits whose function(s) are field-programmable. A circuit is "wired" on a PLD chip by blowing fuses or storing 1, 0 bit patterns. A PLD has a much higher level of integration than an SSI or MSI chip.

Programmable Logic Structures

The PLD schematic has a "regular" repetitive AND/OR structure based on the fact that any switching function is realized as a sum of product terms. The number of terms is seven or eight in the more popular devices. An attractive PLD feature is the multitude of AND gates (e.g., 64), each with a very large number of inputs (e.g., 32). The design of state machines is accommodated by PLDs incorporating D flip-flops. The AND/OR is replaced by AND/XOR for counting-type state machines.

Programmable Logic Devices

The choice of a specific device is influenced by the equations of the function to be implemented. Widely used PLDs are listed in Table 10.2.

Programmable Logic Languages

Effective use of PLDs requires learning one or more of the PLD programming languages available commercially.

Combinational Circuit Designs

Design using PLDs has a high art-form content. This is especially true when one is trying to "fit" equations into a PLD. A twelve-to-one multiplexer design shows how nonstandard logic is implemented. The advantage of the large number of AND gate inputs is emphasized when they are used to implement very fast address decoding. The ease with which complex latch designs are fitted into a PLD is demonstrated.

Sequential Circuit Designs

The state machine design shows how a registered PLD with AND/OR circuits implements all the necessary circuits. The waveform generator example shows how counter equations are modified to suit the XOR circuit. The binary-to-BCD converter deals with some of the PLD pin limitation problems and other design issues.

REFERENCES

Alford, R. C. 1989. *Programmable Logic Designer's Guide*. Indianapolis: H. W. Sams.

Bolton, M. 1989. *Digital Systems Design with Programmable Logic*. Reading, MA: Addison-Wesley.

Broesch, J. D. 1991. *Practical Programmable Circuits: A Guide to PLDs, State Machines, and Microcontrollers*. Orlando, FL: Academic Press.

Burton, V. L. 1990. *The Programmable Logic Device Handbook*. Blue Ridge Summit, PA: TAB.

Couleur, J. F. 1958. BIDEC—A Binary-to-Decimal or Decimal-to-Binary Converter. *IEEE Transactions Electronic Computers* EC-7(6) (December): 313–316.

Lehman, C. L. 1989. *Programmable Logic Design Tools*. Hillsboro, OR: Orcad Systems.

PAL Device Data Book. 1990. Sunnyvale, CA: Advanced Micro Devices.

Programmable Array Logic Handbook. 1988. Sunnyvale, CA: Advanced Micro Devices.

Programmable Logic Data Book. 1990. Dallas: Texas Instruments.

The ProLogic Compiler. 1991. Dallas: Texas Instruments.

PROBLEMS

10.1 *Clocked flip-flops.* Design D, JK, T, and SR flip-flops, each with complementary outputs (Figure P10.1). Add asynchronous clear (*c*) and preset (*p*) inputs to each flip-flop. Let the three inputs clock, clear, and preset be common to all flip-flops. Use a 16R8. *Hint:* use two PLA flip-flops to generate complementary outputs, as shown in the following truth tables.

FIGURE P10.1

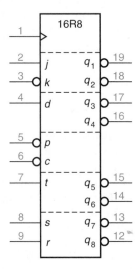

INPUTS				OUTPUTS		INPUTS				OUTPUTS	
$--$	$--$	$++$	$--$	$++$	$++$	$--$	$--$	$++$	$++$	$++$	$++$
p	c	j	k	q_1^+	q_2^+	p	c	s	r	q_7^+	q_8^+
H	L	$-$	$-$	L	H	H	L	$-$	$-$	L	H
L	H	$-$	$-$	H	L	L	H	$-$	$-$	H	L
H	H	L	H	q_1	q_2'	H	H	L	L	q_7	q_8'
H	H	L	L	L	H	H	H	L	H	L	H
H	H	H	H	H	L	H	H	H	L	H	L
H	H	H	L	q_1'	q_2	H	H	H	H	H	L

INPUTS			OUTPUTS		INPUTS			OUTPUTS	
$--$	$--$	$++$	$++$	$++$	$--$	$--$	$++$	$++$	$++$
p	c	d	q_3^+	q_4^+	p	c	t	q_5^+	q_6^+
H	L	$-$	L	H	H	L	$-$	L	H
L	H	$-$	H	L	L	H	$-$	H	L
H	H	L	L	H	H	H	L	q_5	q_6'
H	H	H	H	L	H	H	H	q_5'	q_6

10.2 Design the decoder in Problem 9.14 (Figure P10.2). Use an 18N8.

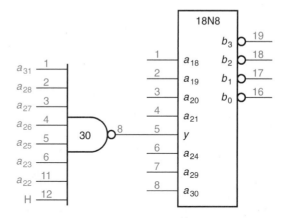

10.3 Design an equivalent to one-half of a 139 decoder, with the following difference (Figure P10.3): replace the 139 enable g with $g = g_{0H}g_{1L} + g_{2L}$. Use a 16L8.

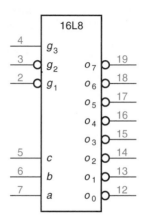

10.4. Design an equivalent to the 138 decoder (Figure P10.4). Use a 16L8.

10.5 Design an equivalent to one-half of a 153 multiplexer, with the following difference (Figure P10.5): replace each data input d_j with $d_j = p_{jH}q_{jH} + r_{jH}s_{jH}$. Use a 20L8.

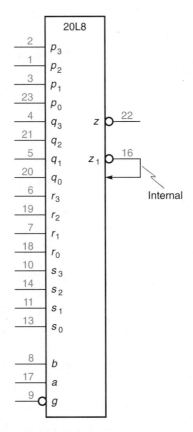

10.6 Design an excess-3-Gray-code-g_{3210}–to–decimal-d_{3210} convertor (Figure P10.6). The g_j are active high, and the d_j are active low. Use a 16L8. Here are the codes in hex:

Input g_{jH} 2 6 7 5 4 C D F E A B 9 8 0 1 3

Output d_{jH} 0 1 2 3 4 5 6 7 8 9 0 0 0 0 0 0

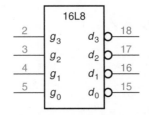

10.7 Design a circuit implementing the following truth table (Figure P10.7). Use a 16L8.

e_1	$i_{76543210}$	a_{210}	g_s	e_0	←assignments
H	– – – – – – – –	HHH	H	H	
L	HHHHHHHH	HHH	H	L	
L	L – – – – – – –	LLL	L	H	
L	HL – – – – – –	LLH	L	H	
L	HHL – – – – –	LHL	L	H	
L	HHHL – – – –	LHH	L	H	
L	HHHHL – – –	HLL	L	H	
L	HHHHHL – –	HLH	L	H	
L	HHHHHHL –	HHL	L	H	
L	HHHHHHHL	HHH	L	H	

FIGURE P10.7

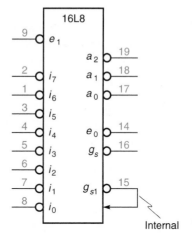

10.8 Design a cascadable four-bit comparator circuit with output $a = b$ (Figure P10.8). Inputs are cascading input e_{in}, bits a_j, and bits b_j ($j = 3, 2, 1, 0$). Use a 16L8. *Hint:* use the not-equals function.

FIGURE P10.8

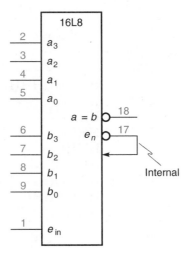

10.9 Design a cascadable four-bit comparator circuit with output $a > b$ (Figure P10.9). Inputs are cascading input g_{in}, bits a_j, and bits b_j ($j = 3, 2, 1, 0$). Use a 16L8.

FIGURE P10.9

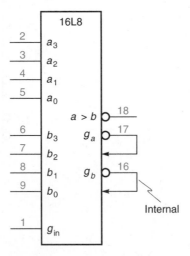

10.10 Design a four-bit-Gray-code–to–four-bit-binary-code converter (Figure P10.10). Use a 16L8.

Input g_j 0 1 3 2 6 7 5 4 C D F E A B 9 8

Output b_j 0 1 2 3 4 5 6 7 8 9 A B C D E F

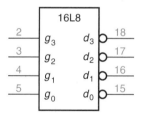

10.11 Design a circuit (Figure P10.11) implementing the following truth table. Use a 16R4.

clr $\overline{}$	s_{10} ++	Function	q_3^+	q_2^+	q_1^+	q_0^+
			⎯⎯⎯⎯⎯⎯⎯ ←assignments			←after one clock pulse
H	HH	load	a_3	a_2	a_1	a_0
H	HL	shift right	p	q_3	q_2	q_1
H	LH	shift left	q_2	q_1	q_0	r
H	LL	hold	q_3	q_2	q_1	q_0
L	––	clear	H	H	H	H

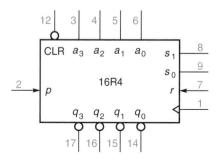

10.12 Design a circuit (Figure P10.12) implementing the following truth table. Use a 20R8.

g load $\overline{}$	h hold $\overline{}$	Function	$q_7^+ q_6^+ q_5^+ q_4^+ q_3^+ q_2^+ q_1^+ q_0^+$ $\overline{}$ ←assignments ←after one clock pulse
L	—	load	a_7 a_6 a_5 a_4 a_3 a_2 a_1 a_0 ←++ for a_j
H	L	hold	q_7 q_6 q_5 q_4 q_3 q_2 q_1 q_0
H	H	shift left	q_6 q_5 q_4 q_3 q_2 q_1 q_0 r ←++ for r

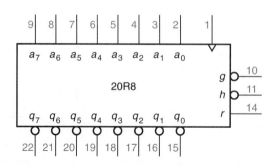

10.13 Design a four-bit *binary* up counter with count control p and r_{co} output (Figure P10.13). Use a 20X4.

FIGURE P10.13

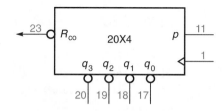

10.14 Design a four-bit *decade* up counter with count control p and r_{co} output (Figure P10.14). Use a 20X4.

FIGURE P10.14

10.15 Design a multifunction four-bit register (Figure P10.15). Use a 20X4.

INPUTS					OUTPUTS			
$\overline{oe}$	$\overline{clk}$	$\overline{clr}$	$\overline{pre}$		$\overline{load}$	d_j^{++}	$\overline{q_j^+}$	FUNCTION
H	–	–	–		–	–		high-z
L	↑	L	–		–	–	H	clear
L	↑	H	L		–	–	L	preset
L	↑	H	H		L	d_j	d_j	load
L	↑	H	H		H	–	q_j	hold

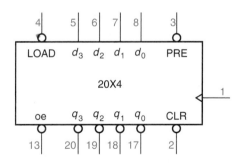

10.16 Design a priority-interrupt encoder that prioritizes eight lines i_0 to i_7 (Figure P10.16). The encoder produces binary 111 for highest priority, i_7, through 000 for lowest priority, i_0. Output f goes low when any i_j line goes high. The outputs are registered. Use a 16R4.

	INPUTS				OUTPUTS	
$\overline{\text{oe}}$	clk	$^{++}$ $i_{76543210}$	FUNCTION		$\overline{f}$	$\overline{q_{210}}$
H	–	– – – – – – – –	disabled high-z			
L	↑	LLLLLLLL	—		H	HHH
L	↑	– – – – – – – H	interrupt$_0$		L	HHH
L	↑	– – – – – – HL	interrupt$_1$		L	HHL
L	↑	– – – – – HLL	interrupt$_2$		L	HLH
L	↑	– – – – HLLL	interrupt$_3$		L	HLL
L	↑	– – – HLLLL	interrupt$_4$		L	LHH
L	↑	– – HLLLLL	interrupt$_5$		L	LHL
L	↑	– HLLLLLL	interrupt$_6$		L	LLH
L	↑	HLLLLLLL	interrupt$_7$		L	LLL

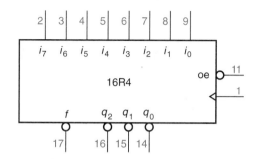

10.17 Design a state machine and output circuits (Figure P10.17b) for the ASM (Figure P10.17a). Use a 16R4.

FIGURE P10.17

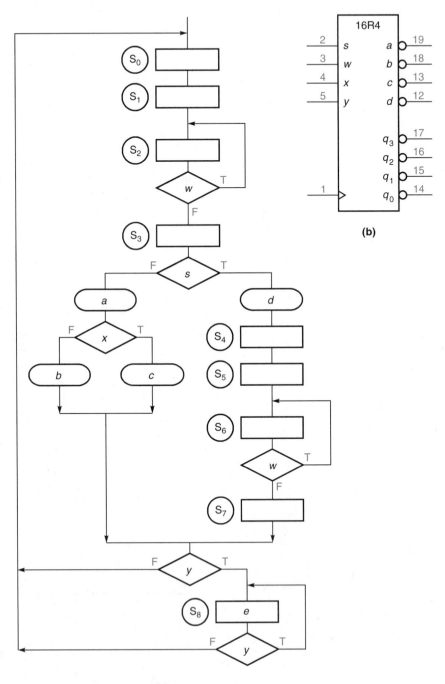

(b)

(a)

10.18. Design a stepper motor controller (Figure P10.18). Each step turns the shaft 7.5°. Control and step truth tables follow. Use a 16R4.

Clock	g	a	b	Function
–	0	–	–	Hold motor in current position
↑	1	1	–	Set outputs to step 1 levels
↑	1	0	0	Step motor clockwise
↑	1	0	1	Step motor counterclockwise

Step	q_3	q_2	q_1	q_0	
1	0	1	0	1	clockwise 1234123 . . .
2	1	0	0	1	counterclockwise 14321432 . . .
3	1	0	1	0	
4	0	1	1	0	
1	0	1	0	1	

FIGURE P10.18

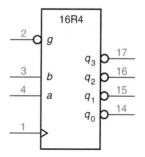

11 ASYNCHRONOUS SEQUENTIAL CIRCUITS

We adapt our knowledge of combinational logic design and synchronous sequential circuit design to analysis and synthesis of asynchronous circuits.

Asynchronous sequential circuit outputs respond to input changes after propagation time delays elapse. There is no clock edge to "wait for," which is basically why these circuits are faster than synchronous circuits. We begin with analysis of well-known asynchronous sequential circuits (RS latch, oscillator, and D flip-flop) in order to establish clearly the nature of the problems treated here and to introduce asynchronous sequential circuit terminology. An asynchronous synthesis method is presented and applied to design a switch debouncer, a D latch, and an edge-triggered flip-flop. The important problem of critical races is discussed, and a method for eliminating such races is presented.

INTRODUCTION

Asynchronous sequential circuits are combinational networks with internal feedback of outputs. This internal feedback of outputs creates latches that provide the memory function required by a sequential circuit. Edge-triggered flip-flops are not used. Asynchronous sequential circuits have the property that outputs respond to changes in one or more inputs as soon as input to output propagation delay times expire.

Designing reliable asynchronous sequential circuits is different from the designing of synchronous circuits because the delays are not controlled by the designer. Given the differences, why bother with asynchronous sequential circuits? First, asynchronous sequential circuits are faster than synchronous sequential circuits. Second, since they exist, we should know about them. Third, since all flip-flops are asynchronous sequential circuits, we should be able to design flip-flops if only to understand them better.

If q_1, q_0 are state variables then the internal state is q_1q_0.

We begin with the analysis of well-known asynchronous sequential circuits; in the process, we introduce the terminology of asynchronous sequential circuits. These definitions are usefully applied to asynchronous sequential functions.

If x, y are input variables then the total state is q_1q_0xy.

Information on past input states is stored in *internal states*. The combinations of input states and internal states are called *total states*. The present output and the next state are functions of the present total state. This parallels synchronous sequential circuit theory (Chapters 6 and 7).

An asynchronous sequential circuit is said to be in a *stable total state* when the next internal state is the same as the present internal state. Stable refers to the fact that nothing is scheduled to happen until an input change occurs. (The clock sets the schedule in a synchronous sequential circuit.) Unstable means the present-state/next-state inequality, with no other input changes, forces action that will equalize them.

Starting from an initial internal state, a sequential function relates output sequences of values to input sequences of values. A sequential function is described by the total states, the output function, and the next-state function.

The design process has two restrictions: (a) an asynchronous sequential circuit must be stable before an input changes, and (b) only one input at a time may change. Designs based on these restrictions are known as *fundamental mode designs*.

11.1 Delays and Reliability

Delays affect reliable asynchronous sequential circuit operation. An example based on an ASM chart illustrates why. The ASM chart for a combinational circuit is straightforward once two facts are recognized: all outputs are conditional, and there is only one state (Figure 11.1). Assume the delays from the x, y inputs (Figures 11.2 and 11.3) to x_{out} and y_{out} are t_{px} and t_{py}, respectively. The signals x_{out} and y_{out} represent the delayed impact of the inputs x and y in the combinational circuit (see Section 5.1.3, page 227).

When t_{px} and t_{py} are equal, the outputs are reliable (Figure 11.2). When t_{px} does not equal t_{py}, unwanted outputs a_2, a_1 appear (Figure 11.3). In this sense the circuit is not reliable, which is why asynchronous sequential circuit design requires extra care.

FIGURE 11.1

Combinational network ASM chart

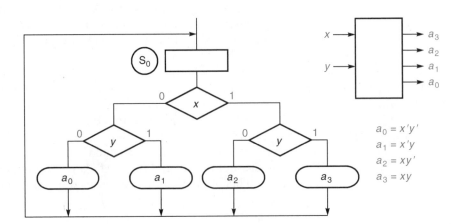

$$a_0 = x'y'$$
$$a_1 = x'y$$
$$a_2 = xy'$$
$$a_3 = xy$$

FIGURE 11.2
Equal delays in a combinational network

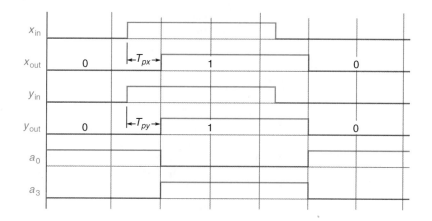

In a synchronous system, the clock decides when action occurs. A clock can ignore unwanted outputs like a_2, a_1 (Figure 11.3), which is why synchronous systems are reliable when inputs meet flip-flop setup and hold time requirements (Section 6.5, page 307).

11.2 Asynchronous Analysis Method

Analysis of the RS latch and the oscillator in what follows is based on the given circuits. The key is cutting the feedback lines and analyzing the resulting combinational circuits. In contrast, the subsequent D

FIGURE 11.3
Unequal delays in a combinational network

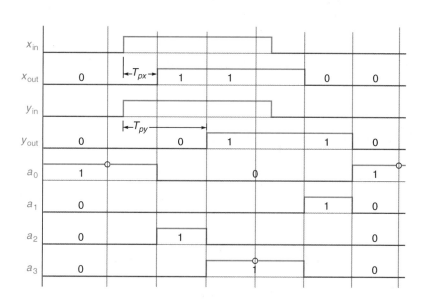

flip-flop analysis proceeds from the given equations. These are not different methods. The apparent difference arises from the choice of starting point in the analytical method. The RS Latch and oscillator analyses start from Step 1 of the method. The D flip-flop analysis starts from Step 2. The method has six major steps:

1. Identify the feedback lines and cut them.
2. Calculate the equations for the next-state outputs q_j^+.
3. Plot the next-state K maps for the q_j^+.
4. Merge the q_j^+ K maps into one composite next-state K map, with state as one coordinate.
5. Identify stable states, and circle them to form the transition map from the composite K map.
6. Make timing diagrams for sequences of input variable transitions.

11.2.1   RS Latch Analysis

Analysis means calculating how outputs of a given circuit respond to input changes. Some inputs are internally generated. They are outputs that have been fed back to provide memory capability (Section 6.1, page 295).

STEP 1   **Identify the feedback lines and cut them.** We note that the latch circuit in Figure 11.4 was a combinational circuit prior to the addition of feedback lines. So we cut the feedback lines to restore the combinational circuit because we already know how to analyze a circuit without feedback. One side of the cut connects to the output; the other side connects to the input. The input side is the feedback signal source. If we think of the output side as the next-state output q^+, then the input side is the present-state input q (Figure 11.4). The present-state inputs q, s, and r determine the next state q^+.

FIGURE 11.4
RS latch circuit with cut feedback line

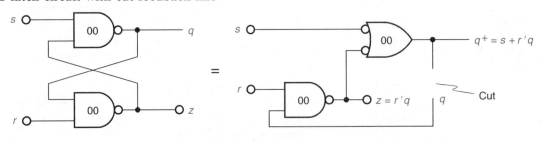

FIGURE 11.5
RS latch K map for
$q^+ = s + r'q$

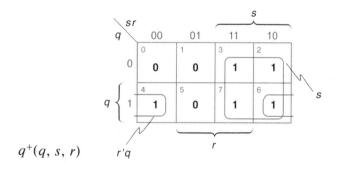

$$q^+(q, s, r)$$

STEP 2   **Calculate the equations for the next-state outputs q_j^+.** Analysis of the RS latch circuit with cut feedback line is straightforward. We calculate the circuit's equation for q^+. In this case there is only one q_j^+ output.

$$q^+ = s + r'q$$

STEP 3   **Plot the next-state K maps for the q_j^+.** We write the RS latch equation to the q^+ K map (Figure 11.5).

STEP 4   **Merge the q_j^+ K maps into one composite next-state K map, with state as one coordinate.** In this case there is only one q_j^+ output. Therefore no merger is necessary, and the composite K map is the same as the K map in Figure 11.5.

STEP 5   **Identify stable states, and circle them to form the transition map from the composite K map.** We observe that if we reconnect the cut feedback wire when $q^+ = q$, the circuit does not know we did that. From the K map we see that $q^+ = q$ in states 0, 1, 4, 6, and 7. These are the stable states. Nothing happens when the feedback line is reconnected. (The next state equals the present state.) We circle the stable states to create the transition map (Figure 11.6). In a moment it will be clear why we say transition.

A state is stable when $q^+ = q$. State 7 is stable because $q^+ = 1$ and $q = 1$.

If q^+ does not equal q when we restore the cut feedback line (states 2, 3, and 5), the circuit reacts to the conflicts created. What happens? The transition map provides the answer. First we note that the reconnection changes only the q input, not inputs s and r (Figure 11.4).

If $q = H$ and $q^+ = L$ then connecting q to q^+ creates a transient that forces a transition.

Let's suppose the latch circuit with the cut feedback line is in state 3, where $qsr = 011$. The map shows for state 3 that $q^+ = 1$, and $q = 0$. They are not equal when we restore the feedback line. The reconnection forces q to transition instantly from 0 to 1 because q^+ is 1. On the map (Figure 11.6) with inputs s, r fixed at $sr = 11$, the only possible

FIGURE 11.6
RS latch transition map for $q^+ = s + r'q$

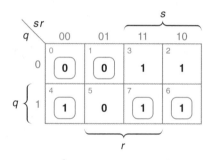

move is in the vertical q direction from state 3 to state 7 as q changes from 0 to 1.

State 7 is a stable state because $q^+ = q$; the circuit remains there. Clearly, state 3 is unstable. Since a circuit cannot remain in an unstable state (why?), it transitions ($\rightarrow$), to stable state 7 in this case. The transition map tells you what happens to unstable states: $s_2 \rightarrow s_6$, $s_5 \rightarrow s_1$, and $s_3 \rightarrow s_7$.

A state is unstable when $q^+ \neq q$. State 3 is unstable because $q^+ = 1$ and $q = 0$.

EXERCISE 11.1   Refer to Figure 11.6. Derive the q output for the following sequence of sr input states:

$$s \quad 0\ 1\ 1\ 0\ 1\ 1\ 0$$
$$r \quad 0\ 0\ 1\ 1\ 1\ 0\ 0$$
$$Answer:\ 0\ 1\ 1\ 0\ 1\ 1\ 1 \qquad \blacksquare$$

STEP 6   **Make timing diagrams for sequences of input variable transitions.** We analyze the consequences of sequential changes in s and r to ascertain the circuit function. The transition map yields the sequence of states when inputs s or r change. Let's suppose that the latch is in s_4 (a stable set state with $q = 1$) and that we want to reset the latch by making r True (L) for an instant. Will the latch reset, and what does "an instant" mean?

Starting from s_4 ($qsr = 100$), changing r from 0 to 1 (Figure 11.7a) forces a transition to s_5 ($qsr = 101$), which is unstable. This forces the circuit to seek a stable state by moving from s_5 to s_1. This vertical move is the only move possible, because s and r are fixed whereas q can change. This is consistent with the equation for q^+ ($q^+ = 0$ when $qsr = 101$). And when q^+ changes from 1 to 0, so does q. However, the map does not tell us what an instant means. For this we need a timing diagram (Figure 11.7b). The timing diagram also shows when moves from square to square are actually made.

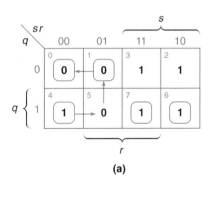

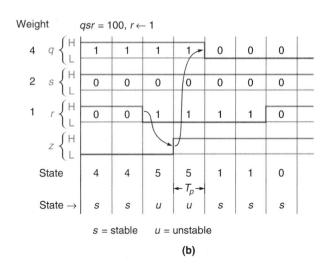

s = stable u = unstable

(b)

FIGURE 11.7
RS latch s_4, s_5, s_1, s_0 sequence of states

An input change changes total state. A reliable change duration must exceed the longest path delay.

(a) A sequence of input states that resets the latch   The RS latch has four nodes, s, r, z, and q (Figure 11.4), so we need four waveforms, starting from state 4 (Figure 11.7b). If we assume the input-to-output propagation time is t_p, then a transition in r ($s_4 \rightarrow s_5$) causes a z transition t_p seconds later. After another t_p seconds, the z transition forces a q transition from 1 to 0 ($s_5 \rightarrow s_1$), which is stable as long as r remains low. When r transitions back to 0, the circuit moves to s_0 ($s_1 \rightarrow s_0$).

FIGURE 11.8
RS latch s_0, s_2, s_6, s_4 sequence of states

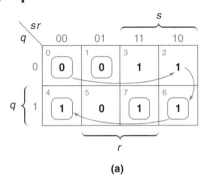

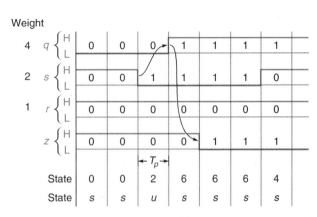

(b)

The timing diagram (Figure 11.7b) is marked off in t_p time units. In the diagram, the r pulse is true for an instant equal to $4t_p$. The instant can be $3t_p$ but not $2t_p$, because a q transition needs to propagate through the first gate during the third t_p after r transitions to 1 in order to hold z high before and after r returns high. Pulsing r resets the latch.

(b) A sequence of input states that sets the latch   This time the transition sequence of states is s_0, s_2, s_6, s_4 (Figure 11.8). Pulsing s sets the latch.

11.2.2   Oscillator Analysis

An oscillator has no stable states. By definition, the output continuously alternates between H and L voltage values. Feedback around an inverter (Figure 11.9) should produce oscillations.

STEP 1   **Identify the feedback lines and cut them.**

STEP 2   **Calculate the equations for the next-state outputs $q_j{}^+$.** After we cut the feedback line in Figure 11.9, we find that

$$q^+ = q' + g'$$

STEP 3   **Plot the next-state K maps for the $q_j{}^+$.**

STEP 4   **Merge the $q_j{}^+$ K maps into one composite next-state K map, with state as one coordinate.**

STEP 5   **Identify stable states, and circle them to form the transition map from the composite K map.** When $g = 0$ (L), the output q^+ is forced high.

A circuit will oscillate for fixed inputs when all states in a transition map column are unstable.

The transition map (Figure 11.10) shows this to be the only stable state $(q^+ = q)$. In the $g = 1$ column all the states are unstable. When $q = 0$, then $q^+ = 1$ and reconnecting the feedback loop creates a conflict. When $q = 1$, then $q^+ = 0$. In both cases, reconnecting the feedback loop creates a conflict because q and q^+ are not equal.

FIGURE 11.9
Oscillator circuit

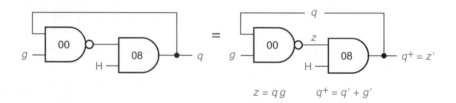

$$z = qg \qquad q^+ = q' + g'$$

FIGURE 11.10
Oscillator transition map

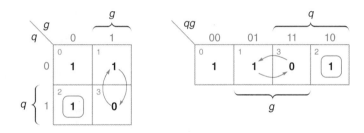

STEP 6 **Make timing diagrams for sequences of input variable transitions.** The timing diagram (Figure 11.11) shows that the circuit responds to g's transitioning to H by oscillating. Figure 11.12 illustrates a practical oscillator circuit. Observe that the waveform is almost a pure sinusoidal wave. The waveform is not a squarewave because the harmonics of the fundamental frequency are not supported by the circuit. (The finite oscilloscope 100-MHz bandwidth also attenuates the harmonics.) Anyone who has studied Fourier series knows that a squarewave is a weighted sum of all the odd harmonics of the fundamental sinusoid (see any text on differential and integral calculus).

EXERCISE 11.2 Refer to Figure 11.10. Derive the q output for the following sequence of g input states:

$$g \quad 0\ 0\ 1\ 1\ 1\ 1\ 1\ 1\ 0\ 0$$
$$\textit{Answer:} \quad 1\ 1\ 0\ 1\ 0\ 1\ 0\ 1\ 1\ 1 \qquad \blacksquare$$

11.2.3 Edge-Triggered D Flip-Flop Analysis

The clocked D latch (Figure 11.13) is derived from the RS latch (see Figure 11.4). A D flip-flop is assembled by using two clocked D latches

FIGURE 11.11
Oscillator waveforms

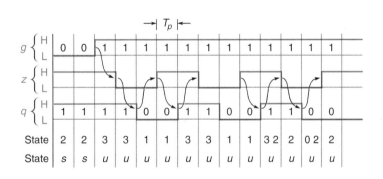

FIGURE 11.12

FIGURE 11.12
Oscillator in practice

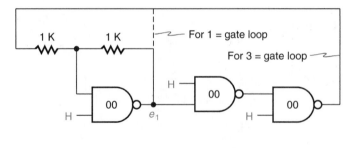

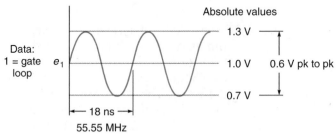

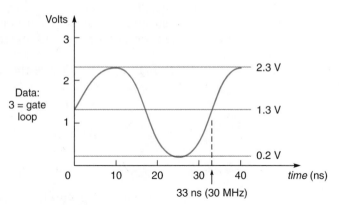

FIGURE 11.13
Clocked D latch

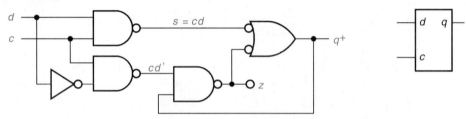

FIGURE 11.14

Edge-triggered D flip-flop

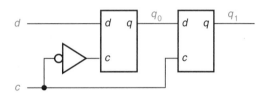

(Figure 11.14). The clocked D latch next-state equation is readily taken from the circuit:

$$q^+ = s + r'q = cd + c'q$$

When c is 0 (L), the latch holds the present q value ($q^+ = q$). When c is 1 (H), the latch takes the value of d ($q^+ = d$). The latch is controlled by the levels of clock c.

One problem with the D latch (see Figure 11.13) is that, if d fluctuates when c is high, so does q^+. In other words, the output can change during the half clock period when c is H and $q^+ = d$. The latch is transparent, as we know from Section 6.2 (page 297). On the other hand, the output of a second latch (call it q_1) cannot change if the first latch's q_0 output is connected to the second latch's d input, and if the first latch is in the hold mode during this half clock period. In this way the idea of using a second latch surfaces (Figure 11.14). Here we derive the equations:

Output follows the input when a circuit is transparent.

$$q^+ = cd + c'q$$

$$q_0^+ = c_0 d_0 + c_0' q_0 \qquad q_1^+ = c_1 d_1 + c_1' q_1$$

We let

$$c_0 = c' \qquad c_1 = c \qquad d_0 = d \qquad d_1 = q_0$$

Then

$$q_0^+ = c'd + cq_0 \qquad q_1^+ = cq_0 + c'q_1$$

When clock is low ($c = 0$), q_0^+ takes the value of d and q_1^+ holds the value $q1$:

$$q_0^+ = d + 0 \quad \cdot \quad q_1^+ = 0 + q_1$$

When clock switches from low to high ($c = 1$), q_0^+ holds the prior value q_0 and q_1^+ takes the value q_0, which is the input to d_1:

$$q_0^+ = 0 + q_0 \qquad q_1^+ = q_0 + 0$$

If d now changes while clock is high, q_0 and q_1 will not change. Observe that q_1 changes value only when clock goes from low to high

FIGURE 11.15

**Edge-triggered D
flip-flop waveforms**

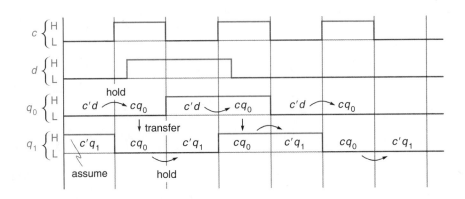

(Figure 11.15). The circuit output responds to the clock edge. This is
the edge-triggered D flip-flop. Now we can proceed with the D Flip-
flop analysis:

STEP 1   **Identify the feedback lines and cut them.** We omit this step because we
start from the equations.

STEP 2   **Calculate the equations for the next-state outputs q_j^+.** The equations
are given:

$$q_0^+ = c'd + cq_0 \qquad q_1^+ = cq_0 + c'q_1$$

STEP 3   **Plot the next-state K maps for the q_j^+.** The present state is q_1q_0, so we
will need a next-state map with q_1q_0 as a coordinate. This means we
need K maps for q_1^+ and q_0^+ with four dimensions, q_1, q_0, c, and d.

FIGURE 11.16

**Edge-triggered D
flip-flop q_0^+ K map**

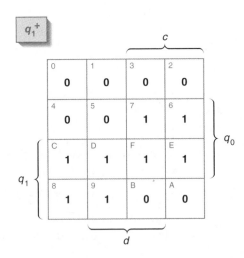

FIGURE 11.17
Edge-triggered D flip-flop q_1^+ K map

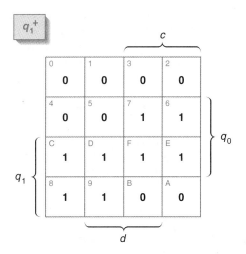

Any function can be mapped onto a higher-dimension map. (Why?) The original functions q_0^+ and q_1^+ map easily (Figures 11.16 and 11.17).

STEP 4   **Merge the q_j^+ K maps into one composite next-state K map, with state as one coordinate.** There are two q_j: q_1 and q_0. Two K maps are merged by making two entries in each square of the composite map. One entry is taken from the q_1 K map, the other from the q_0 map. The map coordinates, q_1, q_0, c, and d, are the same as before (Figure 11.18). Figure 11.18 is a composite map if we ignore the circles. Each row corresponds to one present state of q_1q_0.

FIGURE 11.18
Edge-triggered D flip-flop transition map

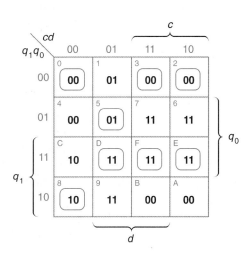

STEP 5   **Identify stable states, and circle them to form the transition map from the composite K map.** The next state represented by $q_1 q_0$ is found in each transition map square (Figure 11.18). The present state for each row is listed at the left of the transition map.

The next state of any present state is stable when the two states are equal. The next states in those squares are circled. The present state is unstable when the next state is different (circles are ignored).

STEP 6   **Make timing diagrams for sequences of input variable transitions.** From Figure 11.15 we deduce the D flip-flop defining equation, $q_1^+ = d$. (If we want to derive this result directly from the equations, we need to eliminate q_0 from the q_1^+ equation. We are stymied because we only have an equation for q_0^+, not for q_0.)

EXERCISE 11.3   Refer to Figure 11.18. Derive the q_1 output for the following sequence of cd input states.

$$c \quad 0\ 0\ 1\ 1\ 0\ 0\ 1\ 1\ 0\ 0\ 1\ 1$$
$$d \quad 1\ 1\ 1\ 0\ 0\ 0\ 0\ 1\ 1\ 1\ 1\ 1$$
$$Answer:\ 0\ 0\ 1\ 1\ 1\ 1\ 0\ 0\ 0\ 0\ 1\ 1 \qquad \blacksquare$$

11.3   Stable and Unstable States

Timing diagrams show that a transient is in progress when the circuit is in an unstable state (Figures 11.7, 11.8, and 11.11). The arrows in the figures emphasize the cause-and-effect relationships. The transition diagram predicts the sequence of states if the time duration of the input change exceeds a lower bound of several t_p propagation time units. This is why the input changes in Figures 11.7 and 11.8 were sustained for more than several t_p (also see Problems 11.1 and 11.2). Perhaps this shows why we need the fundamental mode restriction that a circuit be quiescent when an input changes.

11.4   Asynchronous Synthesis Method

We need a method to show us one way to synthesize asynchronous circuits. The complex synthesis process is assisted by an organized method. The significance of each step in the method outlined next becomes clear after the examples that follow are studied. A practical method requires two assumptions.

a. The system is quiescent prior to any input change.
b. Only one input changes at any instant.

The method for asynchronous synthesis has eight steps:

1. Write the specification.
2. Make a primitive flow table.
3. Make a reduce flow table.
4. Convert the reduced flow table into an unassigned-state map.
5. List the state transitions. Make an adjacency diagram.
6. Assign states, and make a next-state map.
7. Make next-state and output K maps.
8. Translate the K maps into equations, and design the circuits.

In Section 11.5.1 you will find a detailed explanation showing how to implement Steps 4, 5, and 6 for complex problems.

The following switch debouncer example provides a detailed explanation of each step.

11.4.1 Switch Debouncer Synthesis

In earlier chapters we were told to cross-couple two gates when we wanted a debounder circuit. Here is how one is designed.

STEP 1   **Write the debouncer specification.** A break-before-make switch breaks clean when it departs from a terminal and bounces when it makes (hits) the destination terminal. The timing diagram (Figure 11.19) specifies that debouncer circuit output q (in Figure 11.19) does not

FIGURE 11.19
Switch debouncer specification

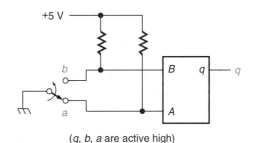

(q, b, a are active high)

State | 2 2 | 3 3 | 5 5 | 7 | 5 | 7 | 5 | 7 | 5 ———→ | 7 7 | 2 2 | 3 | 2 | 3 | 2 | 3 | 2

Each state in a wave-
form set is a stable
state.

have the bounces that occur at the A and B circuit inputs. The typical
specification is a set of waveforms. There is one waveform for each
input and one for each output. A waveform set is in some state between transitions. Waveform transitions are state boundaries. In lieu
of the typical random letter assignments to these states, we prefer to
assign weights to each waveform that are powers of 2. The propensity
for error drops significantly when this is done. Each state is given a
number equal to the sum of the waveform weights in the interval
representing the state. This form of representation makes the method
very tractable as it is executed. (*Note:* The suggested representation
for the states has one exception that is readily accommodated. We
show what the exception is and how to deal with it straightforwardly
in Section 11.6.)

   The key here is to define every possible sequence of states in the
waveform set (Figure 11.19). These are the stable states. In many
cases the physical circuit does not allow some of the possible states to
be entered by the waveform set. This is why states 0, 1, 4, and 6 do not
appear in Figure 11.19. For example, inputs *a* and *b* cannot be at the L
level (states 0 and 4) at the same time unless there is a fault. Such
states are the unstable states—unstable in the sense that if the circuit
enters such a state it cannot remain 'there for physical reasons.

   We assign binary weights 1, 2, and 4 to waveforms *a*, *b*, and *q*,
respectively. We say the waveforms are in some state numbered from
0 to 7 at any point in time. In this case waveforms enter only four
stable states, 2, 3, 5, and 7. Unstable states 0, 1, 4, and 6 are not
entered. State sequences 3, 5 and 7, 2 signal a change in the output
(Figure 11.19). Inputs *b*, *a* enter *ba* input states 1, 3, and 2 but not
state 0.

STEP 2

Primative flow table
horizontal coordinates
are input variables.
Vertical coordinates
are output variables.

Make a debouncer primitive flow table. This table has one column for
each combination of input variables. Input changes are said to be left-
or right-horizontal moves from one input state to another. Two input
variables require four columns (Figure 11.20). There is one row for
each stable state. (The waveforms tell you which states are stable.)
Outputs for each row are specified by the waveforms. The stable-state
number and the row number do not have to coincide.

   We map the timing diagram onto what looks like a three-variable K
map (Figure 11.20). This is the primitive flow table corresponding to
the timing diagram. The column headings are the four possible *ba*
input states. A row represents one stable circuit state, so there are
four rows in this table, corresponding to states 2, 3, 5, and 7. By
definition, every state that appears in the waveforms is a stable state.
The following describes how the rows are filled with entries. All variables are assumed to be active high.

FIGURE 11.20
**Switch debouncer
primitive flow table**

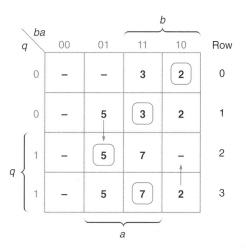

A transition from stable state 2 to stable state 3 is via row 0 unstable state 3. This recognizes that a change causes transients.

State 2 in row 0   Let us suppose that inputs ba are in state 10_2 and that output q is 0 before we throw the switch. This is a stable state we designate with a circled digit 2 ($qba = 010$) in row 0 of the flow table. When the switch arm breaks from A (Figure 11.19), the input state changes to 11_2, which is a horizontal move in the flow table. The circuit is unstable while the circuit responds to the input change. This is why we enter an uncircled 3 in column 11_2 of row 0. Another reason is that only one stable state per row is allowed, by definition. Next we mark the squares in columns 00_2 and 01_2 with dashes to show moves that cannot happen or are not allowed (why?). (Leaving the squares blank may create the perception they were not analyzed.) Unstable state 3 forces a move to row 1.

State 3 in row 1   After the circuit responds, the circuit moves to row 1 stable state 3. The circuit dwells in stable state 3 in row 1 as the arm travels toward B. This may seem to be a strange form of stable state as the arm travels from A to B. Nevertheless, the input is stable and so is the output ($q = 0$). We fill row 1, column 11_2, with a circled 3. When the arm hits B, input state ba changes from 11_2 to 01_2. Since the specification dictates that q switch to 1 when the arm hits B, we enter 5 for unstable state 5 ($qba = 101$) in column 01_2, row 1. While in stable state 3, inputs can change one at a time from 11_2 to 10_2 or 01_2. If the arm happens to fall back to input A, state 10_2 is reentered, which is why a 2 is entered in row 1, column 10_2. The output q is 0 for stable states 2 and 3. We enter a dash in column 00_2 because this column cannot be entered (why?).

State 5 in row 2   After the arm hits B, the circuit responds and q switches from 0 to 1 because that is what has been specified. Stable state 5 ($qba = 101$) is entered in row 2. Bounces after the "make" at terminal B switch the inputs between 01_2 and 11_2. We enter 7 ($qba = 111$) in row 2, column 11_2, and dashes in columns 00_2 and 10_2.

State 7 in row 3   The input change from 01_2 to 11_2 forces a horizontal move in row 2 to the 7 entry. After the circuit responds, the circuit moves to row 3 stable state 7. We enter a stable circled 7 in row 3, column 11_2. If the bouncing continues, the circuit switches between row 2 stable state 5 and row 3 stable state 7 while q remains at value 1. When the switch is thrown back to input A, the output is reset to 0; and as it bounces at A, the circuit switches between row 0 stable state 2 and row 1 stable state 3.

Recapitulation   The table we have just completed is the primitive flow table. Key characteristics of the table are one stable state per row and an output value corresponding to the stable state in the row. Possible unstable states are shown in each row. The unstable states in a row allow transitions to new stable states in different rows to take place when inputs change.

Input changes cause horizontal moves, and feedback changes cause vertical moves on the flow table. Horizontal moves emanate only from stable states. This is consistent with the restriction that no input changes until the circuit is stable (quiescent). Furthermore, there is one column in each row that cannot be reached by perturbing only one variable, because reaching that column requires changing two input variables at the same time. This is consistent with the fundamental mode restriction that only one input changes at a time.

The circled state numbers 2, 3, 5, and 7 of Figure 11.20 are internal states in the primitive flow table. When an input change causes a horizontal move from a stable state, the destination square is an unstable internal state, because transients are ongoing. After transients have expired, a vertical move to a stable internal state in another row takes place.

STEP 3

Mergers eliminate transient states in columns with a stable state.

Make a debouncer reduced flow table. The theoretical basis for reducing the primitive flow table is a significant digression we choose to avoid. If you want to delve into this subject, then research "state reduction by implication table." What follows is an intuitive explanation.

Two present states that have identical outputs and the same next state as the result of an input change are defined as "identical states." One of these two identical states can be eliminated. We apply this concept to pairs of rows because each row has only one stable state.

By definition, identical states have identical outputs. So we select a pair of rows with the same output. Are the next states equal? Any input condition is represented by a column in the row pair. We let the states (in any row pair) in this column be the present state. Any input change selects another column of next states, which have to equal each other. A second input change selects yet another column of next states, which also have to equal each other, and so on. Thus, states in a pair of rows must be equal to each other in every column.

There is one more detail to contend with: a dash means no state number, that is, a "don't care" condition. This implies that a dash can be any state number. The process of combining primitive flow table rows is called *merger*. The rules for merging rows are as follows.

a. Two rows may be merged and thereby replaced by one new row if the pairs of entries in corresponding columns have the same state number, circled (i.e., stable) or not, or at least one dash. In turn the new row may be merged with another row, and so forth.

b. The results of merging all possible entry pairs that are in the same column are:

$$circled\text{-}uncircled \rightarrow circled$$

$$uncircled\text{-}uncircled \rightarrow uncircled$$

$$uncircled\text{-}dash \rightarrow uncircled$$

$$circled\text{-}dash \rightarrow circled$$

$$dash\text{-}dash \rightarrow dash$$

Circled-circled pairs do not occur. (Why?)

In the debouncer's primitive flow table (Figure 11.20), rows 0 and 1 meet the merger criteria, as do rows 2 and 3. Thus the two pairs of rows merge into two single rows. The mergers convert the primitive flow table into the reduced flow table (Figure 11.21a). The reduced flow table emphasizes the fact that output q changes value only on 3, 5 and 7, 2 state transitions.

STEP 4 **Convert the debouncer reduced flow table into an unassigned-state map.** This step is necessary for problems with more than one state variable. We skip this step here because there is only one state variable in this problem.

STEP 5 **List the debouncer state transitions. Make an adjacency diagram.** There being one state variable allows us to skip this step.

FIGURE 11.21
**Switch debouncer
reduced flow table
and output map**

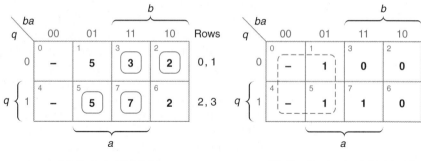

(a) Reduced flow table

(b) Feedback map (which in this
case is also the output map)

STEP 6   **Assign debouncer states, and make a next-state map.** The assignment
is straightforward because there is only one variable q. When there are
several variables, assignment can be a difficult problem.

STEP 7   **Make the debouncer next-state and output K maps.** The reduced flow
table in Figure 11.21a represents in each of its squares a waveform
state from Figure 11.19. The next output q^+ corresponding to each
waveform state is entered into the output q^+ K map (Figure 11.21b).
The output is 0 for stable waveform states 2 and 3, and the output is 1
for stable waveform states 5 and 7. The flow table dashes become
"don't cares" in the q^+ K map. In this case there is no separate output
map because q^+ is the only output.

STEP 8   **Translate the debouncer K maps into equations, and design the circuits.** The two "don't cares" in the debouncer K maps imply four
possible q^+ equations. (There are four ways to assign 0 and 1 to the
two dashes.) There are two cases of interest: both dashes are 0, and
both dashes are 1. Reading the maps yields the equations, and equations are readily transformed into circuits (Figure 11.22).

FIGURE 11.22
Switch debouncer circuits

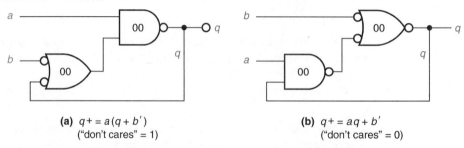

(a) $q+ = a(q + b')$
("don't cares" = 1)

(b) $q+ = aq + b'$
("don't cares" = 0)

EXERCISE 11.4

What is the sequence of input states for the x, c waveform set? What is the sequence of stable states for the q, x, c waveform set?

Answer:

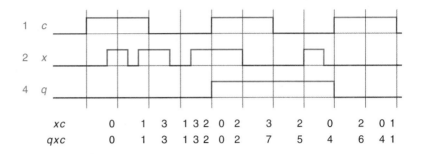

xc	0	1	3	1 3 2	0 2		3	2	0	2	0 1
qxc	0	1	3	1 3 2	0 2		7	5	4	6	4 1

EXERCISE 11.5

Refer to Exercise 11.4. What is the primitive flow table for variables c, x, and q?

Answer:

q \ xc	00	01	11	10
	⓪	1	–	2
	0	①	3	–
0	–	1	③	2
	0	–	7	②
	④	1	–	6
	4	⑤	7	–
1	–	5	⑦	6
	4	–	7	⑥

Columns 01, 11, 10 span x; columns 01 and 11 span c.

EXERCISE 11.6   Refer to Exercise 11.5. Derive the reduced flow table and make a set of state assignments.

Answer:

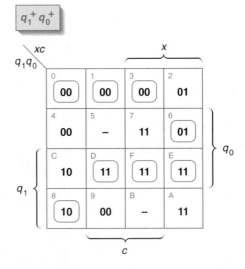

EXERCISE 11.7   Refer to Exercise 11.6. Derive the next-state and output K maps.

Answer:

$q_1^+ q_0^+$

EXERCISE 11.8   Refer to Exercise 11.7. Derive the next-state K maps for q_1^+ and q_0^+. Derive the equations for q_1^+ and q_0^+.

Answer:

$$q_1^+ = cq_0 + (x + c')q_1 + q_1q_0 \qquad q_0^+ = cq_0 + xc' + xq_0$$

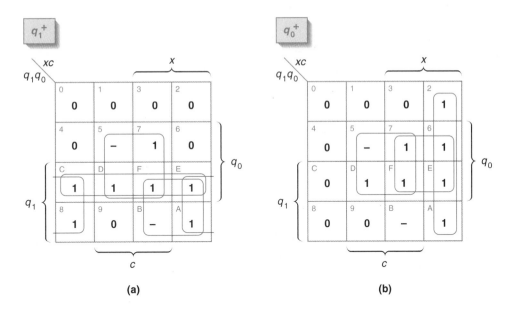

(a)   (b)

(*Note:* the $q_1 q_0$ and $x q_0$ terms eliminate hazards.)   ∎

11.4.2 D Latch Synthesis

A specification shows desired behavior (stable states). Perturbations show misbehavior.

A latch is transparent or holds the present value. These two modes of operation are controlled by an input sensitive to high or low level. The input c can be at either level for any length of time. Although the c input often is called "clock," no clock period is involved, so don't let the clock label mislead you.

STEP 1   **Write the specification.** We want the latch to take the value at the d input when clock is high and to capture the value at the instant clock goes low, and while clock is low to hold the captured value. Three waveforms define latch behavior: clock, d input, and q output (**Figure 11.23**). The q, d, and c waveforms are given binary weights 4, 2, and 1, respectively. In this way we can say, for example, that the waveforms are in state 6. Perturbations of various states are shown separately (Figure 11.24) in order not to clutter the main waveforms.

STEP 2   **Make a primitive flow table.** Three variables imply eight possible stable states in eight rows in the flow table. This time our description building the flow table can be very cryptic, because we already have the debouncer experience. Keep looking at the waveforms (Figure 11.23) as you follow the text. In the following, the notation "Seq$_{ij}$"

FIGURE 11.23
D latch specification

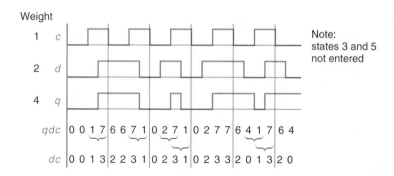

Note:
states 3 and 5
not entered

means "sequence from input state i to input state j when inputs change." Entries are made in Figure 11.25a.

State 0: $q^+ = 0$ when $qdc = 000$. Enter circled 0 in row 0, col 00.

  Seq_{01}: q^+ stays at 0.   Enter 1 in row 0, col 01.

  Seq_{02}: q^+ stays at 0.   Enter 2 in row 0, col 10.

State 1: $q^+ = 0$ when $qdc = 001$. Enter circled 1 in row 1, col 01.

  Seq_{10}: q^+ stays at 0.   Enter 0 in row 1, col 00.

  Seq_{13}: q^+ changes to 1.   Enter 7 in row 1, col 11.

State 2: $q^+ = 0$ when $qdc = 010$. Enter circled 2 in row 2, col 10.

  Seq_{20}: q^+ stays at 0.   Enter 0 in row 2, col 00.

  Seq_{23}: q^+ changes to 1.   Enter 7 in row 2, col 11.

State 3: Enter four dashes, because the waveforms do not enter this state.

FIGURE 11.24
**D latch perturbed c
and d waveforms**

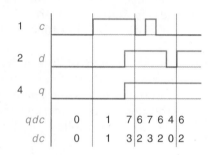

FIGURE 11.25
D latch flow tables

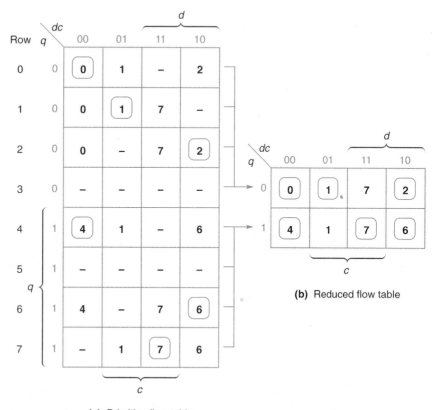

(a) Primitive flow table

(b) Reduced flow table

State 4: $q^+ = 1$ when $qdc = 100$. Enter circled 4 in row 4, col 00.

    Seq_{02}: q^+ stays at 1.   Enter 6 in row 4, col 10.

    Seq_{01}: q^+ changes to 0.   Enter 1 in row 4, col 01.

State 5: Enter four dashes, because the waveforms do not enter this state.

State 6: $q^+ = 1$ when $qdc = 110$. Enter circled 6 in row 6, col 10.

    Seq_{20}: q^+ stays at 1.   Enter 4 in row 6, col 00.

    Seq_{23}: q^+ stays at 1.   Enter 7 in row 6, col 11.

State 7: $q^+ = 1$ when $qdc = 111$. Enter circled 7 in row 7, col 11.

    Seq_{31}: q^+ changes to 0.   Enter 1 in row 7, col 01.

    Seq_{32}: q^+ stays at 1.   Enter 6 in row 7, col 10.

FIGURE 11.26

D latch next-state and output K map

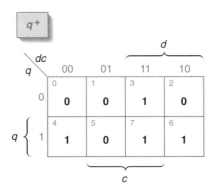

STEP 3   **Make a reduced flow table.** Clearly, rows 0, 1, 2, and 3 merge into one row, and rows 4, 5, 6, and 7 merge into another row (Figure 11.25b).

STEP 4   **Convert the reduced flow table into an unassigned-state map.** Omit.

STEP 5   **List the state transitions. Make an adjacency diagram.** Omit.

STEP 6   **Assign states, and make a next-state map.** In this case, one variable q makes the assignment automatic.

STEP 7   **Make next-state and output K maps.** The state numbers in the reduced flow table are replaced by output values corresponding to the state numbers. This process creates the K map for the circuit (Figure 11.26).

STEP 8   **Translate the K maps into equations, and design the circuit.** The circuit K map yields the equation for the clocked D latch:

$$q^+ = cd + c'q$$

The circuit is found in Figure 11.13.

11.4.3   D Flip-Flop Synthesis

A circuit is a latch when the circuit responds to clock input levels. A circuit is an edge-triggered flip-flop when the circuit responds to the clock input transitions or edges (see Section 6.5, page 307). Here is how we synthesize an edge-triggered flip-flop using only AND and OR gates.

STEP 1   **Write the specification.** The three variables involved with a D flip-flop are output q, input d, and clock c. The defining equation $q^+ = d$ does

FIGURE 11.27
**D flip-flop
specification**

q	d	c	q+
0	0	0	0
0	0	1	0
0	1	0	0
0	1	1	1
1	0	0	1
1	0	1	0
1	1	0	1
1	1	1	1

$\begin{smallmatrix}0\\1\end{smallmatrix}$ = edge triggering

$\_\!\!\Gamma$ stores d input

Defining equation: $q^+ = d = d(c + c')$

A clock edge captures
and holds the D input
value until the next
clock edge.

not reveal the clock. This is true of all clocked flip-flop defining equations. We can make the role of the clock explicit by drawing up a three-variable truth table from the equation $q^+ = d(c + c')$. We improvise by drawing arrows to define edge triggering (Figure 11.27). However, this format does not provide the visual assist to the next step that waveforms do.

We draw q, d, c waveforms implementing the equation $q^+ = d$ (Figure 11.28). The waveforms are complete, in the sense that all qdc states 0 to 7 are shown. Perturbed waveforms are shown in Figure 11.29. Two perturbations are not allowed (see x's in Figure 11.29) because inputs c and d are changing simultaneously at those times.

STEP 2   **Make a primitive flow table.** Eight states require eight primitive flow table rows with circled stable states 0 to 7 (Figure 11.30a). We enter these stable states in their rows and appropriate columns. In any row, we then perturb the inputs (Figure 11.30a) to move horizontally to any other column. All of the states reached by these moves are, by definition, unstable. All other states in any row are unstable because only one stable state per row is allowed. Furthermore, in each row there is one column that cannot be reached by perturbing only one variable,

FIGURE 11.28
**D flip-flop
waveforms**

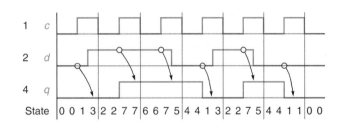

State 0 0 1 3 2 2 7 7 6 6 7 5 4 4 1 3 2 2 7 5 4 4 1 1 0 0

FIGURE 11.29
**D flip-flop
perturbed
waveforms**

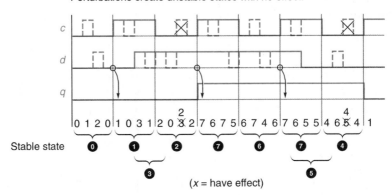

Perturbations create unstable states with no effect.

Stable state

(x = have effect)

because to reach that column would require changing two input variables at the same time, which violates one of our restrictions. We enter a dash in those inaccessible squares (Figure 11.30a).

STEP 3   **Make a reduced flow table.** The merger criteria are clearly met, and rows are merged in two ways out of a possible four (Figure 11.30b). This time we have four rows in the reduced tables. Four rows imply a need for two state variables q_1, q_0.

STEP 4   **Convert the reduced flow table into an unassigned-state map.** Omit.

STEP 5   **List the state transitions. Make an adjacency diagram.** Omit.

STEP 6   **Assign states, and make a next-state map.** An adjacency diagram (Figure 11.31) explicitly relates how moves from row to row are made as inputs change one at a time in any sequence. (The adjacency diagram is defined in upcoming Section 11.5.1.) The state number is defined by the value of two state variables $q_1 q_0$. If we assign 00 to state a, then a move to state b changes only one state variable if we assign 01 to state b. For the same reason we assign 10 to state c. And, as if by magic, it is okay to assign 11 to state d.

STEP 7   **Make next-state and output K maps.** Given the assigned states, the next-state K map (Figure 11.32) is constructed from the upper merged flow table in Figure 11.30. The three stable states in row a of the merged flow table have the same next state output, which is equal to present state a. We enter 00_2 in the corresponding next-state K map squares, because row a was assigned state 00_2.

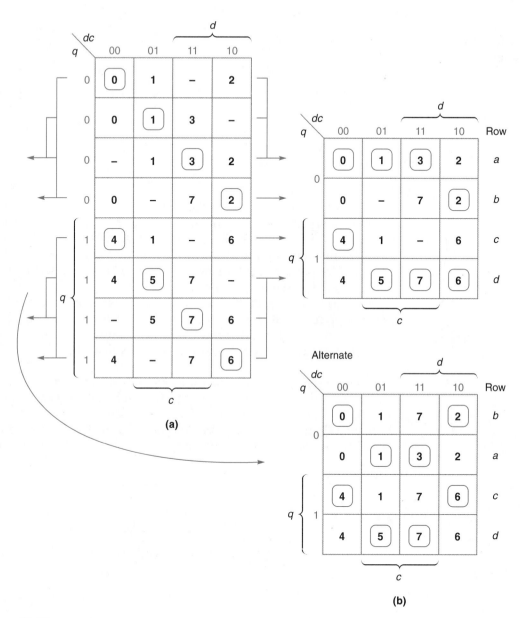

FIGURE 11.30
D flip-flop flow tables

FIGURE 11.31
D flip-flop state assignment

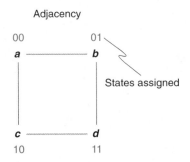

We enter 01_2 in the fourth column of row a, because the corresponding merged flow table square entry is a 2. The 2 corresponds to stable state 2 in row b, which was assigned state 01. In the same way, we determine and enter next-state values in the next three rows. Note that we have put row 2 after row 3 in order to have "adjacent" rows.

Only state variable q_1 is an output; it corresponds to the single output q in Figure 11.29. The state assignments have been chosen so that the state variable q_1 is the same as output variable q. These assignments are not mandatory; other assignments are possible. However, in those cases output q would be a function of q_1 and q_0.

STEP 8   **Translate the K maps into equations, and design the circuit.** The trick here is to create one next-state map for each state variable (Figure 11.33). This is simply a matter of copying from the composite next-state map (Figure 11.32). The equations in Figure 11.34 are read from the K maps. Here is a point that is not straightforward. First we note

FIGURE 11.32
D flip-flop next-state K map

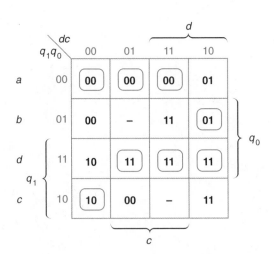

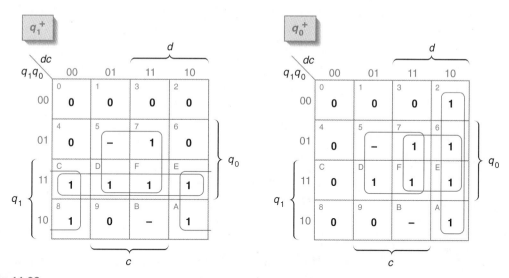

FIGURE 11.33
D flip-flop $q_1^+ q_0^+$
K maps

FIGURE 11.34
D flip-flop circuit in
two parts

$$q_1^+ = q_1 q_0 + q_1 c' + q_0 c$$

$$q_1^+ = q_1(q_0 + c') + q_0 c$$

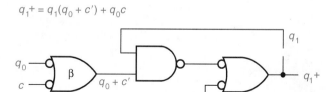

$$q_0^+ = q_0 c + q_0 d + dc'$$

$$q_0^+ = q_0 c + d(q_0 + c')$$

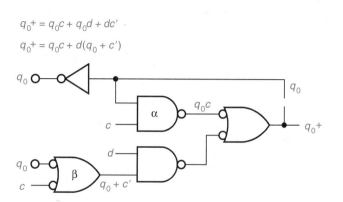

FIGURE 11.35
D flip-flop circuit

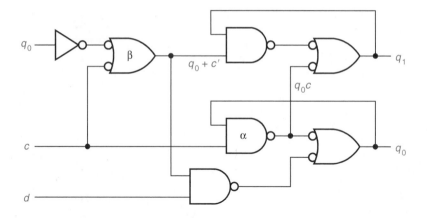

that q_0c is a term common to the $q_1{}^+$ and $q_0{}^+$ equations. Then we focus on the remaining terms and observe that $q_0 + c'$ is a common factor, which is why we recast the equations as shown in Figure 11.34. The circuits follow. The physically realized common term and factor are marked α and β, respectively.

We could use excess gates and pronounce the D flip-flop circuit to be as shown in Figure 11.34. However, we merge the two circuits, saving two gates (Figure 11.35). A variation of this circuit can be found in the TTL data book with asynchronous set and reset added. This is logic family member 74.

11.5   Races

A *race* occurs whenever a state transition requires the change of two or more of the state variables simultaneously. The race is between the variables to see which one changes first. The crucial question is whether or not the race is critical. An example in Figure 11.36 tells the story. Races are analogous to hazards, in the sense a hazard has the *potential* to produce a glitch. The glitch does not necessarily appear. A race has the potential to move to a next state that is incorrect. This can happen when the race is critical. A race is rendered harmless when assigned states make it noncritical. This process may require the addition of new states, necessitating additional variables.

A critical race may end in the wrong state.

Let's suppose the state machine is resting in the total state $q_2q_1q_0i_1i_0 = 01110$ (Figure 11.36). When i_1i_0 switches from 10 to 11, there is a horizontal move left to column 11 in row 2 to unstable total state 01111. It is unstable because the next state is $q_2q_1q_0 = 000$ ($q_2q_1q_0$ must change from 011 to 000). One of three events will occur.

FIGURE 11.36

Reduced flow table fragment with races

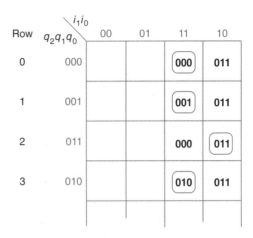

1. If the improbable happens and q_1q_0 both change simultaneously from 11 to 00, the next move is vertically up to stable total state 00011 in row 0. This is the correct final state.
2. If q_0 changes first from 1 to 0, the next move is vertically down to stable total state 01011 in row 3. This is an incorrect final state.
3. If q_1 changes first from 1 to 0, the next move is vertically up to stable total state 00111 in row 1. This is an incorrect final state.

This is a *critical* race, because the final state is incorrect for at least one event.

Now let's assume the state machine is resting in the row 0 total state $q_2q_1q_0i_1i_0 = 00011$. When i_1i_0 switches from 11 to 10, there is a horizontal move right to column 10 in table row 0 (Figure 11.36) to unstable total state 00010. It is unstable because the next state is $q_2q_1q_0 = 011$ ($q_2q_1q_0$ must change from 000 to 011). One of three events will occur.

1. If the improbable happens and q_1q_0 both change simultaneously from 00 to 11, the next move is vertically down to stable total state 01110 in row 2, column 10. This is the correct final state.
2. If q_0 changes first from 0 to 1, the next move is vertically down to unstable total state 00110 in row 1. Then when q_1 changes from 0 to 1, the next move is again vertically down to stable state 01110 in row 2. This is the correct final state.
3. If q_1 changes first from 0 to 1, the next move is vertically down to unstable total state 01010 in row 3. Then when q_0 changes from 0 to 1, the next move is vertically up to stable total state 01110 in row 2. This is the correct final state.

Non-critical races
always end in a correct
state.

> This is a *noncritical* race, because the final state is correct for all events.

11.5.1 Eliminating Races

Critical races can al-
ways be eliminated by
state reassignment that
may require additional
states.

The good news is that critical races can always be eliminated. The keys to success are reassigning states and/or adding states. The adjacency diagram assists in solving critical race problems.

This section shows how to implement Steps 4, 5, and 6 of the asynchronous synthesis method (see Section 11.4) for complex problems.

4. Convert the reduced flow table into an unassigned-state map.
5. List the state transitions. Make an adjacency diagram.
6. Assign states, and make a next-state map.

The *adjacency diagram* consists of points and lines. Each point represents one row (state) of a flow table. Each line joining two points signifies that a transition present in the flow table requires that the states of these two points differ in only one state variable.

If we assign adjacent states in the sequence 00, 11, 01, 10, then we are building in races (Figure 11.37) because the A/B and the C/D assignment pairs require two variables to change state simultaneously. If, however, we assign adjacent states in the sequence 00, 01, 11, 10, then the races are eliminated.

For a more significant problem, let us make up a next-state K map (Figure 11.38) to which states have been assigned. This is a product of synthesis method Step 6 (see Section 11.4). Do races exist? Are they critical? For example, if the system is in stable total state 1101 in row 2 and column 01 and $i_1 i_0$ switches from 01 to 00, then there is a horizontal move left to unstable state 1100; unstable because the next state is 00 whereas the present state is 11. Therefore $q_1 q_0$ must switch from 11 to 00. This is a race between two variables that have to change.

FIGURE 11.37

Adjacency diagram with state assignments

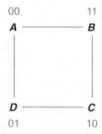

FIGURE 11.38
Next-state K map with two transitions that result in critical races

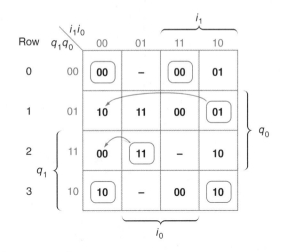

State 00 is a stable state in row 0, where $q_1q_0 = 00$. If q_0 switches first from 1 to 0, the system moves vertically down and locks in stable total state 1000 in row 3. If q_1 switches first, the system moves vertically up one row to unstable total state 0100, where 10 is the next state. This is a new race between two variables that have to change from 01 to 10. From there the next move is up or down, depending on which changes first this time. The races in column 00 are critical regardless of the source of the transition into column 00 (such as from column 10, row 1). How do we reassign states to eliminate these critical races?

STEP 4 **Convert the reduced flow table into an unassigned-state map:** We convert the state numbers entered in the next-state K map (Figure 11.38) into symbols as shown in Figure 11.39. For example, 00 is replaced by

FIGURE 11.39
Unassigned-state map

q_1q_0 \ i_1i_0	00	01	11	10
A	A	–	A	B
B	D	C	A	B
C	A	C	–	D
D	D	–	A	D

FIGURE 11.40
Adjacency diagrams

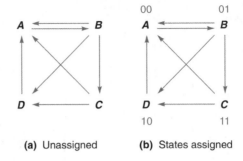

(a) Unassigned   (b) States assigned

the symbol A throughout the map. This is the unassigned-state map that emanates from any Step 4.

STEP 5

The adjacency diagram is a graphical aid for the process that eliminates races.

List the state transitions. Make an adjacency diagram. We list the state-to-state transitions that can occur in each row. We construct an adjacency diagram from the list (Figure 11.40a).

ROW	TRANSITIONS
A	$A \rightarrow B$
B	$B \rightarrow A, B \rightarrow C, B \rightarrow D$
C	$C \rightarrow A, C \rightarrow D$
D	$D \rightarrow A$

If we show the original assignments of 00, 01, 11, and 10 to A, B, C, and D, respectively, we can mark the transitions in bold type where we know critical races exist (Figure 11.40b). This is not a solution.

ROW	TRANSITIONS
00	$00 \rightarrow 01$
01	$01 \rightarrow 00, 01 \rightarrow 11, \mathbf{01 \rightarrow 10}$
11	$\mathbf{11 \rightarrow 00}, 11 \rightarrow 10$
10	$10 \rightarrow 00$

STEP 6

Assign states, and make a next-state map. The solution is to add one *state* variable q_2. This increases the number of states to eight so that we might be able to eliminate the need to change two variables simultaneously. The unit cube is our graphic assistant (Figure 11.41), and the adjacency diagram (Figure 11.40) is our guide.

FIGURE 11.41
Eight-state unit cube as an adjacency diagram

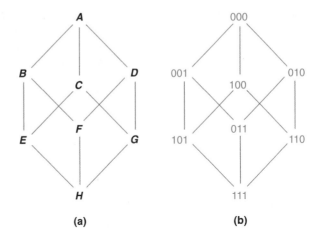

(a)   (b)

We start by making states B, C, and D adjacent to state A in the unit cube (Figure 11.42). Then we assign state numbers to each state, maintaining unit distance between all states (Figure 11.42). (By "distance" we mean the number of different bits in any pair of states.) This guarantees that there are no critical races.

We note that the race pair C → A (11 to 00) is resolved by reassignment (C → A becomes 000 to 100), and the race pair B → D (01 to 10) is resolved by interposing state F (B → D becomes 001 to 011 to 010). We verify that the new assignment set is free of critical races by making a new list of all possible transitions. Now we need to make a revised reduced flow table before we can design a race-free solution.

FIGURE 11.42
Adjacency diagram with assignments

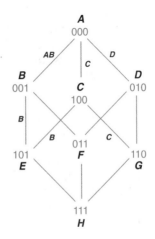

STEP 3 REVISITED   **Making a (revised) reduced flow table.** We start with a new list of state-to-state transitions taken from the adjacency diagram in Figure 11.42. Note that the two critical races are eliminated, and the C → D transition that was not a race will require more propagation time to complete. The latter is an unanticipated cost.

TRANSITIONS

New	Prior
A → B	A → B
B → A	B → A
B → E → C	B → C
B → F → D	B → D*
C → A	C → A*
C → G → D	C → D
D → A	D → A

*Critical races

We use information in the new list and the original reduced flow table to make the revised reduced flow table (Figure 11.43). The A → B, B → A transitions are not changed, so we copy row A from Figure 11.39. The new list shows that B transitions to states A, E, and F. These transitions are represented in new row B. The next step is not unique. We label the next row F and enter D in column 00, because B → F → D.

The D → A transition is not changed, so we copy row D from Figure 11.39. G transitions to D, so we let G be a new row with D entered in column 10.

C transitions to A or G, so we copy row C from Figure 11.39 and change the column 10 D entry to G.

Since we want B → E → C, we add row E, with C entered in column 01.

11.6   An Important Exception

Our suggested representation for the specified states has one exception that is readily accommodated. We next show by example what the exception is and how to deal with it straightforwardly.

The typical specification is a set of waveforms (Figure 11.44). There is one waveform for each input c, g, and one for the output q. In

FIGURE 11.43

Revised reduce flow table

$q_2q_1q_0$		i_1i_0 00	01	11	10
A	000	A	–	A	B
B	001	F	E	A	B
F	011	D			
D	010	D	–	A	D
G	110				D
C	100	A	C	–	G
E	101		C		
H	111				

lieu of the typical random letter assignments to these states, we have been assigning weights that are power of 2 to each waveform. Each state is given a number equal to the sum of the waveform weights in the interval representing the state. This form of representation makes the method very tractable, as we have shown. Our preference for this representation is due to the significant improvement in tractability, however, there is an exception.

The active high waveforms (Figure 11.44) specify that q transitions on state sequences 2, 7 and 7, 2 and 5, 0. In other words, the wave-

FIGURE 11.44

Single-pulser waveform specification

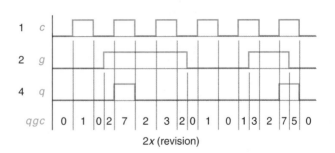

2x (revision)

FIGURE 11.45
**Single-pulser
primitive flow table**

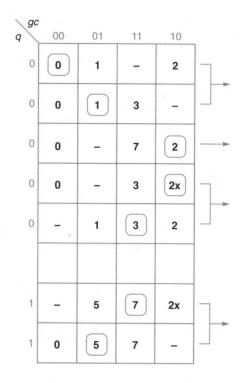

forms specify that only one pulse is generated every time g is asserted, which why the 2, 3 sequence is not a 2, 7 sequence. The 2, 3 sequence is the exception. This 2 is not like the 2 in a 2, 7 sequence. We revise this 2 to 2_x, and construct the primitive flow table accordingly (Figure 11.45). Should we overlook the exception, we will discover that there is a problem when we attempt to construct the primitive flow table when 2_x is a 2. Try it yourself.

SUMMARY

Delays and Reliability
Delays affect reliable asynchronous sequential circuit operation. An example in Figure 11.3 based on an ASM chart illustrates why.

Asynchronous Analysis Method
The RS latch, an oscillator, and the edge-triggered D flip-flop are analyzed. The RS latch and oscillator analyses are based on their circuits. The key is cutting the feedback lines and analyzing the resulting combinational circuits. In contrast, the D flip-flop analysis proceeds from the given equations. In effect, the circuit is known because it is represented by the equations. These are not fundamentally differ-

ent procedures; they differ only in their starting point. Here is our analytical method.

1. Identify the feedback lines and cut them.
2. Calculate the equations for the next-state outputs q_j^+.
3. Plot the next-state K maps for the q_j^+.
4. Merge the q_j^+ K maps into one composite next-state K map, with state as one coordinate.
5. Identify stable states, and circle them to form the transition map from the composite K map.
6. Make timing diagrams for sequences of input variable transitions.

Stable and Unstable States

An asynchronous circuit is in a stable state when the next internal state is the same as the present internal state. The circuit is in an unstable state when the next internal state is different from the present internal state. Timing diagrams show that a transient is in process when the circuit is in an unstable state (Figures 11.7, 11.8, and 11.11). The arrows in the figures emphasize the cause-and-effect relationships.

Asynchronous Synthesis Method

A practical method requires two assumptions (circuits satisfying these assumptions are known as fundamental-mode designs):

a. The system is quiescent prior to any input change.
b. Only one input changes at any instant.

This method has eight steps.

1. Write the specification.
2. Make a primitive flow table.
3. Make a reduced flow table.
4. Convert the reduced flow table into an unassigned-state map.
5. List the state transitions. Make an adjacency diagram.
6. Assign states, and make a next-state map.
7. Make next-state and output K maps.
8. Translate the K maps into equations, and design the circuits.

Races

A race occurs whenever a state transition requires the change of two or more of the state variables simultaneously. The race is between the variables to see which one changes first. The crucial question is whether or not the race is critical. Critical races can always be eliminated, and we show how to do this.

REFERENCES

Breeding, K. J. 1989. *Digital Design Fundamentals*. Englewood Cliffs, N.J.: Prentice-Hall.

Fletcher, W. I. 1980. *An Engineering Approach to Digital Design*. Englewood Cliffs, N.J.: Prentice-Hall.

Huffman, D. A. 1954. Synthesis of Sequential Switching Networks. *Journal of the Franklin Institute* (March, April): 161–190, 275–303.

McClusky, E. J. 1986. *Logic Design Principles*. Englewood Cliffs, N.J.: Prentice-Hall.

Kohavai, Z. 1978. *Switching and Finite Automata Theory,* 2nd ed. New York: McGraw-Hill.

Mead, C., and L. Conway 1980. *Introduction to VSLI Systems*. Reading, Mass.: Addison-Wesley.

Ungar, S. H. 1969. *Asynchronous Sequential Switching Circuits*. New York: Wiley.

PROJECT 19 Demonstration Report

Name ______________________________ SID# ________________ Section _____
(last) *(initials)*

Approved by ______________________ Date ______________________________

Grade on report ______________________

Asynchronous Single-Pulse Generator

1. Design:

_____ primitive flow table
_____ merged flow table
_____ unassigned-state maps
_____ state transitions and adjacency diagram
_____ assigned states and next-state map
_____ next-state and output maps
_____ equations and circuit

2. Build the circuit.

_____ circuit documents

3. Test the circuit.

_____ test data

The following waveforms define an asynchronous single-pulse generator with clock c, pulse generator command g, and output z. If g goes high, then the next time the clock goes high, z is generated as a pulse with duration equal to the time the clock remains high. This time equals one-half of a clock period.

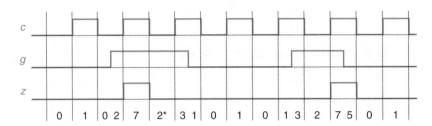

*Note: state 2 is **not** followed by state 7 here.

1. Design the circuit.

   Map the timing diagram into a primitive flow table.
   Make the merger diagram and the merged flow table.
   Make the state assignments, next-state map, and output map.

2. Build the circuit.

3. Test the circuit.

PROBLEMS

11.1 Make the timing diagram analogous to Figure 11.7b for the case when r is:

(a) low for one t_p
(b) low for two t_p
(c) low for three t_p

11.2 Make the timing diagram analogous to Figure 11.8b for the case when s is:

(a) low for one t_p
(b) low for two t_p
(c) low for three t_p

11.3 In Figure 11.7b, the AND and the OR delays both equal t_p. Draw a revised Figure 11.7b for the case when the AND delay is t_p, the OR delay is $1.5t_p$, and r is low for $2t_p$.

11.4 Refer to Figure 11.9. Delete the 08 gate in, and make a timing diagram for, this revised circuit analogous to Figure 11.11.

11.5 Refer to Figure 11.13. Verify that $q^+ = cd + c'q$ for the clocked D latch.

11.6 Refer to Figure 11.21b.

(a) Assign 1 to all dashes and derive the K map.
(b) Assign 0 to all dashes and derive the K map.
(c) Derive equations from the two K maps.

Note: Some of the following problems require analysis of an asynchronous circuit. The asynchronous analysis method of Section 11.2 indicates that solutions to these problems include the following six steps:

1. Identify the feedback lines and cut them.
2. Calculate the equations for the next-state outputs q_j^+.
3. Plot the next-state K maps for the q_j^+.
4. Merge the q_j^+ K maps into one composite next-state K map, with state as one coordinate.
5. Identify stable states, and circle them to form the transition map from the composite K map.
6. Make timing diagrams for sequences of input variable transitions.

11.7 Refer to Figure P11.1. Analyze the 375 bistable latch.

FIGURE P11.1

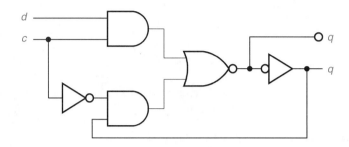

11.8 A latch function $q^+ = sq' + r'q + sr'$, where r is an active low variable and s and q are active high variables. Draw the circuit. Analyze the circuit.

11.9 A latch function $q^+ = sq + rq + sr'$, where s, r, and q are active high variables. Draw the circuit. Analyze the circuit.

11.10 Refer to Figure P11.2. Use s as an input to simplify the analysis. This replaces the R, S_1, S_2 inputs. Analyze the 120 when:

(a) $M = 0$
(b) $M = 1$

11.11 Analyze the circuit represented by the following equations. What is the circuit's function?

$$q_1^+ = c'q_1 + cq_0$$
$$q_0^+ = c'x + cq_0$$

11.12 Analyze the circuit represented by the following equations. What is the circuit's function?

$$q_1^+ = cq_0 + q_1(q_0 + c')$$
$$q_0^+ = cq_0 + d(q_0 + c')$$

11.13 Analyze the circuit represented by the following equations. What is the circuit's function?

$$q_1^+ = q_1q_0 + cq_0 + c'q_1$$
$$q_0^+ = cq_1q_0 + xq_0 + c'x$$

11.14 Analyze the circuit represented by the following equations. What is the circuit's function?

FIGURE P11.2

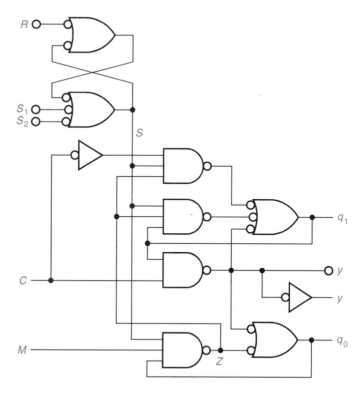

$$q_1{}^+ = gq_1 + cq_0$$

$$q_0{}^+ = gq_1'q_0 + cq_0 + c'gq_1'$$

11.15 Analyze the circuit represented by the following equations. What is the circuit's function? z is the output.

$$q_1{}^+ = cq_0$$

$$q_0{}^+ = cq_0 + xq_0 + c'x$$

$$z = cq_1q_0$$

11.16 Analyze the circuit represented by the following equations. What is the circuit's function?

$$q_2{}^+ = (c' + x)q_2q_1' + cxq_1'q_0 + q_2q_1'q_0$$

$$q_1{}^+ = xq_1q_0' + cxq_1'q_0'$$

$$q_0{}^+ = c'xq_2'q_1' + xq_2'q_1'q_0 + cq_2q_1'q_0$$

Note: In some of the problems that follow, the next-state defining equation is given. If the equation is "edge triggered," the equation implies that one input is a "clock" input. A "clock" input may be aperiodic or periodic. The defining equation is executed on every rising clock edge. The asynchronous synthesis method of Section 11.4 indicates that solutions to these problems include the following eight steps.

1. Write the specification.
2. Make a primitive flow table.
3. Make a reduced flow table.
4. Convert the reduced flow table into an unassigned-state map.
5. List the state transitions. Make an adjacency diagram.
6. Assign states, and make a next-state map.
7. Make next-state and output K maps.
8. Translate the K maps into equations, and design the circuits.

11.17 The toggle function edge-triggered defining equation is $q^+ = q'$. Design a circuit. (Remember that flip-flops are asynchronous circuits.)

11.18 The edge-triggered JK flip-flop defining equation is $q^+ = jq' + k'q$. Design a circuit.

11.19 The edge-triggered T flip-flop defining equation is $q^+ = t \text{ xor } q$. Design a circuit.

11.20 Design the circuit implementing the waveform specification in Figure P11.3.

FIGURE P11.3

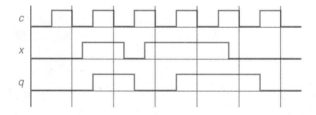

11.21 Design the circuit implementing the waveform specification in Figure P11.4.

FIGURE P11.4

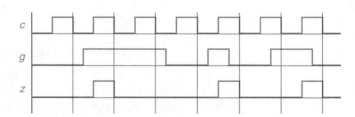

11.22 Design the circuit implementing the waveform specification in Figure P11.5.

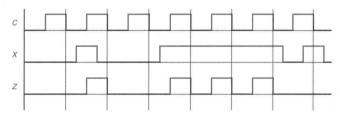

11.23 Design the circuit implementing the waveform specification in Figure P11.6.

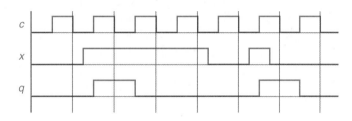

12 PROJECTS

OVERVIEW

We cross the bridge from pencil and paper to hardware. Suddenly an additional set of considerations comes into play, including: the LSTTL databook, pin numbers for inputs and outputs, the use of tools, the use of a multimeter, the use of an oscilloscope, how to avoid shorting 5 volts to an output or to ground, what parts look like, how we know if they are functional, and how the physical layout of parts affects the ease of wiring.

The initial projects are theoretically simple so you can focus on the new and practical considerations. You move on to more complex projects as you gain experience.

INTRODUCTION

The projects supplement the text. These are not fill-in-the-squares projects. They require planning and thinking. In some projects, new material is introduced to challenge you. Projects start with one inverter and work up to projects requiring 12 or more LS chips. The projects are suitable for point-to-point wiring on a solderless breadboard.

The projects can be implemented with standard, low-cost LS TTL integrated circuits and a few discrete parts (Table 12.1). Some projects list in their text required parts not listed in Table 12.1. Section 12.6 discusses software tools.

12.1   One Right Way to Do a Project

STEP 1   **Understand the requirements before you act.** Ask questions if you do not understand. This will save you a lot of time and energy.

STEP 2   **Create the ASM chart required by the project.** The initial "gate" projects do not require an ASM chart.

STEP 3   **Design the circuit, and make an accurate schematic.** Errors in schematics are time eaters. Omissions are also errors. Enter pin numbers to simplify the wiring process.

STEP 4   **Plan the layout of the ICs on the pinboard.** Make a layout drawing. Identify the ICs as U_1, U_2, etc.

STEP 5   **Build the circuit *exactly* per the schematic.**

a. Put a piece of tape on each IC and write the U_j identifier on it.

TABLE 12.1   **Needed Project Parts* and Equipment and Recommended Tools**

QUANTITY†	PART NO.	DESCRIPTION
2	74LS00	quad 2-input NAND
2	74LS02	quad 2-input NOR
3	74LS04	hex inverter
1	74LS05	hex inverter, open collector
2	74LS08	quad 2-input AND
2	74LS10	triple 3-input NAND
3	74LS20	dual 4-input NAND
2	7425	dual 4-input NOR [not LS]
1	74LS30	8-input NAND
2	74LS74	dual D flip-flop
3	74LS85	comparator, 4-bit
2	74LS86	quad XOR
2	74LS109	dual JK flip-flop
2	74LS130	13-input NAND
2	74LS138	decoder, 3 to 8
1	74LS153	mux, dual 4 to 1
1	74LS157	mux, quad 2 to 1
3	74LS163	counter, 4-bit binary up
1	74LS166	8-bit shift register: parallel in, serial out
1	74LS169	counter, 4-bit binary up/down
2	74LS173	4-bit enabled D flip-flop register
1	74LS175	quad D flip-flop
2	74LS194	4-bit bidirectional shift register
2	74LS283	4-bit binary full-adder with look-ahead carry
1		plastic storage box
2		dip switch, 8 single-pole-single-throw
10		red LED
1		10-MHz crystal
1		capacitor, 100 pf
10		capacitor, 0.1 μf
1		potentiometer, 5K ohm
1		resistor, 390 ohm (all resistors 1/4 watt)
10		resistor, 1K ohm
2		resistor, 10K ohm
various		hookup wire, 22 gauge, solid

The following recommended tools will make life in the lab easier.

1		multimeter (volt-ohm-current), 2% analog
1		logic probe
1		IC inserter

TABLE 12.1   **(continued)**

1	IC puller
1	tweezer, fine point
1	wire stripper
1	side cutters, small
1	long-nose pliers, small

* All parts must be suitable for use on a solderless breadboard.
† Quantities shown assume only one project is wired at a time.

b. Line out with a marker pen on a copy of the schematic the wire, the IC, or the discrete component you just installed. This minimizes wiring errors and can be a tremendous assist. Try it.

c. When you can, test the circuit after adding one part but *before* adding the next. Testing as you proceed helps immensely in the debug process and the design-check process.

d. If you detect a design error, take corrective action immediately.

(1) Design the change that corrects the error.

(2) Correct the schematic.

(3) Rewire the circuit.

The goal is to have at *any* instant a schematic that matches the circuit. Otherwise, you risk losing control of your project.

Do not innovate carelessly as you go. If you want to make a change, return to Step 3 and proceed from there. There are *no* short-cuts or minor changes. *All changes are major.* If you find yourself spending hours debugging, you have probably violated one of the preceding rules. Think about this: how can you debug a circuit that is inaccurately documented or carelessly designed?

STEP 6   **Test the completed project.** There is much more to testing than testing as you go. Create a test plan that exercises the project and demonstrates that it works. This is usually a cut-and-try process until you have acquired considerable experience. Now is the time to start.

Make sure your tests include unusual input bit patterns as well as incorrect inputs.

STEP 7   **Write the project report.** In industry, the project report can make or break a project. Do your best.

FIGURE 12.1
**Solderless
breadboard
fragment**

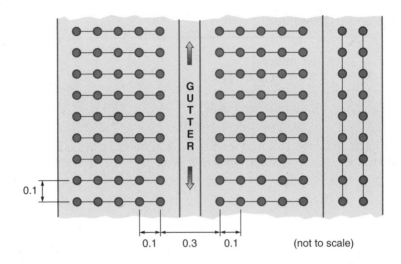

12.2 Solderless Breadboards

Solderless breadboards in their most basic form consist of one or more socket strips mounted on a metal plate. A breadboard fragment is shown in Figure 12.1. A socket strip has two arrays of holes separated by a long channel, or gutter. The holes in the columns parallel and adjacent to the channel are 0.3 inches apart, consistent with standard IC minimum pin row spacing. Each hole array is a two-dimensional array of holes, e.g., 5×59, on 0.1-inch centers. The five holes in each row are shorted together. The rows are not shorted together.

We plug an IC into the board so that the IC straddles the channel and each IC pin plugs into one hole of a row. Then the four other holes in each "pin" row are available to receive wires, which are automatically connected to the IC pin. In this way a circuit is wired from point to point.

Larger ICs have 0.4-inch and 0.6-inch pin row spacing. These are inserted in the same way that the smaller 0.3-inch ICs are inserted; however, the covered-up row holes are not available for point-to-point wiring. Other types of IC packages cannot be used on these breadboards.

Another socket strip (Figure 12.1) is formatted to distribute power and ground. The array has two 1×59 columns. All holes in each column are shorted together. The two columns are not shorted together so that one column can distribute 5 volts and the other ground.

12.3 **Laboratory Equipment**

In addition to the parts and tool kit listed in Table 12.1, you will need the following:

1	100-MHz oscilloscope
1	power supply, 5 volts, 200 mA

12.4 **Project Reports**

Formal project reports usually follow the following plan outline. Informal project reports can follow a project's demonstration sheet. Your instructor may require a different report format.

1. Table of contents
2. List of figures
3. Digital design
    a. *Data path architecture as derived from the specifications*
        The circuit (figures)
    b. *Control ASM chart*
        A short essay on how you created the chart
        The chart (a figure)
    c. *State machine design*
        Truth table of next state and outputs
        Next-state equations
        The circuit (a figure)
    d. *Output circuit design*
        Output equations
        The circuits (figures)
    e. *Timing diagrams*
        A short essay on how you created the diagrams
        The timing diagrams (figures)
4. Test plan: A short essay on how you will test your circuits and why the results will prove the circuits meet the specifications.
5. Demonstrating performance: A short essay on tests you actually made (modifying the test plan as deemed necessary), the data taken, and why the results prove the circuits meet the specifications. Includes any revised figures.
6. Conclusions: A short essay on what you learned from the project.

12.5 **Time Savers**

Useful signal sources are implemented with dip switches, wires, and resistors. The H/L dip switch circuit (Figure 12.2a) generates H and L levels. The latch driven by a toggle switch is a switch debouncer

FIGURE 12.2

FIGURE 12.2
Signal sources using dip switches and wires

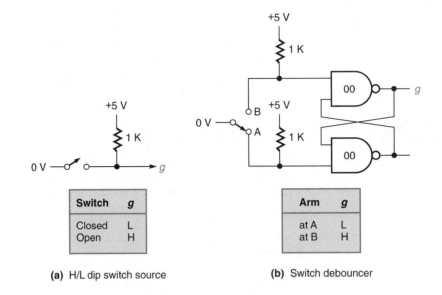

(a) H/L dip switch source

Switch	g
Closed	L
Open	H

(b) Switch debouncer

Arm	g
at A	L
at B	H

(Figure 12.2b). A toggle switch arm is conveniently implemented with a wire.

Not only are schematics drawn with point-to-point wiring tedious to draw, but they are also difficult to analyze (Figure 12.3a). Point-to-point wiring requires us to literally trace wires from point to point. We do not choose to spend our time and energy tracing wires when an attractive alternative is available (Figure 12.3b). There is very little time available for, and even less interest in, tracing point-to-point wiring. Make schematics that can be understood in an instant! That guarantees an audience.

Data books are thick and heavy, and their pages are thin. Looking up chip pinouts over and over again is another waste of time. A simple solution is to take time out to draw pinouts on 3-inch × 5-inch cards (Figure 12.4). You can make a new card as the need arises, thereby avoiding a peak work load. You do this even though software tools are used. The reason is simple: the cards are often-used "instant" information.

12.6

Software Tools

Software tools separate into two basic groups: logic design aids and physical design aids. We next describe these two types of tools and indicate how they can be used to help you implement your projects. (We do not teach how to use any specific tool, simply because there

FIGURE 12.3
**Two ways to draw a
schematic**

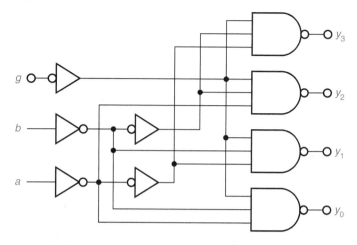

(a) Do you have the time and energy to trace signal paths?

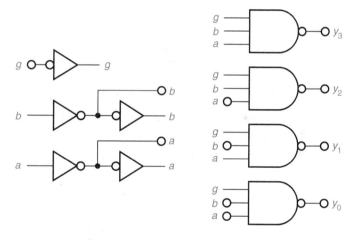

(b) Or do you prefer "instant" tracing?

are too many choices.) Then we point out that project implementation
can usefully employ pencil and paper as well as computer-aided tools.
(A judicious balance may the best solution.) Finally, we point out the
parts used in projects are not limited to TTL parts.

Logic design   LogicAid[1] is a computer-aided logical design assistant.
LogicAid assists you in performing the many functions necessary to

---

[1] LogicAid is a trademark of West Publishing Company.

FIGURE 12.4
IC pinout cards

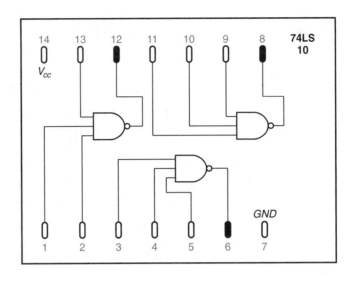

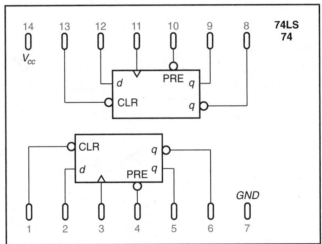

the logic design process. The program is capable of simplifying Boolean functions specified in various ways, such as equations, truth tables, and K maps. Given a state table, LogicAid routines can be used for state reduction, state assignments, state table comparison for equivalence, and generating flip-flop input equations.

Physical design   Computer-aided design (CAD) and computer-aided engineering (CAE) are names for types of physical design software tools that can aid you in your project development. CapFast[2] is one of many such tools. These tools affect project development like calculators influence mathematical calculation. Both augment, if not replace, pencil-and-paper methods. Tools allow you to manage large projects effectively. Without tools, in fact, large projects become impractical except for those few individuals with extraordinary talent. What is large? The apparently small number 20 is large when there are 20 integrated circuits in a circuit designed with pencil-and-paper methods. Here are some of the ways CAD/CAE tools assist you.

Get and place library components.

Select, copy, move, and erase components.

Get and place input, output, ground, and V_{cc} connections.

Draw wires.

Enter a title and your name.

Plot a schematic.

Assign reference designators and component pin numbers.

Perform an electrical rules check.

Perform a logical simulation to verify your equations.

Perform a timing analysis and create timing diagrams.

Create a parts list.

Create a net list of all the point-to-point connections.

The price you pay for this assistance is having to learn new languages. The tools' languages have vocabularies of, say, 200 commands plus the syntax rules. However, the benefits are worth the effort. Without tools, you are probably not realizing your potential as a designer.

Using tools in your projects   You can use computer-aided tools more creatively and more effectively when you truly know and understand pencil-and-paper methods. Learning those methods is the first order of business, which is why we recommend you use both types of methods when you implement the projects.

When you turn to computer-aided methods, the immediate problem that arises is the need to learn the new methods. This is a problem because you have limited time available for implementing each project

---

[2] CapFast is a trademark of Phase Three Logic, Inc.

if they are assignments in a laboratory course. The lack of adequate time is the basic issue. We recommend you learn useful increments. That is easy to say, yet difficult to do. If specific tools are part of your lecture series, the problem is mitigated.

The projects are presented with no mention of tools. Questions relevant to tools are part of the design problem, for example, what tools to use, how to use them, and when to use them. Which methods to use in individual projects is a local decision. Perhaps you start off with pencil-and-paper methods while learning how to use tools in parallel.

Table 12.1 lists TTL parts you can use to implement your projects. Also, we recommend specific parts in some projects if you use TTL parts. However, you do not have to use those parts. For example, you can use PLDs, parts from another logic family, or a mixture. Like methods, which parts to use in individual projects is a local decision. Flexibility is the key to successful and varied project implementation.

12.7 The Projects

Follow the procedure in Section 12.1 until you create a better one.

Answer every question in essay form.

Use a plastic template to draw schematics.

Use mixed-logic.

Show all pins.

Identify inputs.

Identify all lines.

Show part numbers.

Use L and H on schematics. Do *not* use 0 and 1.

Keep bit position numbers in order, e.g., 3210, *not* 3102.

Breadboard circuit exactly per schematic.

Document all changes.

Project standards Project solutions must be visually and logically clear. Readers should not have to decode your schematics nor read your mind. This is an opportunity to communicate. Do not fall into the trap of talking to yourself.

Do not make freehand drawings.

Use template and straightedge.

Use engineering paper or blank paper.

Use H and L—there are no ones and zeros in hardware.

All trick circuits must be explained by a detailed analysis.

Do not use package drawings. Use logical symbols.

APPENDIX
IEEE STANDARD
91-1984

INTRODUCTION

The IEEE Standard 91-1984 is a relatively new symbology standard that covers a multitude of details we choose not to explain. Instead, we concentrate on explaining the IEEE symbols for gates and building blocks used in the text. In this way you will acquire significant knowledge about the standard, since the text covers most, if not all, of the commonly used circuits that happen to be represented by a diverse set of IEEE symbols.

A symbol comprises an outline, or a combination of outlines, plus one or more qualifying symbols. The IEEE symbols we show for standard parts are common to most data books.

Gates

The distinctive-shape symbols (Figure A.1) are the ones we have used throughout the text. The IEEE gate symbols have rectangles for outlines and a qualifying symbol inside the rectangle (Figure A.1). Our mixed-logic bubble (Section 4.3) is replaced by the IEEE half-arrow above the signal line. (The IEEE bubble symbol implies inversion in a positive-logic system.)

The 266 XNOR gate has an open collector output, which is marked with the IEEE open collector qualifier.

JK Flip-Flop

The IEEE symbol for the 109 shows a C1 qualifier at the clock input (Figure A.2). The C symbol identifies a control function. The 1 in combination with the C identifies signals controlled by the C input by marking those signals with prefix 1. (C1 represents a control dependency.) When control input C1 is asserted, all dependent signals perform their function.

The J and K inputs perform their function when the clock is asserted, which is why the J and K inputs have a 1 as a prefix. This is how we know J and K inputs are synchronous.

The IEEE symbol for the 109 JK flip-flop shows the active low PRE and CLR inputs. We know they are asynchronous inputs because they have no prefix 1. They are not controlled by the C1 input.

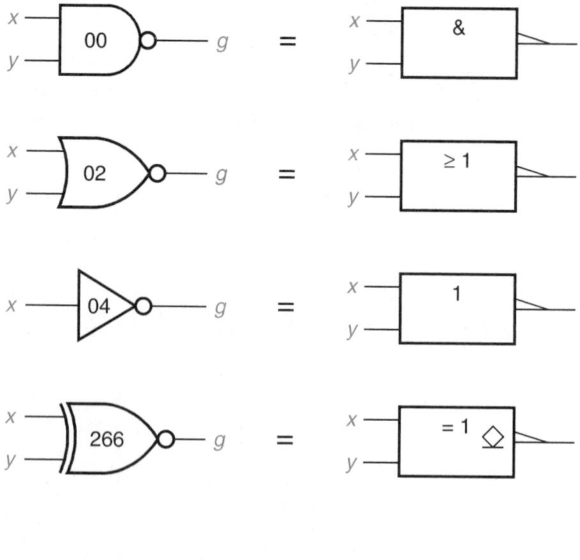

FIGURE A.1

Distinctive-shape symbols and IEEE gate symbols

FIGURE A.2
JK flip-flop IEEE symbol

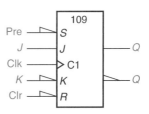

A.3 Decoder

The 138 decoder symbol (Figure A.3) shows EN dependency. The EN dependency enables all output lines, unless EN is qualified by a numerical suffix that also marks only the specific outputs enabled (e.g., EN4 enables all outputs marked with a 4).

All 138 decoder outputs are potentially enabled when EN (without a suffix) is asserted. The qualifier BIN/OCT specifies that binary inputs A, B, and C select one of eight outputs to be asserted. EN is asserted when the term $G_1 G_2 G_3$ is true. Notice how the AND gate is placed inside the 138 outline. Our preference would have been to use a G dependency for the A, B, C inputs, as is done for the 138 demultiplexer symbol discussed next.

The 138 demultiplexer symbol shows G dependency. G dependency performs a selection function. Input lines labeled 0, 1, and 2 represent powers of 2, to show their binary weight. The input-selection group is bracketed, as shown in Figure A.3. The bracket is followed by a G plus a range of numbers from 0 to 7, shown as 0 over 7. The outputs to be selected by the G dependency are marked with the range 0 to 7. This time the term $G_1 G_2 G_3$ is the input that is passed to the selected output.

FIGURE A.3

138 decoder and demultiplexer IEEE symbol

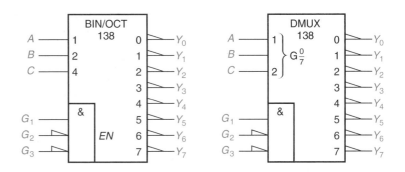

**153 multiplexer
IEEE symbol**

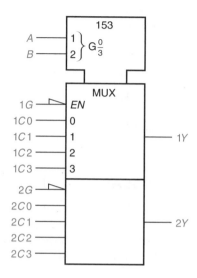

A.4   Multiplexer

The 153 package includes two 4×1 multiplexers, each with individual enables and common selection controls. The common-control block allows us to assemble compound symbols, such as the symbol required to represent the 153 package. The common-control block looks like the handle of a spade attached to the blade, where the blade represents the two multiplexers (Figure A.4).

The G dependency is used because the A, B control inputs select multiplexer inputs. EN dependencies enable individual multiplexers.

A.5   Register

The 173 common-control block controls four enabled D flip-flops (Figure A.5). The down-pointing triangle at a flip-flop output identifies a three-state output. Asserted M and N inputs assert EN, which enables all three-state outputs. The right-pointing triangle identifies outputs with extra drive capability.

Control dependency C1 is asserted when data-enable inputs $G_1 G_2$ and the clock are asserted. Asserted C1 loads the enabled D flip-flops.

Clr input is asynchronous, because the associated R symbol does not have a 1 prefix.

FIGURE A.5
173 register IEEE symbol

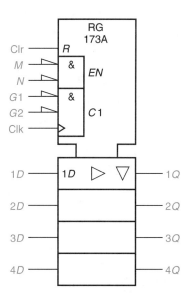

FIGURE A.5
173 register IEEE symbol

A.6   Counter

The CTRDIV16 qualifier specifies that the outline represents a "divide by 16 (binary)" counter (Figure A.6). M dependency marks signals that control mode selection—modes such as clear, load, and count in the 163 counter. M dependency affects inputs the same as does C dependency. When Mj is asserted, inputs marked with j perform their function. M1 loads data into the counter, because the flip-flop inputs are marked with a 1 (Figure A.6). When the load input is not asserted, M2 is asserted to allow the counter to count, because the clock input is marked with a 2+. This is explained in a moment.

The C5 dependency asserts the synchronous counting function marked as 5D in each flip-flop followed by the [n], where $n = 1, 2, 4, 8$. The synchronous clear function is marked as 5, showing it is controlled by the clock input. 5CT = 0 shows that the asserted clear input sets the count to zero.

The solidus (/), such as the one following the C5 notation, separates sets of labels. The clock input has two sets of labels: C5 and 2, 3, 4+. The 2, 3, 4+ label means the counter increments by one (+) whenever the combination M2 *and* G3 *and* G4 is asserted.

The R$_{co}$ output is asserted when qualifier G3 is asserted *and* count equals 15.

FIGURE A.6
**163 counter IEEE
symbol**

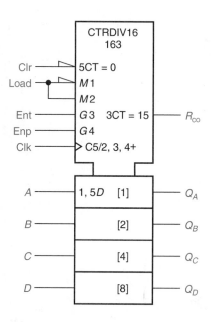

FIGURE A.7
**194 shift register
IEEE symbol**

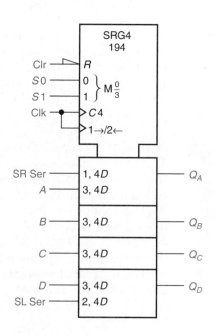

**245 transceiver
IEEE symbol**

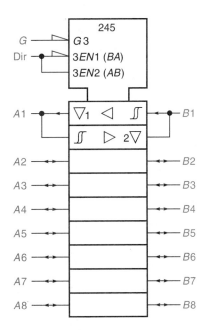

A.7   Shift Register

The SRG4 qualifier specifies that the outline represents a four-bit shift register (Figure A.7). The clr input is qualified by R, which means "reset all bits to 0." The clr input is asynchronous, because it is not controlled by the C4 qualifier.

$S_1 S_0$ inputs define an M dependency, selecting shift register modes 0 to 3. Mode 0 does nothing; mode 1 selects the serial input mode, with bit A as input; mode 2 selects the serial input mode with bit D as input; mode 3 selects the load data in parallel mode.

The clock input shows three control functions. C4 clocks the four bits in all modes. Therefore all modes are synchronous. A parallel action is: clock implements shift right in mode 1 (1→), and shift left in mode 2 (2←).

A.8   Transceiver

The G3 dependency selects the data flow direction BA or AB by marking the dir inputs' EN qualifiers with prefix 3 (Figure A.8). EN1 asserts the A lines as outputs, whereas EN2 asserts the B lines as outputs. The right-pointing triangles identify all outputs with extra drive capability. The hysteresis symbol identifies all inputs with this capability. This function "squares-up" input signals (see Section 2.5).

INDEX

Note: first page number is principal reference.

Mixed Logic: Notation for Active High and Low Variables

Modern, practical digital circuits and data books make extensive use of mixed logic. However, mixed logic notation needs to be simplified.

Data book notation We call a bubble at a gate terminal a "bubble of the first kind." These are the bubbles found in data books. The bubble of the first kind at a gate terminal means merely that the gate input or output is asserted when the voltage is low (L). The absence of a bubble at a gate terminal means the gate input or output is asserted when the voltage is high (H). Furthermore, in data books variables associated with variables are marked with an overbar ($\bar{x}$). The bar does not mean the complement of the variable; it means what the bubble means—the gate input or output is asserted when the voltage is low (L).

Notation for text and schematics No special notation is needed in positive- or negative-logic systems. In positive-logic systems all variables x are understood to be active high, and, in effect, active low variables are represented by the complements x'. In negative-logic systems all variables y are understood to be active low, and, in effect, active high variables are represented by the complements y'.

In mixed-logic systems where variables are active high or low in different parts of the circuit, a special notation is required in order to know by inspection the active level at any circuit node.

One special notation is the use of the subscript "H" for active high variables and the subscript "L" for active low variables. For instance, the variable g is written as g_L at active low nodes and as g_H at active high nodes.

Our preferred notation for mixed logic systems is to mark the circuit nodes of active low variables with a new "bubble of the second kind." The absence of a bubble at a circuit node indicates that the variable at that circuit node is true (logical 1) when the voltage is high (H).

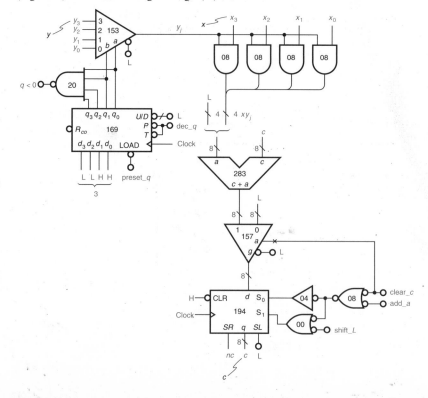